Three-Dimensional Turbulent Boundary Layer

International Union of Theoretical and Applied Mechanics

Three-Dimensional Turbulent Boundary Layers

Symposium, Berlin, Germany,
March 29 – April 1, 1982

Editors
H. H. Fernholz and **E. Krause**

With 288 Figures

Springer-Verlag
Berlin Heidelberg New York 1982

Prof. Dr.-Ing. H. H. Fernholz

Hermann-Föttinger-Institut für
Thermo- und Fluiddynamik der
Technischen Universität Berlin
1000 Berlin 12

Prof. E. Krause, Ph. D.

Aerodynamisches Institut der
Rheinisch-Westfälischen Hochschule Aachen
5100 Aachen

ISBN-13:978-3-642-81889-9 e-ISBN-13:978-3-642-81887-5
DOI: 10.1007/978-3-642-81887-5

Library of Congress Cataloging in Publication Data
Main entry under title:
Three-dimensional turbulent boundary layers.
At head of title: International Union of Theoretical and Applied Mechanics.
Bibliography: p.
Includes index.
1. Turbulent boundary layer--Congresses.
I. Fernholz, H. (Hans), 1935- . II. Krause, E. (Egon), 1933-
III. International Union of Theoretical and Applied Mechanics.
TA 357. T 48 1982 620. 1'064 82-10546

2061/3020 5 4 3 2 1 0

Scientific Committee

G. N. Abramovich, USSR
P. Bradshaw, UK
G. Drougge, Sweden
H. H. Fernholz, FRG (Co-Chairman)
S. J. Kline, USA
E. Krause, FRG (Co-Chairman)
Y. Senoo, Japan

Local Organizing Committee

A. Behlow
H. Berger
A. Dahm
H. H. Fernholz
A. Leutz

Symposium Sponsors

International Union of Theoretical and Applied Mechanics (IUTAM)

Technische Universität Berlin

Ausstellungs-Messe-Kongress-GmbH (AMK Berlin)

Messerschmitt-Bölkow-Blohm (MBB) München

Eternit AG Berlin

Preface

The IUTAM Symposium on Three-dimensional Turbulent Boundary Layers was suggested by the Gesellschaft für Angewandte Mathematik (GAMM) and sponsored by the International Union of theoretical and Applied Mechanics. The symposium was organized by H.H. Fernholz (Hermann-Föttinger-Institut für Thermo- und Fluiddynamik der Technischen Universität Berlin) and E. Krause (Aerodynamisches Institut der RWTH Aachen). After two successful Euromech Colloquia on the same topic in Berlin 1972 and Trondheim 1975 the organizers felt that another meeting should be convened, this time with participants from inside and outside Europe.
The aim of the symposium has been to bring together scientists who are actively engaged in boundary layer research, both experimental and theoretical. The scope of the meeting encompassed incompressible and compressible three-dimensional turbulent boundary layers. Special emphasis was laid on economical calculation methods, on measurements of fluctuating quantities and on measuring techniques designed for and applied successfully to three-dimensional boundary layers.
From among thirty-four papers submitted for presentation, twenty-six contributions of twenty-five minutes each were selected by the European members of the Scientific Committee. Furthermore there were four invited lectures of forty-five minutes. Short discussions were held directly after each presentation with a long discussion period at the end of each day. The final discussion on the last day of the symposium was recorded on tape and is presented in a slightly shortened version as the last contribution in this volume.

We hope that the meeting has been useful in demonstrating the state of the art and in giving participants a chance to talk to each other and to exchange ideas. The symposium has shown again that there exist few experiments which contain sufficient information for improving turbulence models. Calculation methods agreed either with measurements of mean velocity or of turbu-

lence qualitatively but were not able to show consistent agreement with both. This was probably due to rather simple turbulence models and not to numerics, and the opinion was prevailing among participants that the structure of turbulence in three-dimensional boundary layers is not yet well understood.
The organisation of any conference is time consuming and needs assistance from many sources. We are grateful therefore to all the members of the Scientific Committee and the local Organizing Committee and to the Hermann-Föttinger-Institute who have assisted us in the preparations for and the running of the symposium. Thanks are due to our sponsors who have generously helped by providing parts of travel grants for fifteen participants. Last but not least we wish to thank the participants for all the trouble they have taken to prepare their manuscripts.

We are indebted to Mr. von Hagen and the editorial staff of Springer Verlag for valuable assistance in preparing these proceedings.

Berlin, Mai 1982

H.H. Fernholz and E. Krause

Participants

Aihara, Y.	JAP	Department of Aeronautics, Faculty of Engineering, University of Tokyo, Bunkyo-Ku, Tokyo
Arnal, D.	F	ONERA/CERT, 2, Avenue E. Belin, 31055 Toulouse Cedex
Berg, van den, B.	NL	National Lucht- en Ruimtevaartlab, Anthony Fokkerweg 2, 1059 CM Amsterdam
Bogdonoff, S.M.	USA	Department of Mechanical Aerospace Engineering, D-214 Engineering Quadrangle, Princeton University, Princeton, N.J. 08544
Broberg, L.	S	Division of Mechanics, Chalmers University of Technology, 412 96 Göteborg
Carstensen, C.	FRG	Institut für Schiffs- u. Meerestechnik, Technische Universität Berlin, Straße des 17. Juni 135, 1000 Berlin 12
Cousteix, J.	F	ONERA/CERT, 2, Avenue E. Belin, 31055 Toulouse Cedex
Dallmann, U.	FRG	DFVLR, Bunsenstr. 10, 3400 Göttingen
Dengel, P.	FRG	Hermann-Föttinger-Institut für Thermo- und Fluiddynamik, Technische Universität Berlin, Straße des 17. Juni 135, 1000 Berlin 12
Drougge, G.	S	Aerodynamics Department, The Aeronautical Research Institute of Sweden, P.O. Box 11021, 16111 Bromma
Dumas, R.	F	I.M.S.T., 12, Avenue du Général Leclerc, 13003 Marseille
Elsholz, E.	FRG	VFW Fokker G.m.b.H., Postfach 107845, 2800 Bremen
Fannelop, T.K.	N	Division of Aero and Gasdynamics, NTH, 7034 Trondheim-NTH
Ferguson, A.M.	UK	Glasgow University, Hydrodynamics Laboratory, Acre Road, Glasgow G20 OTL
Fernholz, H.H.	FRG	Hermann-Föttinger-Institut für Thermo- und Fluiddynamik, Technische Universität Berlin, Straße des 17. Juni 135, 1000 Berlin 12

Gessner, F.B.	USA	University of Washington, Department of Mechanical Engineering, FU-10, Seattle, Wash. 98195
Gibson, M.M.	UK	Imperial College of Science and Technology, Dept. of Mechanical Engineering, Exhibition Road, London SW7 2BX
Grandke, T.	FRG	Siemens AG, Forschungslaboratorien, Abt. FL PLAS, G.-Scharowsky-Str. 2, 8520 Erlangen
Hirschel, E.H.	FRG	MBB Flugzeuge, Postfach 801160, 8000 München 80
Hoekstra, M.	NL	MARIN Maritime Research Institute, 2, Haagesteeg, 6700 AA Wageningen
Honami, S.	JAP	Science University of Tokyo, Faculty of Engineering, Dept. of Mechanical Engineering, Kagurazaka, Shinjuku-Ku, Tokyo 162
Hoving, L.	S	FFA The Aeronautical Research Institute of Sweden, Aerodynamics Department, P.O.Box 11021, 161 11 Bromma
Hsieh, T.	USA	Applied Mathematics Branch, R44, Naval Surface Weapons Center, White Oak, Silver Spring, MD 20910
Humphreys, D.A.	S	FFA The Aeronautical Research Institute of Sweden, Aerodynamics Department, P.O.Box 11021, 161 11 Bromma
Ishii, J.	JAP	Science University of Tokyo, Dept. of Mechanical Engineering, Kagurazaka, Shinjuku-Ku, Tokyo 162
Joubert, P.N.	AUS	Department of Mechanical Engineering, University of Melbourne, Parkville, Victoria, 3052
Krause, E.	FRG	Aerodynamisches Institut der Technischen Hochschule Aachen, Wüllnerstr. zw. 5-7, 5100 Aachen
Kux, J.	FRG	Institut für Schiffbau, Universität Hamburg, Lämmersieth 90, 2000 Hamburg 60
Lakshminarayana,B.	USA	The Pennsylvania State University, 153 Hammond Building, University Park, Penna 16802
Larsson, L.	S	Swedish Maritime Research Centre, SSPA, Box 42001, 400 22 Göteborg

Lindhout, J.P.F.	NL	National Aerospace Laboratory NLR, Anthony Fokkerweg 2, 1059 CM Amsterdam
Löfdahl, L.	S	Swedish Maritime Research Centre, SSPA, Box 42001, 400 22 Göteborg
Miyata, M.	JAP	Department Mechanical Engineering, Yamanashi University, Kofu
Nakamura, I.	JAP	Nagoya University, Dept. of Mechanical Engineering, Furocho, Chikusaku, Nagoya
Nakkasyan, A.	CH	Ecole Polytechnique Fédérale de Lausanne, 1015 Lausanne
Okuno, T.	JAP	Department of Naval Architecture, College of Engineering, University of Osaka Prefecture, Mozu, Umemachi, Sakai, Osaka, 591
Patel, V.C.	USA	The University of Iowa, Institute of Hydraulic Research, Iowa City, Iowa 52242
Pierce, F.J.	USA	Virginia Polytechnic Institute and State University, College of Engineering, Dept. of Mechanical Engineering, Blacksburg, VA 24061
Prahlad, T.S.	IN	Aerodynamics Division, Vikram Sarabhai Space Center, Trivandrum - 695 022
Pu, Q.	PRC	Research Associate, Institute of Mechanics, Chinese Academy of Sciences, Zhong Guan Cun, Beijing
Ryhming, I.L.	CH	Ecole Polytechnique Fédérale de Lausanne, Départment de Mécanique, Laboratoire de mécanique des fluides, ME (Ecublens), 1015 Lausanne
Sarohia, S.	UK	Queen Mary College, University of London, Dept. of Aeronautical Engineering, Mile End Road, London E1 4NS
Schneider, G.R.	FRG	DFVLR-AVA-SM-TS, Bunsenstr. 10, 3400 Göttingen
Senoo, Y.	JAP	Research Institute of Industrial Science, Kyushu University, Fukuoka 812
Smith, P.D.	UK	Royal Aircraft Establishment, AE6 Division, Bedford MK41 6AE

Steverding, B.	UK	Aeronautics & Mechanics Branch, European Research Office, United States Army, 223 Old Marylebone Road, London NW1 5TH
Stock, H.W.	FRG	Dornier G.m.b.H., Postfach 1420, 7990 Friedrichshafen
Sutton, E.P.	UK	University Engineering Department, Trumpington Street, Cambridge CB2 1PZ
Tai, T.C.	USA	Code 1606, David Taylor Naval Ship Research and Development Center, Bethesda, Md. 20084
Teng, H.Y.	USA	Princeton University, Princeton, N.J. 08540
Vagt, J.-D.	FRG	Hermann-Föttinger-Institut für Thermo- und Fluiddynamik, Technische Universität Berlin, Straße des 17. Juni 135, 1000 Berlin 12
Wang, K.C.	USA	Department of Aerospace Engineering and Engineering Mechanics, San Diego State University, San Diego, CA 92182
Woods, J.D.	FRG	Institut für Meereskunde, Regionale Ozeanographie, Universität Kiel, Düsternbrooker Weg 20, 2300 Kiel 1
Yamada, H.	JAP	Research Associate, Yamaguchi University, Dept. of Mechanical Engineering, Tokiwadai, Ube
Young, A.D.	UK	Queen Mary College, University of London, Dept. of Aeronautical Engineering, Mile End Road, London E1 4NS
Zhu, Z.	FRG	DFVLR-AVA-SM-TS, Bunsenstr. 10, 3400 Göttingen

Contents

Session 1

Chairman: H.H. Fernholz

B. van den Berg — Some Notes on Three-Dimensional Turbulent Boundary Layer Data and Turbulence Modelling 1

J.-D. Vagt — Experimental Techniques in Three-Dimensional Turbulent Boundary Layers 19

Session 2

Chairman: F.B. Gessner

F.J. Pierce, J.E. McAllister — Measurements in a Pressure-Driven and a Shear-Driven Three-Dimensional Turbulent Boundary Layer 44

T. Okuno, Y. Himeno — Turbulence Measurements in Ship-Model Flow 55

L. Löfdahl, L. Larsson — Measurements of Reynolds-Stress Profiles in the Stern Region of a Ship Model 66

Session 3

Chairman: E.P. Sutton

F.B. Gessner, S.H. Arterberry — A Method of Flow Field Determination for Moderately Skewed, Three-Dimensional Flows 79

K.C. Wang — New Developments in Open Separation ... 94

J. Ishii, S. Honami — Unsteady Structures of a Separated Three-Dimensional Turbulent Boundary Layer 106

T.C. Tai — Effect of Crossflow on the Vortex-Layer-Type Three-Dimensional Flow Separation 117

Session 4

Chairman: G. Drougge

S. Sarohia, A.D. Young — Wind Tunnel Investigations of some Three-Dimensional Separated Turbulent Boundary Layers 126

E.P. Sutton — Experiments on a Flow With Swept Separation and Reattachment of a Boundary Layer 138

Session 5

Chairman: B. van den Berg

Y. Senoo — Three-Dimensional Boundary Layers in Turbomachines 149

B. Lakshminarayana, T.R. Govindan, C. Hah — Experimental Study of the Boundary Layer on a Turbomachinery Rotor Blade 165

I. Nakamura, S. Yamashita — Boundary Layers on Bodies of Revolution Spinning in Axial Flows 177

L. Fulachier, E. Arzoumanian, R. Dumas — Effect on a Developed Turbulent Boundary Layer of a Sudden Local Wall Motion 188

Session 6

Chairman: Y. Senoo

I. Nakamura, M. Miyata, T. Kushida, Y. Kagiya — Some Measurements in the Intermittent Region of a Turbulent Boundary Layer along a Corner 199

Y. Aihara, H. Koyama — Three-Dimensional Turbulent Boundary Layer along a Concave Wall 210

H. Osaka, H. Yamada, I. Nakamura — Three-Dimensional Structure of the Turbulent Wake behind an Intersecting Cruciform Circular Cylinder

Session 7

Chairman: A.D. Young

J.D. Woods — The Three-Dimensional Turbulent Boundary Layer under the Sea Surface *not available at the day of printing*

S.M. Bogdonoff — Three-Dimensional Shock Wave/Turbulent Boundary Layer Interactions: Some Observations and Conjectures 232

T. Hsieh — An Investigation of Surface Flow Pattern and Pressure Distribution for Viscous, Sonic Flow over Hemisphere-Cylinder at Incidence 243

Session 8

Chairman: A.D. Young

Q. Pu — Cross-Hatching - An Interaction between Shock and Turbulent Boundary Layer 254

Session 9

Chairman: T. Fannelop

P.D. Smith — The Numerical Computation of Three-Dimensional Boundary Layers 265

J. Cousteix — Integral Method and Turbulence Models Applied to Three-Dimensional Boundary Layers 286

J.P.F. Lindhout, A.C. de Bruin, J.M. Gooden — A Method for Three-Dimensional Boundary Layer Calculations on Arbitrary Bodies: Some Results on Aircraft Wings and Engine Cowls 298

Session 10

Chairman: J. Cousteix

U.R. Müller — Comparison of Three-Dimensional Turbulent Boundary-Layer Calculations with Experiment (presented by E. Krause) 309

I.L. Ryhming, T.K. Fannelop — A Three-Dimensional Law-of-the-Wall Including Skewness and Roughness Effects 325

D. Arnal, J. Cousteix — Numerical Study of Corner Flows 343

Session 11

Chairman: E. Krause

E.H. Hirschel — Three-Dimensional Boundary-Layer Calculations in Design Aerodynamics ...353

H.W. Stock, W. Seibert — Turbulent Boundary Layers on Three-Dimensional Configurations 366

H.H. Fernholz — An Evaluation of the Final Discussion of the IUTAM Symposium on Three-Dimensional Boundary Layers 381

Some Notes on Three-Dimensional Turbulent Boundary Layer Data and Turbulence Modelling

B. van den Berg

National Aerospace Laboratory NLR, The Netherlands

Summary

The main results of the three-dimensional turbulent boundary layer experiments carried out up to now are discussed. The emphasis is on the direction of the measured turbulent shear stress, which generally differs from the direction of the velocity gradient. In a number of cases (but not all cases) the behaviour can be understood when assuming a large time lag in the downstream development of the turbulence. Turbulence modelling is generally based on the turbulent transport equations. It will be shown that these equations may give large differences between turbulent shear stress and velocity gradient direction, when assuming a negative correlation between the fluctuating static pressure and the excess kinetic energy due to the turbulence.

Introduction

Incompressible three-dimensional turbulent boundary layers will be discussed here. Three-dimensional boundary layers are defined as thin shear layers adjacent to walls, in which the derivatives normal to the wall are large. The amount of work performed in this field, both theoretical and empirical, has much increased during the last decade. A large number of three-dimensional turbulent boundary layer experiments have been carried out now. An extensive review was published in 1976 by Johnston (ref. 1). In this paper only those experiments have been considered, which include the measurement of the two important turbulent shear stress components parallel to the wall. This is because the relation with turbulence modelling is more direct in these cases. The selection of the experiments employed for the analysis in the present paper, is mainly based on a recent evaluation of the experiments as test cases for calculation methods. This evaluation was performed for the Stanford Conference on "Complex turbulent flows" (ref. 2)[x].

(x) The evaluation of three-dimensional turbulent boundary layer experiments was carried by D.A. Humphreys and the present author, and was the starting point of this paper. Thanks are due to Dave Humphreys for his part in that work and, therefore, to his contribution to this paper.

The experimental results will not be discussed in great detail here. Emphasis will be laid on some specific features of the three-dimensional turbulent boundary layer. Actually most attention will be paid to the measured direction of the turbulent shear stress. This does not mean that it is believed that the turbulent shear stress direction is more important for the prediction of the flow development than the shear stress magnitude. Both play an equivalent role, of course.

Following the experimental data analysis, the turbulent transport equations will be studied. These equations form the basis of most turbulence models. One of the major problems is the modelling of the so-called pressure-strain term in these equations. Attention will be focussed on this term and its probable properties in three-dimensional boundary layers. A new model for the pressure-strain term will be proposed. The purpose is to explain the measurement results obtained in the three-dimensional turbulent boundary layers.

Experimental data

In three-dimensional boundary layers both the velocity magnitude, Q, and direction, β, vary with the wall distance, Y. At the boundary layer edge $Q = Q_e$, $\beta = 0$ and $Y = \delta$. The angle α is the external flow angle relative to the fixed x-coordinate. Figure 1 shows a typical polar velocity plot and illustrates the various definitions used here.

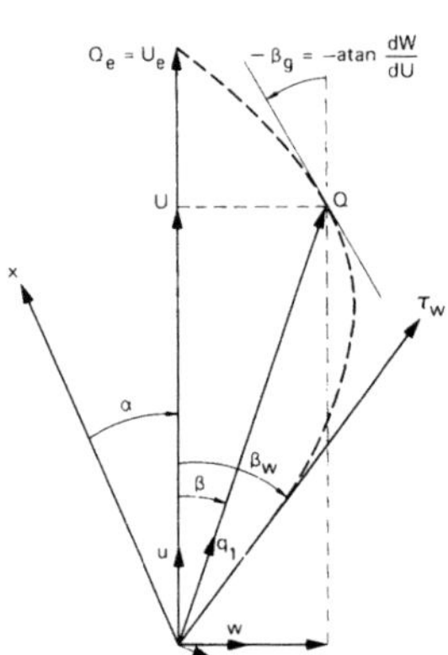

Fig. 1 Polar plot illustrating the symbols employed

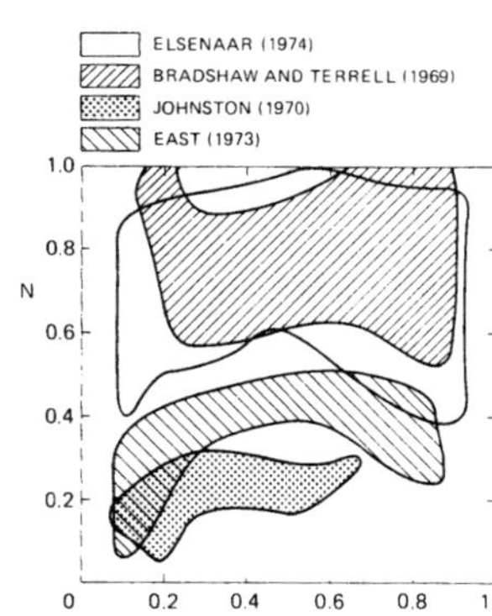

Fig. 2 Ratio of local streamwise to local cross-flow eddy viscosity N. From Johnston (1976)

The turbulent velocity fluctuations may be defined in the external flow coordinate system, u, v, w, or in the local flow coordinate system, q_1, q_2, q_3, where $v = q_2$ is the velocity fluctuation normal to the wall. The turbulent shear stress angle β_τ = atan $(\overline{wv}/\overline{uv})$. In laminar boundary layers the shear stress angle coincides with the velocity gradient angle, $\beta_g = (\delta W/\delta Y)/(\delta U/\delta Y) = dW/dU$. One of the more striking measurement results in three-dimensional turbulent boundary layers is that these angles do not generally coincide. This result can be expressed in a non-isotropy ratio $N = \nu_{e_3} / \nu_{e_1}$, where ν_{e_1} and ν_{e_3} represent the eddy viscosity parallel and normal to the local flow direction respectively:

$$(-\overline{q_\alpha q_2}) = \nu_{e_\alpha} \quad \delta Q_\alpha/\delta Y \qquad\qquad \alpha = 1\ ,\ 3$$

When $\beta_\tau = \beta_g$, then $N = 1$. The values found in some experiments are indicated in figure 2 (from ref. 1). It is evident from this graph that in all these cases $N < 1$, sometimes $N << 1$, which implies a comparatively small turbulent shear stress in cross-flow direction.

It is not only the measured shear stress direction, which deviates from expections based on simple assumptions. Also the turbulent shear stress magnitude may differ considerably from what one would expect on the basis of available knowledge about two-dimensional turbulent boundary layers. As an example the data from the Van den Berg/Elsenaar experiment (ref. 3 and 4) will be considered. This is an infinite swept wing flow, where the boundary layer develops from a two-dimensional zero-pressure-gradient layer to three-dimensional separation. Figure 3 shows the wall streamlines visualized by oil flow and the positions of the measuring stations. An eddy viscosity, ν_e, can be calculated from the magnitude of the measured turbulent shear stresses and velocity gradients. The results are plotted in figure 4. The eddy viscosity values have been non-dimensionalized in the usual way with the velocity at the edge and the displacement thickness (here $\delta^* = \int(1 - U/Q_e)dy$). The value normally assumed for two-dimensional boundary layers is also indicated in the figure. The measured eddy viscosities in the three-dimensional boundary layer are seen to be substantially smaller than that value. This means that the turbulent shear stress magnitude is comparatively small in the three-dimensional boundary layer.

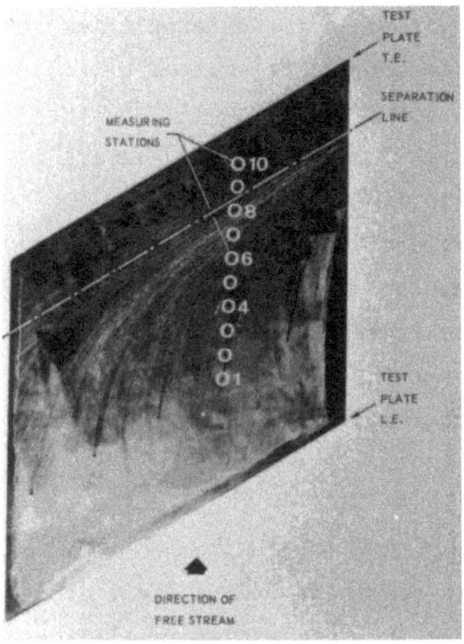

Fig. 3 Van den Berg/Elsenaar experiment. Oil flow pattern and measuring stations

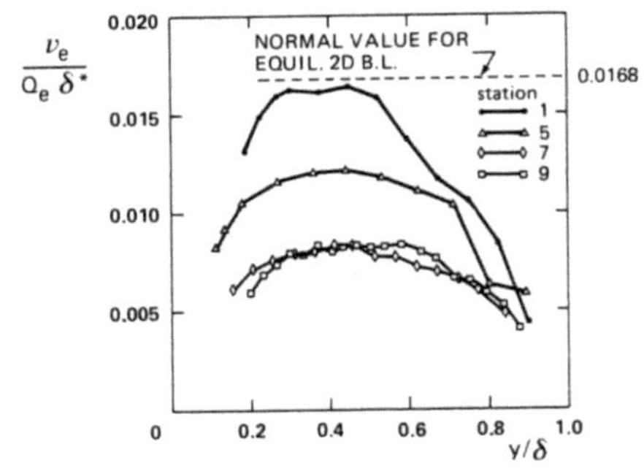

Fig. 4 Van den Berg/Elsenaar experiment. Eddy viscosities deduced from test data

Before drawing definitive conclusions from these experimental results, the accuracy of the turbulent stress measurements has to be discussed. In all experiments the turbulence data were obtained from hot-wire measurements. When performing an error analysis for such measurements, one generally ends up with a large possible error for the turbulence quantities (e.g. ref. 4). The probable error may be expected to be essentially less, but uncertainty about the attained data accuracy remains. However, a check is possible, as redundant information is in principle available when both mean velocities and shear stresses are measured. Such a check has been carried out for the Van den Berg/Elsenaar experiment (ref. 5). The outcome is shown in figure 5

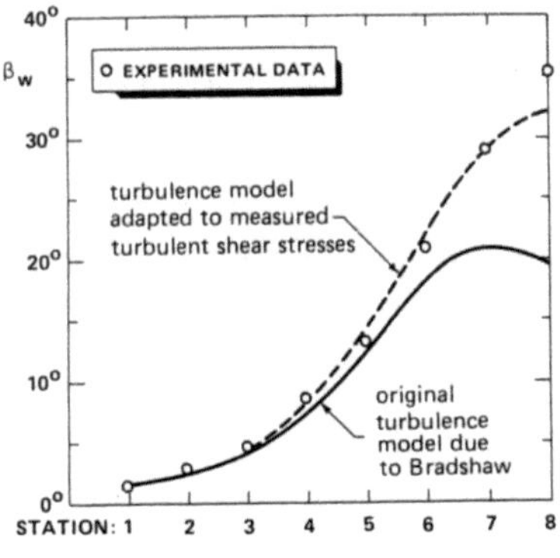

Fig. 5 Van den Berg/Elsenaar experiment. Calculation results with original turbulence model and with model adapted to test data

for the wall angle, β_w. Calculations were carried out using a three-dimensional version of Bradshaw's turbulence model. Figure 5 shows that the results deviate substantially from the experimental data. Subsequently the original turbulence model was so adapted that the measured turbulent shear stress directions and magnitudes were reproduced in the calculations. It is seen that in that case the experimental wall angles are well predicted. The same holds for the other mean flow quantities. It is concluded that the accuracy of the turbulent shear stress measurements, at least in this experiment, is satisfactory.

In the further analysis of the experimental results the role of history effects will be stressed. It is well known that turbulent flows only very slowly forget their initial conditions. The possible importance of history effects for the three-dimensional turbulent boundary layer development was noted earlier (e.g. ref. 1). It will be investigated more extensively here.

First the Van den Berg/Elsenaar experiment will be considered again. Figure 6 gives the direction of the velocity, β, the velocity gradient, β_g, and turbulent shear stress, β_τ, for a number of measuring stations. As mentioned earlier, the velocity gradient and shear stress direction do not generally coincide. The notion that they should approximately coincide comes from the assumption that the production of turbulent shear stress is in the direction of the velocity gradient and the dissipation is in the direction of the shear stress,

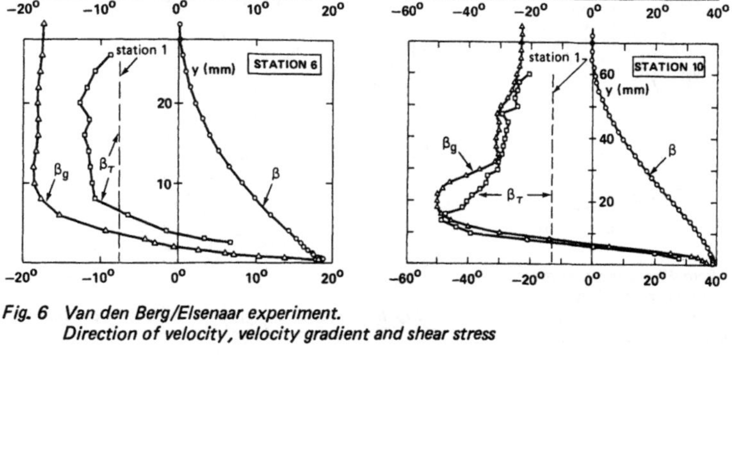

Fig. 6 Van den Berg/Elsenaar experiment.
Direction of velocity, velocity gradient and shear stress

and that these are the mechanisms dominating the turbulence development. At the initial station 1, the boundary layer is nearly two-dimensional: $\beta_\tau \simeq \beta_g \simeq \beta$. In figure 6 the shear stress direction at station 1 is indicated in the plots for the other stations (note that the direction is given relative to the external flow direction at the station considered). It appears that at station 4 the turbulent shear stress direction in the outer region of the boundary layer differs scarcely from that at station 1. This suggests that the turbulent boundary layer in the outer region is not yet affected by the fact that the flow has become three-dimensional. Down-stream of station 4 the shear stress direction gradually changes, becoming nearly equal to the velocity gradient direction at station 10, i.e. behind the three-dimensional separation line (figure 3).

In principle history effects should be studied along streamlines. For an infinite swept wing (with sweep angle Λ) it is easy to prove that for a streamline the following quantity remains constant:

$$Y^* = (1/Q_{ref} \cos\Lambda) \int_0 Q\cos(\Lambda + \alpha + \beta)\, dy$$

A plot against Y^* is shown in figure 7 for the angles relative to the fixed x-coordinate. In figure 8 the turbulent shear stress magnitude, $|\tau|$, has been plotted against Y^* for station 1, 4 and 7. It appears that also the shear stress magnitude only changes slowly. At station 7 the turbulent shear stress magnitude is much nearer the value at station 1 than the value expected for a corresponding equilibrium boundary layer. These results can all be explained very well by assuming that history effects play a dominating role in the development of the turbulence.

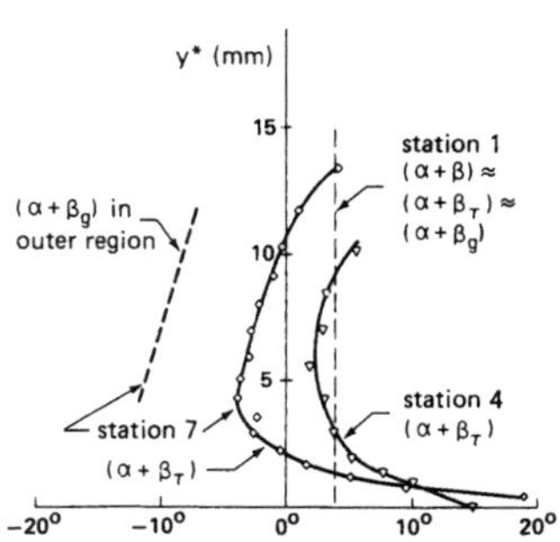

Fig. 7 Van den Berg/Elsenaar experiment. Variation along streamlines (y^* = constant) of shear stress direction relative to fixed reference axis

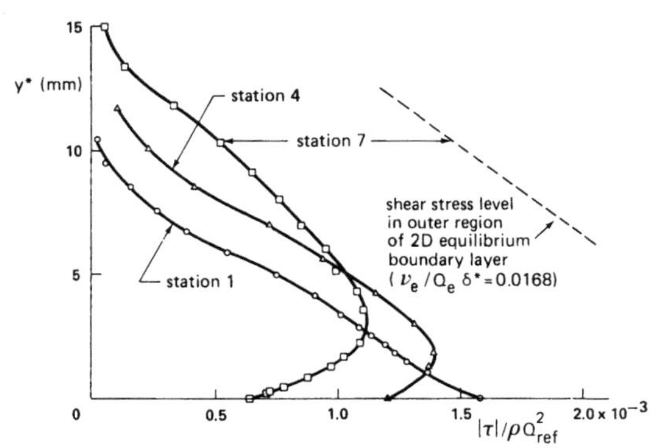

Fig. 8 Variation along streamlines (y^* = constant) of shear stress magnitude

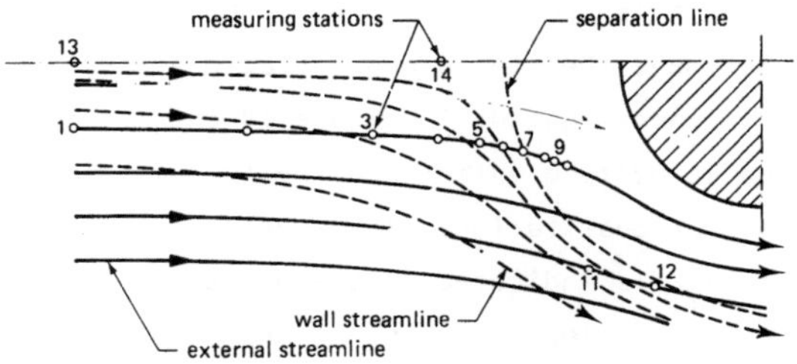

Fig. 9 Dechow/Felsch experiment. Streamlines and measuring stations

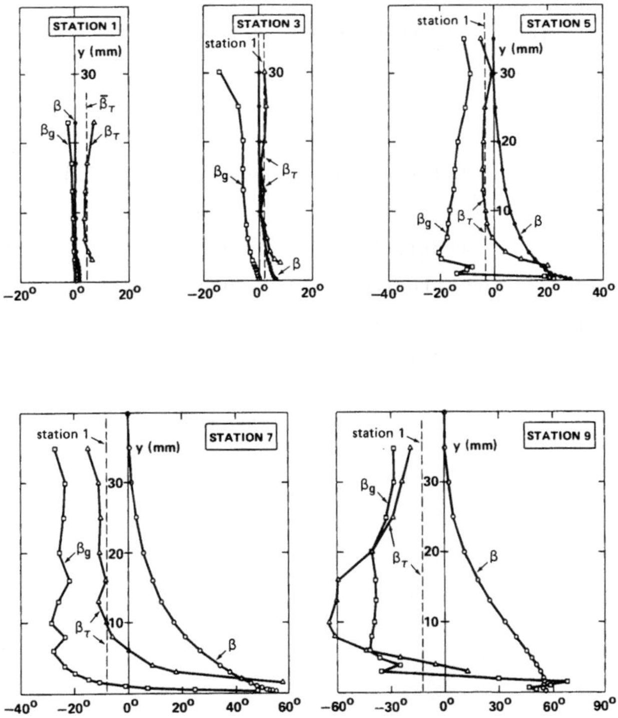

Fig. 10 Dechow/Felsch experiment. Direction of velocity, velocity gradient and shear stress

An even more striking example is provided by the Dechow/Felsch experiment (ref. 6, 7). In this experiment the flow along a flat plate upstream of a cylinder (see figure 9) is investigated. In figure 10 the measured direction of the velocity, the velocity gradient and the turbulent shear stress has been plotted again at various stations. The average shear stress angle at the initial station 1, $\overline{\beta}_{\tau}$, is indicated in the plots for the other stations (relative to the external flow direction at that station). It is evident from figure 10 that up to station 7, i.e. up to the separation line, scarcely any change in turbulent shear stress direction occurs in the outer region of the boundary layer. The reason that history effects are so dominant here is the short region actually covered by the whole flow development: the distance between station 1 and 7 is less than 20 initial boundary layer thicknesses, while significant three-dimensionality only develops downstream of station 3. For comparison: in the Van den Berg/Elsenaar experiment the distance between the initial station and separation is about 60 initial boundary layer thicknesses.

Johnston's experiment of a turbulent boundary layer on a flat plate in front of a swept step (ref. 8) also concerns a flow developing over a very short distance (∿ 5 initial boundary layer thicknesses). As seen in figure 11, not much change in the turbulent shear stress direction is found again in the boundary layer outer region between the initial station and the one close in front of the step.

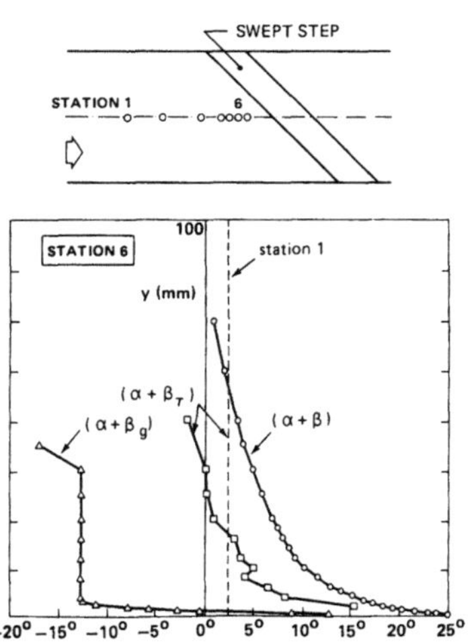

Fig. 11 Johnston experiment.
Direction of velocity, velocity gradient and shear stress

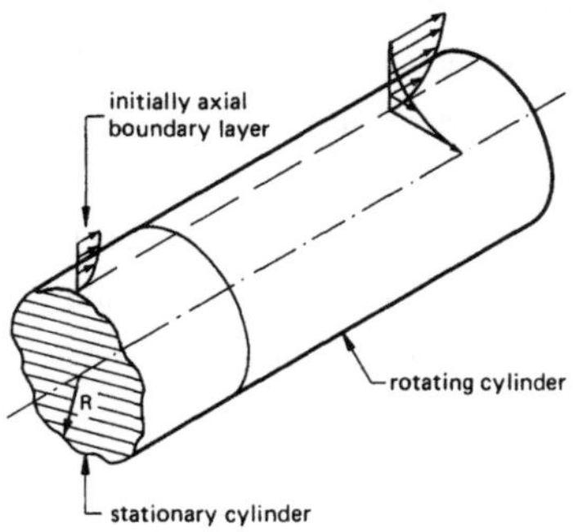

Fig. 12 Rotating cylinder experiment. (Lohmann, Bissonette et al, Fulachier et al)

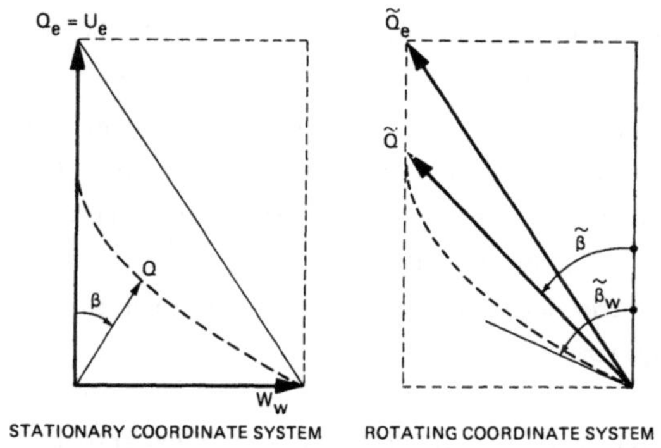

Fig. 13 Polar plot in stationary and rotating coordinate system

Up to now pressure-driven flows have been considered, in which large history effects for the turbulence are more likely. In shear-driven flows things may be different. A typical case is the flow along a rotating cylinder with a stationary front section, as sketched in figure 12. The initially two-dimensional boundary layer is skewed by the abrupt transverse wall motion. The effect of the transverse wall motion is evident first in a thin layer near the wall, which grows downstream. The flow may be viewed from a stationary coordinate system or from a coordinate system attached to the rotating cylinder, as illustrated in figure 13. In the rotating coordinate system the boundary layer is initially three-dimensional and decays to a two-dimensional boundary layer. The flow has been studied by several investigators (ref. 9, 10, 11). Figure 14 gives the measured direction of velocity, velocity gradient and turbulent shear stress in one of the ex-

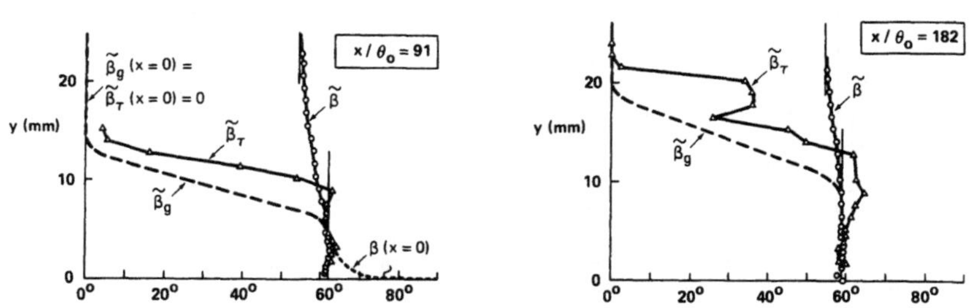

Fig. 14 Lohmann rotating cylinder experiment ($W_w/U_e = 1.41$). Direction of velocity, velocity gradient and shear stress

periments (Lohmann, ref.9). The velocity gradient and shear stress direction are seen to differ again, but here the difference can not be explained by history effects: the turbulent shear stress angle does not lag but leads the velocity gradient angle. In figure 14 the flow is viewed from the rotating and not the stationary coordinate system. This affects the velocity angles, but not the other angles. Note that this means that the magnitude of the non-isotropy ratio N, defined earlier, depends on the coordinate system used.

Figure 15 and 16 show two other experiments, in which it was found that the turbulent shear stress angle leads instead of lags the velocity gradient angle. Both are pressure-driven flows in contrast with the case just discussed. The Müller/Krause experiment (ref. 12, 13) is probably affected by the crude way of tripping the boundary layer, which appears to have led to an uncommon high turbulent shear stress level at the initial station 1. This does not explain, however, why the history effects found in similar experiments (Van den Berg/Elsenaar, Johnston) are not observed here. The Fernholz/Vagt experiment (ref. 14, 15) concerns a boundary layer, which first approaches separation and then only becomes three-dimensional. It may be that so close to separation other mechanisms (extra rates of strain) are important and affect the turbulent shear stress angle. A shear stress angle lead instead of lag was also found by Pierce and Ezweke in a similar situation close to separation (ref. 1).

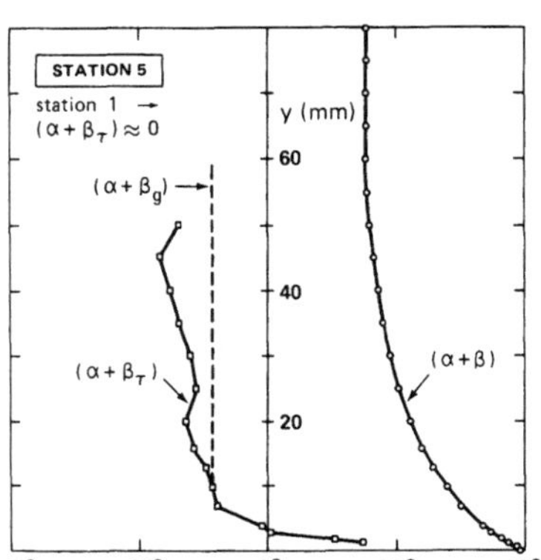

Fig. 15 Müller/Krause experiment. Direction of velocity, velocity gradient and shear stress

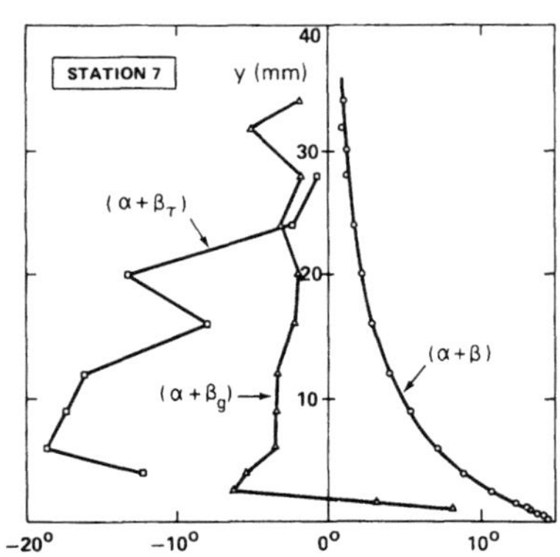

Fig. 16 Fernholz/Vagt experiment. Direction of velocity, velocity gradient and shear stress

It is concluded from the foregoing that in a number of cases there is strong evidence that history effects dominate the turbulence development in the outer region of the boundary layer. However, sufficient exceptions were found to avoid the suggestion that this is the only important mechanism in three-dimensional turbulent boundary layers.

Turbulence modelling

The purpose of experiments is to provide information to check and improve calculation methods and particularly the empirical assumptions made, which are here usually contained in the turbulence model. The basis of most turbulence models are the turbulence transport equation. Such equations may be written for all turbulent stress components, but attention will be focussed here on the equations for the shear stress components parallel to the wall, $\overline{q_1 q_3}$ and $\overline{q_3 q_2}$. For incompressible three-dimensional thin shear layers they read:

$$\frac{D}{Dt}(-\overline{q_\alpha q_2}) = P_{\alpha,2} - \phi_{\alpha,2} + \Delta_{\alpha,2} \qquad \alpha = 1,3$$

$$P_{\alpha,2} = \overline{q_2^2}\,\frac{\delta Q_\alpha}{\delta x_2} \qquad \text{(production term)}$$

$$\phi_{\alpha,2} = \overline{\frac{p}{\rho}\,\frac{\delta q_\alpha}{\delta x_2}} + \overline{\frac{p}{\rho}\,\frac{\delta q_2}{\delta x_\alpha}} \qquad \text{(pressure-strain term)}$$

$$\Delta_{\alpha,2} = \nu\,\frac{\delta^2}{\delta x_2^2}(-\overline{q_\alpha q_2}) + \frac{\delta}{\delta x_2}(\overline{q_\alpha q_2^2}) + \frac{\delta}{\delta x_2}(\overline{\frac{p}{\rho} q_2}) \qquad \text{(diffusion term)}$$

Following Lumley (ref. 16) one may write alternatively:

$$\phi_{\alpha,2} = -\overline{\frac{q_\alpha}{\rho}\,\frac{\delta p}{\delta x_2}} - \overline{\frac{q_2}{\rho}\,\frac{\delta p}{\delta x_\alpha}}$$

$$\Delta_{\alpha,2} = \nu\,\frac{\delta^2}{\delta x_2^2}(-\overline{q_\alpha q_2}) + \frac{\delta}{\delta x_2}(\overline{q_\alpha q_2^2})$$

The latter expression for $\Delta_{\alpha,2}$ may be preferable as it corresponds directly with what one expects for the transport of a transferable quantity in a turbulent flow: $\overline{q_2\,(-q_\alpha q_2)}$. The main reason, however, to introduce here the alternative expressions is that the one for the pressure-strain term $\phi_{\alpha,2}$ will appear convenient later in the analysis.

The term at the left-hand side of the transport equation is the advection term. This is the term which should represent what was called here the history effects. It can be shown, however, that in the thin shear layer approximation assumed here, the advection term (and the diffusion term) are higher-order terms. The advection term in itself can therefore hardly explain the large history effects found in experiment. The two first-order terms are the production term, $P_{\alpha,2}$, and the pressure-strain term, $\phi_{\alpha,2}$. It is the latter term which poses the problems. It involves a correlation of turbulence quantities, which is difficult to grasp, and also direct measurements of the term have not yet been successfully performed. In the following attention will be focussed on this term. The purpose will be here to explain the experimental findings in three-dimensional turbulent boundary layers from the features of the pressure-strain term, $\phi_{\alpha,2}$.

Conventionally the first expression given for $\phi_{\alpha,2}$ is used as the starting point for further analysis. The explicit appearance of the pressure, p, in the correlation is eliminated by taking the divergence of the equations of motion, which leads to a Poisson equation (see e.g. ref. 17). The pressure-strain term can then be rewritten as a volume integral with two terms, one containing only turbulent velocity fluctuations and one containing also the mean velocity gradients. Accordingly one generally distinguishes two contributions to the pressure strain term:

$$\phi_{\alpha,2} = \phi_{\alpha,2}^{(1)} + \phi_{\alpha,2}^{(2)}$$

which will be called here the turbulent strain component and the mean strain component respectively. For the modelling of the turbulent strain component one nearly exclusively has adopted Rotta's early proposal (ref. 17) which reads for the shear stress $(-\overline{q_\alpha q_2})$:

$$\phi_{\alpha,2}^{(1)} \sim (-\overline{q_\alpha q_2})$$

This represents the dissipation in the shear stress transport equations. For the modelling of the mean strain component there is more variety in proposed forms. The most common are the so-called quasi-isotropic form and the following more simple expression:

$$\phi_{\alpha,2}^{(2)} = \gamma\, \overline{q_2^2}\, \frac{\delta Q}{\delta x_2} \qquad (\gamma = \text{constant})$$

Actually this is also the dominant term of the quasi-isotropic form (ref. 18).

The mean strain component as modelled above represents a negative production. The resultant production is still in the direction of the velocity gradient:

$$(1 - \gamma)\ \overline{q_2^2}\ \frac{\delta Q_\alpha}{\delta x_2}$$

The dissipation is in the direction of the shear stress. Consequently the shear stress and velocity gradient direction must coincide, when neglecting the (in principle) small effects of the higher-order terms. One should expect that the same holds approximately for the quasi-isotropic form for $\phi_{\alpha,2}^{(2)}$, when the expression used here is indeed its dominant term. Consequently it seems not likely that with these models large differences between turbulent shear stress and velocity gradient direction will be predicted.

Speculating on the correlation between the turbulent velocity fluctuations and pressure fluctuations, it seems natural to assume that when the momentary kinetic energy is high, the potential energy stored in the pressure will be low in general. Of course, the Bernoulli equation, which expresses the constancy of these energies in inviscid irrotational flows, does not hold here, but it is likely that still a pressure increase will be associated relative often with a velocity decrease (see ref. 19 for a more detailed discussion). The momentary kinetic energy may be written:

$$\tfrac{1}{2}\rho\,(Q_i + q_i)^2 \simeq \tfrac{1}{2}\rho Q_i^{\,2} + \rho Q_i q_i$$

$i = 1, 2, 3$ (summation convention)

The excess kinetic energy due to turbulence, contained in the last term, will be assumed here to be anti-correlated on the average with the pressure. This leads to the assumption:

$$p = -\beta\rho Q_i q_i \qquad (\beta = \text{constant},\ 0 < \beta < 1)$$

To study the consequenses of this assumption for the pressure-strain term, it is convenient to employ the alternative expression given for this term at the beginning of this section. Substitution of the proposed relation for p then leads straightforward to the following equation:

$$\phi_{\alpha,2} = \beta Q_i \left\{ \overline{q_\alpha \frac{\delta q_i}{\delta x_2}} + \overline{q_2 \frac{\delta q_i}{\delta x_\alpha}} \right\} + \beta \left\{ \overline{q_\alpha q_1} \frac{\delta Q_1}{\delta x_2} + \overline{q_\alpha q_3} \frac{\delta Q_3}{\delta x_2} \right\} \tag{1}$$

The first term may be identified as the turbulent strain component $\phi_{\alpha,2}$ and the second term as the mean strain component $\phi_{\alpha,2}^{(2)}$. Again the mean strain component represents a negative production. The resultant production is written in full below for the shear stress components $(-\overline{q_1 q_2})$ and $(-\overline{q_3 q_2})$.

$(-\overline{q_1 q_2})$:

$$(\overline{q_2^2} - \beta\overline{q_1^2}) \frac{\delta Q_1}{\delta x_2} - \beta\overline{q_1 q_3} \frac{\delta Q_3}{\delta x_2}$$

$(-\overline{q_3 q_2})$:

$$(\overline{q_2^2} - \beta\overline{q_3^2}) \frac{\delta Q_3}{\delta x_2} - \beta\overline{q_1 q_3} \frac{\delta Q_1}{\delta x_2}$$

In two-dimensional boundary layers the resultant production is identical to the result obtained with the common $\phi_{\alpha,2}^{(2)}$-model when $\beta\overline{q_1^2} = \gamma\overline{q_2^2}$ (flow in x_1-direction). Due to the last term, however, essential changes occur in three-dimensional boundary layers. From experiment it is known (e.g. ref. 4) that generally $\overline{q_1 q_3} > 0$ when $\delta Q_3/\delta x_2 > 0$. In that case the last term leads to a reduced production in three-dimensional boundary layers. In weakly three-dimensional flows, where $\delta Q_3/\delta x_2 << \delta Q_1/\delta x_2$, particularly the production of the cross-flow component $(-\overline{q_3 q_2})$ is affected, since the last term is relatively large then. The new model for $\phi_{\alpha,2}^{(2)}$ apparently can lead to a production of shear stress in a direction significantly different from the velocity gradient direction and so to significantly different shear stress and velocity gradient angles. Note that these differences are brought about by first-order terms in the transport equations.

As argued above, the last terms in the expressions for the resultant production tend to lead to a comparatively small cross-flow component of the turbulent shear stress, which implies a shear stress angle, β_τ, between the flow angle, β, and the velocity gradient angle, β_g. The actual situation is more complicated, however, as the first terms differ also since normally $\overline{q_1^2} \neq \overline{q_3^2}$. In a more detailed analysis it was found that the direction and magnitude of the resultant shear stress production depends strongly on the relative magnitude of $\overline{q_1^2}$, $\overline{q_2^2}$, $\overline{q_3^2}$ and $\overline{q_1 q_3}$. Considering the fact that the result depends so much on the balance of various terms in the transport equations, it is conceivable that the role of the advection terms, i.e. the history effects, becomes disproportional important. A direct conclusion about the role of history effects does not follow, however, from the present considerations.

No attempt has been made here to apply the proposed expression for $\phi_{\alpha,2}^{(2)}$ in a new turbulence model. Such a model should in any case comprise also accurate predictions of the other stress components, $\overline{q_1^2}$, etc., and consequently would become complicated. Furthermore it is believed that the proposed expression for $\phi_{\alpha,2}^{(2)}$ represents one (rather obvious) feature of the pressure fluctuations, but that reality is much more complex. The aim has been here to make it clear that, when considering the transport equations for the turbulent shear stress, large differences in the direction of the shear stress and velocity gradient can be expected on the basis of the proposed expression for $\phi_{\alpha,2}^{(2)}$.

Up to now attention was focussed on the turbulent transport equations. Turbulence modelling may also be studied from a more simple phenomenological point of view. Looking at turbulent boundary layers visualized for instance by smoke as in figure 17, the presence of large eddies is clearly evident.

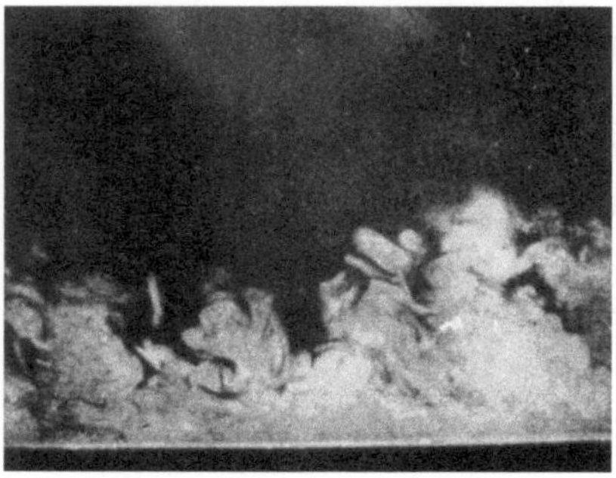

Fig. 17 Turbulent boundary layer flow visualized by means of smoke

It is generally accepted that these large eddies play an important role in the turbulent flow development and also that they contribute essentially to the turbulent shear stress. The life time of an eddy is proportional to its size. The important role of large eddies therefore implies large history effects. Furthermore the skewing of the mean velocity profiles in three-dimensional boundary layers will affect particularly the structure of the large eddies. There is a growing experimental evidence that the large eddy structures show a fair amount of coherence. One may wonder what will happen with these coherent structures, when the boundary layer becomes three-dimensional. Consequently, considering the role of large eddies, both large history effects and changes in turbulence structure due to the three- dimensionality of the flow should be expected.

The turbulent flow has been described above from two view points, which seem more different than they should be. The turbulence transport equations are basically exact, but they describe only local flow properties. The relation with large eddies is difficult to make. The role of the large eddies enters into the solution of the equations via the boundary conditions at the wall and boundary layer edge. Consequently knowledge about large eddy structures can not be easily translated in better assumptions for the appropriate terms in the transport equations.

Concluding remarks

Experimental data indicate that both the turbulent shear stress magnitude and direction are considerably affected by the three-dimensionality of the flow. A more detailed analysis of the data suggests that in a number of experiments history effects dominate the turbulence development in the outer region of the boundary layer.

An investigation of possible features of the pressure-strain term in the turbulence transport equations has led to the conclusion that the large effects found in three-dimensional boundary layers can be explained, when assuming for the modelling of that term a negative correlation between the fluctuating static pressure and kinetic energy.

For further progress it seems essential that the physics of the flow is better understood. This means that in future three-dimensional turbulent boundary layer experiments one should aim at more information about the turbulence than that contained in the turbulent stress tensor alone.

References

1. Johnston, J.P.: Experimental studies in three-dimensional turbulent boundary layers. Stanford Un., Thermosc. Div. Report MD-34 (1976).

2. Humphrey, D.A.; van den Berg, B.: Evaluation of flow 0250: the three-dimensional turbulent boundary layer. In "Complex turbulent flows; comparison of computation and experiments". The 1980-1981 AFSOR-HTTM Stanford Conference.

3. van den Berg, B.; Elsenaar, A.: "Measurements in a three-dimensional incompressible turbulent boundary layer under infinite swept wing conditions". NLR TR 72092 U (1972).

4. Elsenaar, A.E.; Boelsma, S.H.: "Measurements of the Reynolds stress tensor in a three-dimensional turbulent boundary layer under infinite swept wing conditions"; NLR TR 74095 U (1974).

5. Elsenaar, A.; van den Berg, B.; Lindhout, J.P.F.: "Three-dimensional separation of an incompressible turbulent boundary layer on an infinite swept wing". AGARD Conf. Proc. No. 168 (1975).

6. Dechow, R.: "Mittlere Geschwindigkeit und Reynoldsscher Spannungstensor in der dreidimensionalen turbulenten Wandgrenzschicht vor einem stehendem Zylinder". Un. Karlsruhe, Strömungsmechanik und Strömungsmaschinen, Heft 21 (1977).

7. Dechow, R.; Felsch, K.O.: "Measurements of the mean velocity and of the Reynolds stress tensor in a three-dimensional turbulent boundary layer induced by a cylinder standing on a flat wall". Proc. 1977 Turb. Shear Flow Symp., Pennsylvania.

8. Johnston, J.P.: "Measurements in a three-dimensional turbulent boundary layer induced by a swept forward-facing step". J. Fluid Mech., Vol. 42, p. 823 (1970).

9. Lohmann, R.P.: "The response of a developed turbulent boundary layer to local transverse surface motion". J. Fluids Eng., Vol. 98, p. 354 (1979).

10. Fulachier, L.; Arzoumanian, E.; Dumas, R.: "Effect of a sudden discontinuity of the strain at the wall of a rotating cylinder; close-to-wall region of the boundary layer". IUTAM Symp. "Three-dimensional turbulent boundary layers", Berlin, 1982.

11. Bissonette, L.R.; Mellor, G.L;: "Experiments on the behaviour of an axisymmetric turbulent boundary layer with a sudden circumferential strain". J. Fluid Mech., Vol. 63, p. 369 (1974).

12. Müller, U.: "Messung von Reynoldsschen Spannungen und zeitlich gemittelten Geswindigkeiten in einer dreidimensionalen Grenzschicht mit nichtverschwindenden Druckgradienten". Thesis, Un. Aachen (1979).

13. Müller, U.; Krause, E.: "Measurements of mean velocities and Reynolds stresses in an incompressible three-dimensional turbulent boundary layer". Proc. 1979 Turb. Shear Flow Symp., London.

14. Fernholz, H.H.; Vagt, J.D.; Dziomba, B.; Dengel, P.: "Messungen in einer dreidimensionalen turbulenten Wandgrenzschicht an einer gekrümmten Wand: Tabellarsiche Darstellung der Meszwerte". Techn. Un. Berlin, HFI Institutsbericht 01/78 (1980).

15. Fernholz, H.H.; Vagt, J.D.: "Turbulence measurements in an adverse-pressure-gradient three-dimensional turbulent boundary layer along a circular cylinder". J. Fluid Mech., Vol. 111, p. 233 (1981).

16. Lumley, J.L.: "Pressure-strain correlation". Physics of Fluid, Vol. 18, p. 750 (1975).

17. Rotta, J.: "Statistische Theorie nichthomogener Turbulenz". Zeitschrift für Physik, Vol. 129, p. 37 (1951).

18. Launder, B.E.; Reece, G.J.; Rodi, W.: "Progress in the development of a Reynolds-stress turbulence closure". J. Fluid Mech., Vol. 68, p. 537 (1975).

19. van den Berg, B.: "The pressure-strain term in the turbulence transport equations". NLR Report to be published.

Experimental Techniques in Three-Dimensional Turbulent Boundary Layers

J.-D. VAGT

Hermann-Föttinger-Institut für Thermo- und Fluiddynamik
Technische Universität Berlin
Straße des 17. Juni 135, D-1000 Berlin 12

Summary

Various measuring techniques in three-dimensional turbulent boundary layer research are reviewed, and problems of uncertainty and interaction between probes and flow are discussed. Methods for mean and fluctuating quantities in low speed experimental programs are presented, and influence parameters on readings in the highly turbulent near wall region and possible "errors" are given.

Introduction

It is not the aim of this paper to list all experiments in three-dimensional turbulent boundary layers. For this there are at least three reasons: (1) there are extensive reviews, see for example Johnston (1976), (2) not all experiments are relevant for a discussion of experimental techniques, (3) not in all experiments Reynolds stresses have been measured and can thus be discussed. In 1976 Johnston listed nine "full" experiments and seven more were discussed by Fernholz and Vagt (1981). All of them are experiments with a free-stream velocity lower than 100 m/s (see tables 1 and 2). These tables show the measuring techniques used for mean flow data, Reynolds stresses and wall shear stress. Before we begin with the discussion of mean flow measurement techniques we should look at a sketch of a three-dimensional boundary layer (fig. 1*).

At each wall location three quantities define the mean flow field of a three-dimensional boundary layer:

1. the velocity vector at the edge of the boundary layer $\vec{\bar{U}}_e$ and the pressure gradient,
2. the mean velocity profiles,
3. the wall shear stress in magnitude and direction.

In fig. 2 details are given concerning the coordinate system and the Reynolds stresses in the x,z-plane (see footnote).

---

* Two denotations are given. The mean velocity inside the boundary layer is $\bar{c}$ or $\bar{u}$, for example.

Measurements of flow direction

In the years between 1969 and 1975 mean flow measurements were often performed with pitot-tubes and yaw sensing tubes or combinations of them as two- or three-hole impact tubes, which were also used as pitch-sensing tubes. These yaw and/or pitch sensing tubes were two- or three-hole impact tubes soldered side by side and cut off at the tip with a 70° apex angle, for example Johnston (1970) and East (1974). In general the tip angle varies between 100° and 70°. Typical dimensions of three tube heads in the plane of the tubes range between 1.4 and 2 mm.

One should also mention the cylindrical yaw-tube (Pitot-cylinder) with three holes along the circumference which was used in preliminary tests by Van den Berg and Elsenaar (1972), but it does not seem to be very accurate. Large interference problems especially close to the wall must be expected and indeed were found by the above authors. We think that problems of aerodynamic interference are the main reason for Johnston's phrase still being true today that "good mean velocity data are rare in the wall region ($y^+ \leq 100$)".

Two procedures are possible to determine the direction of the flow within the boundary layer. The first one is to measure the actual pressure difference and to calculate the angle using a calibration chart (fig. 3).

The second one is to rotate the probe at each profile point until zero pressure difference is obtained across the two slanted tubes. This "null-reading" method has been used by the majority of experimenters, for example by Elsenaar and Boelsma (1974), Johnston (1970), Dechow (1977), East and Hoxey (1969), and Vagt and Fernholz (1979).

The first method has the advantage that no rotating device is needed but that a calibration chart has to be developed. The main disadvantage probably is that the flow field may be more strongly disturbed when the tube is not aligned with the local flow velocity vector. This method is therefore only to be recommended if the streamline curvature is weak.

Although the second method promises to be more accurate it seems to be wise to check the results with different probe configurations or better, with completely different methods.

Vagt and Fernholz (1979) compared the signals or results of two hot-wire configurations and of a hot-wire with a cobra probe. The flow field was highly turbulent under the influence of a high pressure gradient and has a strong curvature of the streamlines. The near wall measurements of the flow angle were not only hampered by probe size and response time but also because the cobra probe interfered with the curved flow. The two techniques gave good agreement in the outer region of the boundary layer but differed by more than 30 % in the near wall region (fig. 4). One may deduce from the cobra probe measurements that the alignment of the velocity vector is constant close to the wall whereas the hot-wire shows a monotonic increase of the flow angle towards the wall. It is perhaps interesting to note that interference effects even between hot-wire probes and the flow in the near wall region of a boundary layer can become very large (fig. 5). The flow angle was measured with a single normal hot-wire probe, the stem of which was inclined at 5° to the wall in one case (Johnston's arrangement 1970) and perpendicular in the other. The former probe gives flow angles which show a similar behaviour to that of the cobra probe in fig. 4. Our conclusion is that the stems of both cobra probe and inclined hot-wire probe affect the near-wall flow so strongly, that these measurements are not representative of the actual flow which has a much higher deflection angle.

One should always remember: pressure driven three-dimensional boundary layers are driven by the pressure field imposed on the flow. If this pressure field is disturbed by inserting probes with stems and holders, the flow field *may* react by a deflection of the streamlines due to the probe. This effect may not be so severe in shear driven three-dimensional boundary layers or where the curvature is weak.

Usually, other effects which have rarely been corrected are those of displacement and turbulence, perhaps because these corrections are still a matter of speculation (Chue (1975) and Becker and Brown (1974)). It should be mentioned that yaw meter probes are made of round and flattened tubes, for example Cousteix, Pailhas and Michel (1980).

Nissiriou and Pilatis used a five tube yaw meter. These probes require time-consuming calibration and complicated traverse devices and are larger than two or three-hole yaw or pitch meters. Thus larger aerodynamic interference effects must be expected.

A clever device is the single slanted total head probe used by Sarohia and Young (1982) which has to be rotated around its axis to find the direction of flow.

How accurately can the direction be determined with the above mentioned yaw meters. (The hot-wire method is discussed lateron.)

The sensitivity of the probe depends on the angle between the two chamfered side tubes. It is the larger the smaller the angle. For a 70° probe tip the sensitivity is about $0.04 \cdot \frac{1}{2}\rho\,\bar{U}^2$ per degree. Thus for an airflow under normal atmospheric conditions with $\bar{U} = 10$ m/s this amounts to

0.24 mm water column per degree. To measure 0.5° within the boundary layer the manometer or the transducer must have a resolution of at least 0.1 mm of water. But even then it appears to be rather optimistic to resolve down to 0.5°. It should be mentioned that in the near wall region the turbulence level is usually high and large fluctuations may occur. Thus the readings will be affected by these fluctuations and perhaps viscous dampers or, in the case of a transducer, electronic averaging will be necessary. But even then (Chue (1975)) the uncertainty of the measured yaw or pitch angle due to displacement, Reynolds number and wall proximity effects is probably larger than 1° (without any interference problems).

If a highly-sensitive pressure-transducer or a micromanometer is available readings within ± 0.2° should be possible at the edge of the boundary layer. Krogstad (1979) claimed to be accurate even within ± 0.1°.

There is one extreme "error" estimate by Winkelmann and Melnik (1976). They give an error of 0.1° and 0.2°, respectively for the yaw angle within the boundary layer and on the wall using a specially constructed liquid micromanometer and a microscope to observe the meniscus. Nearly 10 minutes were required to take the measurement near the wall. The question remains how stable the flow conditions were in this experiment.

Flow angle measurements can also be made with single wires or x-wire arrays. Because of the smaller dimensions of the hot wires it is possible to perform measurements closer to the wall ($y \leq 0.3$ mm). Ideally, the x-wire should have a better flow-angle response (Delleur (1966)), but in the case of a boundary layer flow a single straight hot-wire is obviously superior. Moussa and Eskinazi (1975) measured the mean flow direction by using a rotatable slanted wire with the probe stem trailing downstream. They calibrated the probe for all "possible" angles and prepared detailed charts including the flow angles as functions of four mean voltages obtained at different rotation angles of the probe. Bissonnette and Mellor (1974) used a straight hot-wire with the sensor mounted perpendicular to the stem axis. The probe was attached to a driving mechanism which rotated the stem very slowly. Making use of the directional properties of the wires which must be symmetrical, the direction can be calculated from traces on an x,y-plotter for each wall distance. This technique should be applicable even for wall distances smaller than 0.3 mm if no other nonsymmetrical effects occur. Basicly, this method is the same as was used by Rogers and Head (1969), Francis and Pierce (1967) or more recently, by Müller (1979), Hebbar and Melnik (1976), Fernholz and Vagt (1981) and (1979). Differences lie in the evaluation of the flow angle and the probe stem arrangement which turned out to be rather critical as far as interference effects are concerned. Such effects may be responsible for the "existence" of collateral near-wall flow as found by some authors.

The procedure to determine the flow angle is as follows: at a fixed distance from the wall the straight wire is rotated to the "left" and to the "right" until the same value is obtained (by integrating). The flow angle is then evaluated by means of a protractor (bisector method).

Bissonnette and Mellor (1974) and Pierce and Ezekwe (1975) evaluated the flow angles using a continuously rotating hot-wire or hot-film ($\sim$1 degree s^{-1}). Using two sample traces in both directions around the mean flow direction, Bissonnette and Mellor (1974) drew an average curve between the two traces of the filtered signal, which was then used to evaluate the mean flow direction. This technique may have problems with the time constant. Looking at their sample traces it is surprising that firstly the curves are quite smooth and secondly not very flat. The reason may be that the turbulence levels at the measuring station is not very high and that the measurements were not carried out close to the wall. (In this case it is difficult to recalculate the turbulence data into the measuring coordinate system, so no conclusion can be drawn.) An evaluation is possible, however, in the experiment of Pierce and Ezekwe (1975) where the turbulence level $\left(\overline{(u'^2)}^{1/2}/\bar{U}\right)$ is only 20 % for the plane of symmetry near the rear plate and for the first points measured close to the wall.

Very close to the wall where the velocity is very small, Vagt and Fernholz (1979) published sample traces (fig. 6) which look very different from those of Bissonnette and Mellor. The traces are quite flat and very "noisy" although they were filtered. Because of the noise a somewhat different technique had to be applied which made use of the minima of the traces. Very near to the wall the signals were amplified to obtain clear maxima and minima of the curves. For practical application of this method the hot-wires must be inspected carefully under a microscope in order to make sure that the wires are straight, even under operating conditions, parallel to the wall and that the connections between the prong tips and the wire are carefully smoothed and polished after the soldering process (perhaps it is not wise to use commercially available probes, or one should calibrate them very carefully when using this technique). Of course, all single measuring components like the x,y-plotter have to be calibrated, too.

Delleur (1966) stated that single and x-wires may be used to determine the flow direction with an accuracy better than 0.05° as long as the turbulence intensity is below 12 %. Usually the turbulence level is larger than 12 % in the flow region close to the wall which is the most interesting one. From that point of view it is interesting to list the statements of some authors:

Müller, U.	1979	accuracy: ± 0.2° - 0.5° , y > 0.2 mm, wire
Klinksiek, W.F. Pierce, F.J.	1970	uncertainty: ± 0.5°, near floor, film, ± 0.5° boundary layer edge
Francis, G.P. Pierce, F.J.	1967	uncertainty: ± 0.2° overall, wire
Hebbar, K.S. Melnik, W.L.	1976	overall, experimental error: ± 0.125° but ± 0.5° close to the wall, wire
Fernholz, H.H. Vagt, J.-D.	1981	repeatability: ± 1° in the vicinity of the wall, uncertainty: ± 2° at the wall, wire

From that list two important features can be deduced. The first one is that the authors use different expressions for the "reliability" of their data. Following Dean (1953) "A c c u r a c y is the measure of the reliability of an estimate in stating the true value including the effects of accidental errors, systematic errors and mistakes". From that, without going into details, it seems to me to be nearly impossible to give a measure for the accuracy, especially when only o n e observer employs only o n e instrument for a measurement.

Again, following Dean, a n e r r o r is "the difference between an observed and the true value of a variable". The true value is almost always unknown. What then do Hebbar and Melnik (1976) mean with their statement of 0.125° "overall experimental error" in a flow field with a turbulence level of up to 45 % at a wall distance of 0.11 mm (port 1)?

R e p e a t a b i l i t y is related to precision, which is a measure of the accidental errors and mistakes ("personal errors") in a set of measurements by the same observer using the same technique and the same instruments.

U n c e r t a i n t y is "a possible value of the error, or what one thinks the error might be". This is perhaps the only value the experimenter can give. Can we, however, really believe the very low uncertainty values given in many papers or do they stem from an imaginary competition to be better than the neighbour?

The second feature of the above list is that Fernholz and Vagt gave rather high values for repeatability and uncertainty. The reason for this are probably temperature and aerodynamic interference effects. Fig. 7 shows two sample traces, one far away from the wall and the other very close to the wall. A temperature variation of 1 °C during rotation will influence the readings quite differently because of the different shapes of the angle characteristic of a straight hot wire.

The aerodynamic interference effect was discussed at the beginning. The probe used by Francis and Pierce (1967) seems to produce very little disturbances due to aerodynamic interference because of the curved shape of the stem and the prongs (see also Larsen (1974)) but no control experiments were made and so some doubts remain. As far as one can learn from the literature such control tests are not very common. But they are necessary to give a fair estimate of the uncertainty.

Positioning at the wall

The position of the wall relative to a reference point of the traversing mechanism is easily found when pitot tubes or similar probes are traversed. The probe is moved towards the wall such that the lower lip touches the surface without any deflection of the probe body. Many techniques are available which are not difficult to use. However, these techniques cannot be employed for hot-wire traverses because of the danger to destroy the probe. Some of the available techniques were summarized by Vagt (1979) but it should be mentioned that problems still occur when one wishes to measure the distance between the probe and the wall. This is not the case for those workers who assume their first measuring points to lie on the sublayer law or/and on the law of the wall.

One method of measuring the distance from the wall is the "mirror image" method. It can be used if the surface is sufficiently smooth and if there is enough space to mount the optics for the observer. Another method was described by Orlando (1974) who used a "wall stop" soldered to the probe stem to prevent the wire from being accidentally damaged by the wall. The distance of the wire from the wall (when the wall stop makes contact) was measured by an optical comparator.

Vagt and Fernholz (1979) discussed a calibration technique where wire deflections after wall contact were measured.

Pozzerini (1976) applied an electrical method which made use of the sudden drop in cold resistance when the probe tips touch the electrically conducting wall.

Krogstad (1979) used a more elaborate method, a combined electro-optical method.

Reported uncertainties are ± 0.03 mm (Klinksiek and Pierce (1970)), ± 0.05 mm (Müller (1979)), ± 0.05 mm (Pierce and Ezekwe (1975)), ± 0.01 mm (Krogstad (1979)), ± 0.02 mm (van den Berg and Elsenaar (1972)), ± 0.01 mm (Vagt and Fernholz (1979)).

All these uncertainty values may be "right" or "wrong". This is very difficult to judge unless the authors discuss the details of the hot-wire, especially, whether the wire is straight under operation condition or not. Unfortunately published papers often contain long sections on data reduction procedures but give few details about measuring techniques.

Velocity measurements

We now discuss methods to measure the magnitude of the mean velocity vector or its components in a reference system. Most often the magnitude itself is measured with a probe aligned in the direction of the mean flow*.

Pitot-tube traverses have been made by nearly all the experimenters working in three-dimensional boundary layers and often three-tube probes were used for measuring the flow direction and the magnitude (see Chue (1975) for various effects on the readings).

Static pressure measurements were made by a disk-static probe, for example in the case of the Johnston-step-flow where a strong transverse pressure gradient was present, and in the case of Dechow (1977) with a cylindrical body on a plate. The latter author discovered fairly large transverse pressure gradients near the cylinder, too. Since this is not in accordance with the results of East and Hoxey (1969) in a very similar experiment but with a different probe, interference problems may have been the cause. Apparently, Dechow's disk probe had a diameter of 6 mm thus rectifying yaw angles across that length in a boundary layer profile. A detailed discussion of static probe interference effects was also given by East (1974). More recently Vagt and Fernholz (1976) presented measurements of static pressure across the boundary layer on the wall of a circular cylinder with flow parallel to its axis and showed that wall-interference effects exceeded corrections due to turbulence by almost an order of magnitude.

---

* Sometimes there is made a distinction between "mean direction of flow" and "mean flow direction" which was introduced by Rose (1962). See also Hebbar and Melnik (1976) and Vagt and Fernholz (1979).

Thus, many experimenters believe in constant pressure distributions across the boundary layer (except in the case of flow in front of a step) and assume the static pressure to be that measured at the wall tapping.

We have seen that accurate measurements of mean velocity in the region below $y^+ = 20$ are not so easy. Impact tubes of about 0.5 mm diameter are too large for the near wall region and, of course, they are subject to large errors due to displacement effects (a shift of the effective center). As Johnston (1976) pointed out we still have no study of the combined effects of velocity gradient and yaw-angle gradient.

The only other instrument is then the hot wire by which data were obtained down to $y^+ = 2$ or less. The state of the art is well known and surveys were given by Strickert (1974), Vagt (1979), and Blackwelder (1981). But before details are discussed, a quotation by Papavergos and Hedley (1979) should be mentioned:

"Although it is believed that hot-wire (film) anemometer measurements may be taken by semi-skilled personnel, the interpretation of the hot-wire measurements for the mean and turbulent velocity components as well as their statistical properties requires detailed and correct knowledge of its sensitivity."

This is even more important for successful measurements in three-dimensional boundary layers which require great skill and considerably more patience than those in two-dimensional turbulent boundary layers (see also Johnston (1976)).

In most cases today, the constant temperature method is preferred. Since the instrument has to be calibrated, the ambient conditions should be as stable as possible. If there are temperature changes in the fluid, corrections must be applied. But it is perhaps not a good idea to spend a lot of time on developing corrections for contamination due to dust. Here a good filter is the only answer or successive cleaning of the wire.

Mean velocity measurements should be made with single normal wires which must be straight under operation conditions. Straight and well aligned wires under operation conditions are an essential taking measurements in the near wall region. It is not obvious that this rule is heeded in all boundary layer experiments and I would like to recall my earlier remarks (see Dahm and Vagt (1976), Vagt and Dahm (1979)). Hot wires not carefully made m a y well have been a reason for the scatter in flow angle measurement discussed by van den Berg and Elsenaar (1972) below values of $y^+ = 20$.

Another very annoying problem is the additional heat loss due to wall proximity, which depends on several parameters such as Reynolds number, wall material, probe design, acceleration and others (Vagt (1979)). The additional heat loss leads one to overestimate the velocity in the near wall region. In three-dimensional boundary layers this effect was discussed by Hebbar and Melnik (1976) who developed a correction using two-dimensional flow field measurement very close to the wall (down to $y^+ \approx 0.5$, which was approximately 0.01 mm wall distance). To achieve this any vibration of the wall must have been excluded and the wires must have been absolutely straight under operation conditions.

Another problem arises if one takes the mean value of the voltage signal. In the highly turbulent region close to the wall this signal does not only contain information about the mean velocity but also about the turbulent quantities which are added up in different ways. If that is included in a correction chart then not only the same wall material must be used but also the turbulence must be in the same state in the calibration and in the experiment. As a guide the mean velocity given by a single normal hot-wire is overestimated by roughly half the squared transverse turbulent intensity normal to the wire axis. This rule is applicable up to 50 % in longitudinal intensity. It was also suggested by Tutu and Chevray (1975) assuming an initial joint normal probability density function. The additional effect on mean velocity readings produced by turbulence may be smaller if the actual joint p.d.f. in the flow is skewed in the proper direction. It may be larger if the opposite is the case.

It is rather astonishing that the problem of the turbulence effects on hot-wire readings does not seem to exist for many experimenters. But even when it is discussed no corrections are applied, perhaps for two reasons. The first is that the turbulence level is below 20 % (Elsenaar and Boelsma (1974) $y > 2.5$ mm*) or smaller than 30 % (Dechow (1977)). In these cases it was assumed that this level is low enough to be ignored. The second one is that the experimenter feels unable to give corrections because of the very high turbulence level in the flow under investigation. In the mid range the problem seems to have been completely overlooked.

It is often a hard task or impossible to recalculate given data into the measuring co-ordinate system in order to estimate local turbulence levels, to discuss uncertainties or other relevant topics. Furthermore, the presentation of normalized data with unknown normalizing quantities is the best way to prevent a reviewer from discussing measuring techniques and related

* But the mean velocity was measured down to $y = 0.15$ mm.

problems later on.

It is also very rare to find uncertainty values of hot-wire data. Although van den Berg and Elsenaar (1972) discussed this very important topic they did not state their uncertainty. The same is true with Müller (1979), Pierce and Ezekwe (1975) and Fernholz and Vagt (1981) for example. The latter authors only gave an estimate of the turbulence effect, which is of about 10 % near the wall. It must be admitted, however, that it is very difficult to calculate the uncertainty since too many various and contradictory aspects have to be taken into account.

Under these circumstances it is surprising to find Hebbar and Melnik (1976) state an "overall experimental error" of $\pm$ 0.5 % and $\pm$ 1 % for the magnitude of mean flow velocity close to the wall. It should be emphasized again that the turbulence level may have been as high as 45 % (port 1; y = 0.11 mm) in their experiment. The maximum "error" of 2 % given by Dechow (1977) seems to be also rather optimistic. Klinksiek and Pierce (1970) estimated an uncertainty of $\pm$ 4 % for near wall velocity values and $\pm$ 1.5 % at the boundary layer edge. Both values are realistic in my opinion.

Reynolds-stress measurements

Reynolds-stress measurement have been mainly performed by means of hot wires, in a few cases by hot films or a split-film sensor.

Johnston (1976) defined three different methods (I, II and III) to obtain the turbulent stresses by means of hot-wire probes. The first one employs normal and slanted wires with continuous rotation, originally proposed by Fujita and Kovasznay (1968), developed by Bissonnette and Mellor (1974) or Bissonnette (1970), and used by several other experimenters in somewhat modified forms. As already mentioned the short averaging time used for the rotating sensors may cause problems in flows with low frequency variations. The three Reynolds stresses $\overline{u'^2}$, $\overline{w'^2}$ and $\overline{u'w'}$ (see fig. 2; ε very small) are measured by a single normal (straight) wire in the streamwise coordinate system, and the remaining three components $\overline{v'^2}$, $\overline{u'v'}$ and $\overline{v'w'}$ by a rotated slanted wire (see fig. 8). The voltage signals were read from traces on the x,y-plotter and then inserted into a set of simultaneous equations to compute the shear and normal stresses. This was done by a least-square method. (It was obviously hoped that the "accuracy" was improved by that procedure.)

Johnston (1976) pointed out that with this technique interference problems may occur when the slanted wire is in different angular positions. Therefore a modified slanted wire was used vor $\overline{v'w'}$-measurements in the investigation of Fernholz and Vagt (1981) with special bent prongs (see also Vagt and Fernholz (1979)).

This latter method was checked against a technique with fixed x-arrays (method II) by Pierce and Ezekwe (1976). It was found that the rotating sensor-method shows more scatter, and that there is a tendency to produce an occasional specious stress value, especially in the slanted wire output. Nevertheless, agreement between the two methods was judged good. But figure 4 of Pierce and Ezekwe shows discrepancies of 10 % - 15 % in their shear stress and normal stress measurements.

Method II as defined by Johnston (1976) is the most conventional one. It employs normal, slanted and/or x-wires which are aligned in the premeasured flow direction and where the stem and prongs trail downstream. This technique has been used by Müller (1979), Swamy et al. (1978), Dechow (1977), and Elsenaar and Boelsma (1974), for example.

Müller (1979) used very small probes to perform measurements close to the wall. But unfortunately, it is not easy to determine the sensitivities of these probes. Thus very time-consuming calibrations were necessary, especially for the longitudinal cooling effect. Furthermore, a more complex response equation had to be used which could take into account pitch effects of the probe. The probe-holder was of the gooseneck type and similar to those of Johnston (1970) and Dechow (1977).

The method used by Elsenaar & Boelsma (1974) lies between method I and II because the probe holder is protruding through the wall with a perpendicular probe body. The probe tip is located 25 mm in front of the axis of rotation, that is only five diameters from the holder. A compromise had to be made between the probe length and the change of the measuring position due to rotation of the whole assembly when the boundary layer was traversed. It was claimed that the interference effects were small for a probe with a length of 25 mm and that a deviation from the measuring position of at most 5 mm was acceptable.

The third method is the three-wire method. It was used by Zimmerman (1975). The instantaneous signals from the three wires can be resolved into a mean velocity vector and three fluctuation signals as long as the instantaneous direction of flow lies inside the 70° apex angle of the cone formed by the wires. If "point"-measurements are intended the use of such a probe may be dubious because of its rather large size. "Near wall" measurements in a boundary layer flow are naturally much easier to perform if the boundary layer thickness is large. Recently, Frota and Moffat (1979) discussed a triaxial probe in detail. But again no general conclusions interesting for boundary layer flows could be drawn, because the turbulence level of the flow was rather low. In highly turbulent flows large interference effects at the wires may be present due to the prongs.

The uncertainty values sometimes given in published papers seem to be quite strange. It is hard to believe that the rms-value of the longitudinal fluctuation can be measured with an "overall experimental error" of $\pm$ 2 % in regions where the longitudinal intensity is about 40 % - 50 % (single normal wire technique). According to Tutu and Chevray (1975) and Bradbury (1976) the turbulence effect alone produces an error of about 5 % - 9 % in this intensity range. The error is even greater, apart from spatial integration problems, for x-wire arrays.

Furthermore, Fernholz and Vagt (1981) have shown experimentally that the u-wire signal varies with a change in the active wire length. When the wire length was changed from 0.8 mm to 2.8 mm, the intensity signal was changed by - 6 %, although each wire had been calibrated at the same mean temperature. Elsenaar and Boelsma emphasized this problem too by giving "wire length bars" in some of their figures. In their discussion of calibration and data evaluation procedures, they state that the "error" in all measurements of the turbulent stresses (normalized with the friction velocity) is 10 % to the measured value or 0.1 absolute, whichever is the largest. It should be mentioned again that the turbulence level in their flow was quite low, rarely exceeding 20 %.

Fernholz and Vagt (1981) obtained their results with an uncertainty of about 10 %, except for the component $\overline{v'w'}$, but for a flow with a higher turbulence level. The term $\overline{v'w'}$, measured with a slanted wire in different angular positions could only be measured with an "error" (uncertainty) of about $\pm$ 25 %. Perhaps the uncertainty very close to the wall for the other two shear stress components is of the same order because of rectification effects on slanted and x-wires. At an assumed true level of only 30 % in $\sqrt{\overline{u'^2}}/\bar{U}$ the error in $\overline{u'v'}$ according to Tutu and Chevray may be of the order of - 20 %.

Perhaps this magnitude of error is more easily accepted when knowing that comparisons between x-wire and single wire measurements show differences up to 20 %* for dimensionless longitudinal intensity values or up to 45 %* for dimensionless shear stress values $\overline{u'v'}$, and when this result is "marked" good (Cousteix, Pailhas and Michel (1980)).

---

* based on lower values

Very large corrections were introduced by Bissonnette and Mellor (1974) for near wall $\overline{v'w'}$ measurements because of considerable variation of flow angle over the wire length when using the slanted wire technique, although tests in a pipe flow showed that Bissonnette and Mellor's technique permits the determination of stresses to an "accuracy" better than 5 % of the local turbulent energy per unit mass.

Very recently Löfdahl (1982) estimated uncertainties of 10 % for $\overline{v'w'}$- and $\overline{u'v'}$-shear stresses in a thick boundary layer near the stern of a ship model with local turbulent intensities smaller than 20 %. These values seem to be realistic. Due to the "straight" probe measurements the uncertainty of $\overline{u'w'}$-stresses was given at only 4 %.

It should be pointed out again that in flows with a turbulence level as high as 20 % no hot-wire gives necessarily the right answer. In fig. 9 a comparison (Dengel and Vagt, to be published) is shown between measurements carried out by the pulsed-wire technique* (Bradbury and Castro (1971)) and a single normal wire. Differences in the normal stress of about 6 % at a turbulence level of about 20 % measured by the hot-wire and about 12 % at a turbulence level of 32 % may occur. These "errors" are much larger than predicted by Tutu and Chevray (1975). They clearly depend on the p.d.f. and, of course, on the amount of negative instantaneous velocities which cause "rectification" effects. Taking this into account any attempt to improve hot-wire measurements by using higher order correlations in the series expansion may give even worse results. Furthermore, evaluation procedures which avoid series expansion (evaluation of squared signals) are not better (see Bartenwerfer (1981)) either.

Wall shear stress

The last topic to be discussed is how to measure wall shear stress in a three-dimensional turbulent boundary layer. Because of the skewed profile magnitude and direction must be measured. It can be seen from the previous section that it is difficult to measure accurately the flow direction very close to the wall. Even if this were possible the angle of the limiting wall stream line would still be different. The extrapolation procedure discussed by Johnston (1976) may lead to erroneous values for the wall angle. Measuring devices protruding from the wall such as Stanton tubes (Van den Berg and Elsenaar (1972)) or sublayer fences (Vagt and Fernholz (1973)) respond

* The use of that instrument is not restricted to low turbulence flows.

to an average value measured across their height. The only instrument, apart from an omni-directional balance, which can give the direction of the limiting streamline,at least ideally,is the heat meter, a hot-wire lying in the wall. But all these instruments are not easy to use.

Although Van den Berg (1972) was not convinced of his oil flow pattern recognition he repeated in a later experiment (1975) wall angle measurement up to $\pm$ 0.5 degrees which is very encouraging. Unfortunately fig. 21 in the investigation of Dechow (1977) shows a large discrepancy between heat meter and oil-flow pattern data. Surprisingly the oil-flow result is in good agreement with the flow angle at $y = 0.2$ mm. Very recently Pierce and McAllister (1980) compared their shearmeter and oil streak measurements and found excellent agreement.

It is F. J. Pierce and his co-workers to whom practically all experimenters in three-dimensional turbulent boundary layer research must be grateful when measuring wall shear stress by means of indirect measuring methods. All these devices rely on two-dimensional calibrations. If an early comparison between balance and indirect devices by Pierce and Krommenhoek (1968) had not been confirmed by Pierce and McAllister (1980), the employment of Preston-tubes, fences and Clauser-charts would still have no firm basis. But it seems that we have to live with an uncertainty up to 10 % even now. Further study of the latter author's paper is strongly recommended.

Conclusion

It seems that we now have about 15 fully documented experiments in the field of three-dimensional turbulent boundary layers. The question remains whether these experiments are accurate enough for testing calculation procedures. There are still extreme difficulties taking measurements very close to the wall. This is especially true for some Reynolds stresses. Improvements, as far as I can see, are only possible in long range experimental programs. That is basicly the same result given by Johnston (1976) eight years ago.

References

Bartenwerfer, M.: Zur Analyse von Hitzdrahtsignalen turbulenter Strömungen. Thesis, TU Berlin, 1981.

Becker, H.A.; Brown, A.P.G.: Response of Pitot probes in turbulent streams. J. Fluid Mech. 62 (1974) 85-114.

Berg, B. van den; Elsenaar, A.: Measurements in a three-dimensional incompressible turbulent boundary layer in an adverse pressure gradient under infinite swept wind conditions. National Aerospace Lab. NLR TR 72092 U (1972).

Berg, B. van den: A three-dimensional law of the wall for turbulent shear flow. J. Fluid Mech. 70 (1975) 149-160.

Bissonnette, L.R.: An experimental study of the development of a three-dimensional turbulent boundary layer under rapidly changing rate of strain. Ph.D. dissertation, Princeton University, 1970.

Bissonnette, L.R.; Mellor, G.L.: Experiments on the behaviour of an axisymmetric turbulent boundary layer with a sudden circumferential strain. J. Fluid Mech. 63 (1974), 2, 369-413.

Blackwelder, R.F.: Hot-wire and hot-film anemometers. Methods of Exp. Physics: Fluid Dyn., Edited by R.J. Emrich, Vol. 18, Part 17, Academic Press, 1981.

Bradbury, L.J.S.: Measurements with a pulsed wire and hot-wire anemometer in the highly turbulent wake of a normal flat plate. J. Fluid Mech. 77 (1976), 3, 473.

Bradbury, L.J.S.; Castro, I.P.: A pulsed wire technique for velocity measurements in highly turbulent flows. J. Fluid Mech. 49 (1971) 657.

Bradshaw, P.; Terell: The response of a turbulent boundary layer on an infinite swept wing to the sudden removal of pressure gradient. NPL Aero Rept. 1305, 1969.

Chandrasekhara Swamy, N.V.; Gowda Lakshmana, B.H.; Lakshminath, V.R.: Turbulence measurements in the three-dimensional boundary layer on a yawed flat plate at incidence. Z.f.W. 5 (1978) 159.

Chue, S.H.: Pressure probes for fluid measurement. Prog. Aerospace Sci., Vol. 16 (1975) No. 2, 147-223.

Cousteix, J.; Pailhas, G.; Michel, R.: Measurements of mean velocity and Reynolds stress tensor within a wake of a swept wing. Rapport Techn. OA 41/2259 AYD (DERAT 25/5004 DY), 1980.

Dahm, A.; Vagt, J.D.: Herstellung interferenzarmer Hitzdrahtsonden. Internal Report at the Hermann-Föttinger-Institut, Techn. Univ. Berlin, 1976.

Dechow, R.: Mittlere Geschwindigkeit und Reynoldsscher Spannungstensor in der dreidimensionalen turbulenten Wandgrenzschicht vor einem stehenden Zylinder. Diss. Univ. Karlsruhe. Strömungsmaschinen und Strömungsmechanik, Heft 21 (1977).

Delleur, J.W.: Flow direction measurement by hot-wire anemometry. J. of the Engineering Mech. Div., Proc. American Soc. of Civil Eng.92 (1966) 45-71.

Dengel, P.; Vagt, J.D.: A comparison between pulsed-wire and hot-wire measurements in a highly turbulent boundary layer. Hermann-Föttinger-Institut, TU Berlin (1982) to be published.

East, L.F.: Measurements of the three-dimensional incompressible turbulent boundary layer induced on the surface of a slender delta wing by leading edge vortex. Royal Aircraft Establishment, TR 73141, 1974.

East, L.F.; Hoxey, R.P.: Low speed three-dimensional turbulent boundary layer data. RAE Tech. Rep. 69041, 1969.

Elsenaar, A.; Boelsma, S.H.: Measurements of the Reynolds stress tensor in a three-dimensional turbulent boundary layer under infinite swept wing conditions. National Aerospace Lab. NLR TR 74095 U, 1974.

Fernholz, H.H.; Vagt, J.D.: Turbulence measurements in an adverse pressure gradient three-dimensional turbulent boundary layer along a circular cylinder. J. Fluid Mech. 111 (1981) 233-269.

Francis, G.P.; Pierce, F.J.: An experimental study of skewed turbulent boundary layers in low speed flow. J. Basic Eng., Trans. ASME (1967) 597-608.

Frota, M.; Moffat, R.J.: Instrumentation for the measurement of turbulence components in a three-dimensional flow field. AFOSR-TR-79-0957, Dept. Mech. Eng., Stanford Univ., 1979.

Fujita, H.; Kovasznay, S.G.: Measurements of Reynolds stress by a single rotated hot-wire anemometer. Rev. Sci. Instr. 39 (1968) 1351.

Hebbar, K.S.; Melnik, W.L.: Measurements in the near-wall region of a relaxing three-dimensional low speed turbulent air boundary layer. Tech. Rep. AE-71-1, Univ. Maryland, 1976.

Johnston, J.P.: Measurements in a three-dimensional turbulent boundary layer induced by a swept, forward facing step. J. Fluid Mech. 42 (1970) 823-844.

Johnston, J.P.: Experimental studies in three-dimensional turbulent boundary layers. Stanford Univ., Thermosciences Div. Rep. MD-34, 1976.

Klinksiek, W.F.; Pierce, F.J.: Simultaneous lateral skewing in a three-dimensional turbulent boundary layer flow. J. Basic Eng. (1970) 83-92.

Krogstad, P.-A.: Investigation of a three-dimensional turbulent boundary layer driven by simple two-dimensional potential flow. Thesis, Norwegian Inst. of Technology, Trondheim, 1979.

Larsson, L.: Boundary layers of ships. Part III: An experimental investigation of the turbulent boundary layer on a ship model. Rep. No. 74/7, Dep. of Appl. Thermo and Fluid Dyn., Chalmers Univ. of Techn., 1974.

Löfdahl, L.: Measurements of the Reynolds stress tensor in the thick three-dimensional boundary layer near the stern of a ship model. Doktorsavhandling vid Chalmers Tekniska Högskola, 1982.

Lohmann, R.P.: The response of a developed turbulent boundary layer to local transverse surface motion. ASME Paper 76-FE-3, 1976. Also Ph.D. thesis, University of Connecticut, 1973.

Moussa, Z.M.; Eskinazi, S.: Directional mean flow measurements using a single inclined hot-wire. Phys. of Fluids 18 (1975), No. 3, 298-305.

Müller, U.: Messung von Reynoldsschen Spannungen und zeitlich gemittelten Geschwindigkeiten in einer dreidimensionalen Grenzschicht mit nichtverschwindendem Druckgradienten. Dissertation RWTH Aachen, Fak. Maschinenwesen, 1979.

Nissiriou, M.I.; Pilatis, N.: Development of three-dimensional turbulent boundary layers in incompressible flow with separation. Bristol Univ. Dept. Aero & Mech. Eng.

Orlando, A.: Turbulent transport of heat and momentum in a boundary layer subject to deceleration, suction and variable wall temperature. Ph.D. thesis, Univ. A Harbor, Michigan, 1974.

Papavergos, P.G.; Hedley, A.B.: A simple practical method for establishing turbulence characteristics by means of a single 45° slant hot-wire probe in a field of known mean flow direction. J. Phys. E: Sci. Instr. 12 (1979) 761.

Pierce, F.J.; Krommenhoek, D.H.: Wall shear stress diagnostics in three-dimensional turbulent boundary layers. Virginia Polytechnic Inst. Interim Technical Rep. No. 2, 1968.

Pierce, J.F.; Duerson, S.H.: Measurements of the Reynolds-stress tensor in a three-dimensional turbulent boundary layer. Int. TR No. 4, 1974, AROD Project 6858E, Durham.

Pierce, F.J.; Ezekwe, C.I.: Turbulent stress tensors in a three-dimensional boundary layer. Virginia Polytechn. Inst. Rep. E-75-1, 1975.

Pierce, F.J.; Ezekwe, C.I.: Comparison of Reynolds stress diagnostics by fixed and rotating probes. AIAA Journ. 14, 3 (1976), 412.

Pierce, F.J.; McAllister, J.E.: Near wall similarity in a pressure driven three-dimensional turbulent boundary layer. Virginia Polytechnic Institute & State Univ. Blacksburg, VPI-E-80.32, 1980.

Pozzorini, R.: Das turbulente Strömungsfeld in einem langen Kreiskegel-Diffusor. Dr.-Ing. Thesis, E.T.H. Zürich, 1976.

Rogers, B.K.; Head, M.R.: Measurement of three-dimensional boundary layers. Aeron. J. of the Royal Aeron. Soc. 73 (1969) 796-798.

Rose, W.G.: Corrections to average measurements in unsteady flow. Proc. ASME Symp. on Measurements in Unsteady Flow. Worcester, Mass. (1962) 85-89.

Sarohia, S.; Young, A.D.: Wind tunnel investigations of some three-dimensional separated turbulent boundary layers. IUTAM Symposium on Three-Dimensional Turbulent Boundary Layers, Berlin, 1982.

Strickert, H.: Hitzdraht- und Hitzfilmanemometrie. VEB-Verlag Technik Berlin, 1974.

Tutu, N.K.; Chevray, R.: Cross-wire anemometry in high intensity turbulence. J. Fluid Mech. 71 (1975) 785.

Vagt, J.D.; Fernholz, H.H.: Wall interference effects of static pressure probes in an incompressible turbulent boundary layer. Aeron. Quarterly (1976) 176-184.

Vagt, J.D.: Hot-wire probes in low speed flow. Progr. Aerospace Sci. 18 (1979) 271-323.

Vagt, J.D.; Dahm, A.: Notes on the production and application of hot-wire probes in the fluids laboratory. Translation of the Europ. Space Agency of German Report DLR-FB 77-16, 1979.

Vagt, J.D.; Fernholz, H.H.: A discussion of probe effects and improved measuring techniques in the near-wall region of an incompressible three-dimensional turbulent boundary layer. AGARD Conf. Proceedings 271, 1979.

Winkelmann, A.E.; Melnik, W.L.: An experimental investigation of a two and a three-dimensional low speed turbulent boundary layer. Univ. Maryland, Tech. Rep. No. AE-76-2, 1976.

Zimmerman, D.R.: An experimental investigation of a three-dimensional boundary layer. Purdue Univ., Ph.D., 1975.

Authors	Source and type of cross flow	Number of 3D profiles	Measurement method for Reynolds stress data (see text for description of methods)	Technique for mean field data
Bradshaw & Terrell (1969)	constant pressure yawed flow decaying to 2D; infinite yawed flat wall	5	Method II - hot wire probes of normal, slant and X-array to obtain $\overline{c_i c_j}$	pitot and twin tube (cobra type) probes for Q and γ
Johnston (1970)	strong $\partial p/\partial x$, growing from developed 2D layer; infinite swept step on wall	3	Method II - same as above except $\overline{q_i q_j}$ data obtained and reduced to $\overline{c_i c_j}$	same as above plus disk static probe to obtain p(y), and 2-hole cobra probe held on edge to obtain $\beta(y)$
Bissonnette & Mellor (1970,1974)	transverse wall motion, growing from 2D layer on cylinder	16	Method I - hot wires of normal and slant types with probe and prong axes normal to wall to ohtain $\overline{c_i c_j}$, data given as $\overline{u^2}$, $\overline{v^2}$ (ref. system)	C and γ by hot wire probes
East (1973)	slightly favorable $\partial p/\partial x$, cross over profiles; delta wing at angle of attack	8	$(-\rho\overline{uv})$ and $(-\rho\overline{wv})$ from mean equations of motion using mean field data	3-hole impact tube (cobra type) for Q and γ
Lohmann (1973,1976)	see Bissonnette & Mellor	9	Method I - similar to Bissonnette and Mellor except probe stems held from downstream	C and γ by hot wire probes
Pierce & Duerson (1974)	layer on end wall of constant area rectangular curved duct	7	Method II - with cylindrical hot film probes of X-array to obtain $\overline{c_i c_j}$. Data reduced to $\overline{u^2}$, $\overline{v^2}$, $\overline{w^2}$, $\overline{uv}$, $\overline{wv}$, $\overline{wu}$ format (ref. system)	C and γ by hot film probes
Elsenaar & Boelsma (1974)	adverse chordwise pressure gradient due to 35° yawed wing to generate 'infinite' swept flow on flat wall with separation	10	Method II - hot wire probe of X-array to obtain $\overline{c_i c_j}$	Q, γ and β by 3-hole impact tube (cobra type)
Pierce & Ezekwe (1975)	strong $\partial p/\partial x$ impinging flow growing from developed 2D layer over flat wall	14	Method I - with cylindrical film sensors on probes	ditto Bissonnette and Mellor but with film sensors
Zimmerman & Abbott (1975)	constant pressure, slightly yawed profile decying to 2D on flat yawed wall	11	Method III - instantaneous u, v, w by triaxial, 3-wire probe and analogue computer Data: $\overline{u^2}$, $\overline{v^2}$, $\overline{w^2}$, $\overline{uv}$, $\overline{wv}$, $\overline{wu}$, $\overline{k}$, $\overline{kv}$ (ref.system)	Q and γ by 3-hole impact probe (cobra type)

Table 1

Authors	Source and type of cross flow	Number of 3-D profiles	Measurement method for Reynolds stress data and listing of data	Technique for mean field data and skin friction measurements	Maximum flow angle*). Measuring method	y^+_{min}
Dechow, R. (1977)	Cylindrical body normal to the wall. Pressure driven flow.	14	Quarter square method, single and X-wires, stem parallel to the wall. Method II $\overline{u'^2}, \overline{v'^2}, \overline{w'^2}, \overline{u'v'}, \overline{u'w'}, \overline{v'w'}$;	Normal single hot wire. Preston probe; direction of skin friction with hot film probe.	60° X-wire.	
Hebbar & Melnik (1976, 1978)	Yawed flow decaying to 2-D; flat wall	9	Single rotated hot-wire probe. Probe protruding out of the wall. $\overline{u'^2}, \overline{w'^2}, \overline{u'w'}$; Method I no continuous rotation.	Normal hot wire. Preston probe; sublayer fence; hot film.	22° Bisector method	0.31
Chandrasekhara Swamy, N.V. et al. (1978)	20° yawed flat plate at two angles of incidence. Zero pressure gradient	more than 7	Normal single and X-wire probe. Probe stem parallel or 3° inclined towards the wall. Method II $\overline{u'^2}, \overline{v'^2}, \overline{w'^2}, \overline{u'v'}, \overline{u'w'}, \overline{v'w'}$.	Normal single hot-wire, Pitot probe. Preston probe.	12° 3-hole yaw probe	~ 30
Ezekwe, C.I., Pierce, F.J., McAllister, J.E. (1978)	Free-jet over a flat wall impinging against a back wall. OFF and on plane of symmetry.	4	Rotating hot-film technique, straight and slanted hot-films. Method I $\overline{u'^2}, \overline{v'^2}, \overline{w'^2}, \overline{u'v'}, \overline{u'w'}, \overline{v'w'}$.	Straight hot film. No skin friction data	- Rotating hot film	-
De Grande, G. Hirsch, Ch. (1978)	60° curved duct flow decaying along a straight extension; flat wall.	140	Single rotated hot wire; single normal and slanted wire. Probe protruding out of the wall. Method I $\overline{u'^2}, \overline{v'^2}, \overline{w'^2}, \overline{u'v'}, \overline{u'w'}, \overline{v'w'}$.	Normal single hot-wire. 3-hole probe. No skin friction measurements.	28° 3-hole probe Rotating hot-wire	< 10 estimate
Müller & Krause (1979); Müller (1979)	Adverse chordwise pressure gradient due to inclined side wall and upper wall	21	Single normal, slanted wire, X-probe. Stem parallel to the wall. Method II $\overline{u'^2}, \overline{v'^2}, \overline{w'^2}, \overline{u'v'}, \overline{u'w'}, \overline{v'w'}$,	Normal single hot-wire Preston probe	53° Bisector method	-
Fernholz & Vagt (1980)	Adverse pressure gradient, pressure driven circumferentially due to inclined back wall.	24	Single normal, slanted and X-wire. Probe stem normal to the wall. $\overline{u'^2}, \overline{v'^2}, \overline{w'^2}, \overline{u'v'}, \overline{u'w'}, \overline{v'w'}$, u'-spectra. Method I, with and without contin. rot.	Normal single hot-wire. Preston probe, surface fence.	31° modified rotating hot-wire.	4

*) against free stream direction

Table 2

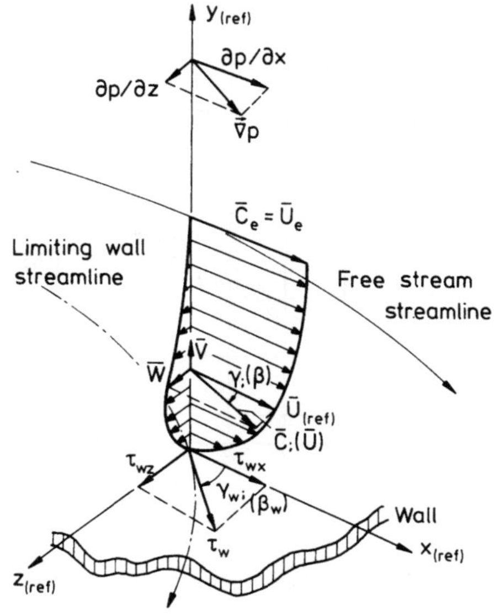

Fig. 1. Skewed boundary layer profile, local free-stream coordinates, mean velocity and wall shear stress components. Johnston (1976)

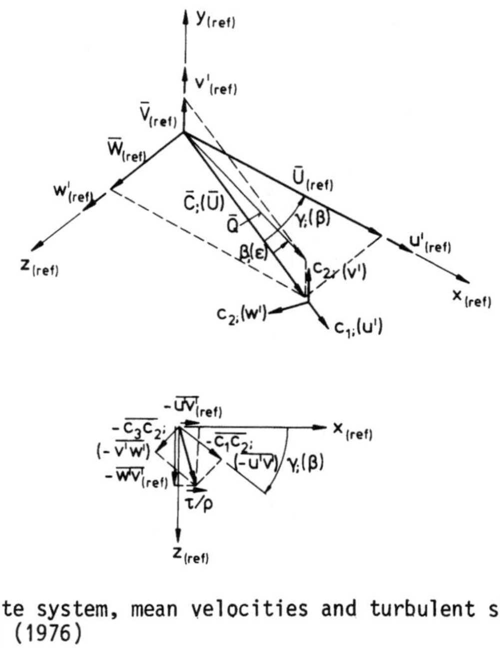

Fig. 2. Coordinate system, mean velocities and turbulent shear stresses. Johnston (1976)

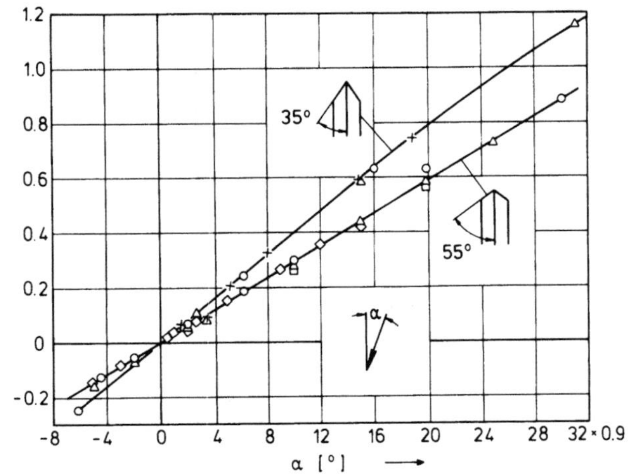

Fig. 3. Yaw meter characteristic

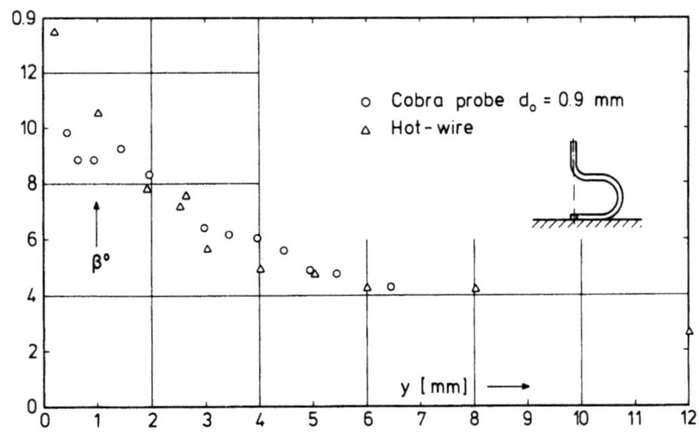

Fig. 4. Flow angle measurement with hot-wire and cobra probe in a three-dimensional turbulent boundary layer. Vagt, Fernholz (1979)

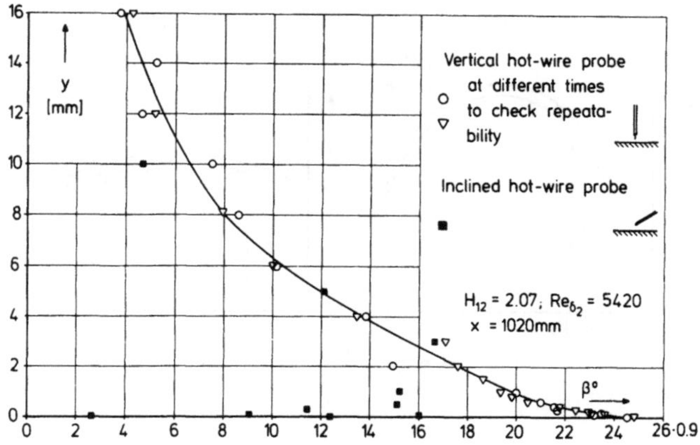

Fig. 5. Flow angle measurement with normal and inclined hot-wire probe in a three-dimensional turbulent boundary layer. Vagt, Fernholz (1979)

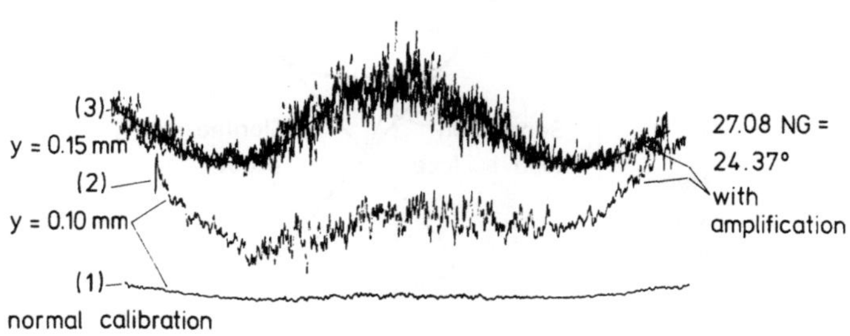

Fig. 6. Flow-angle response characteristic of hot-wire near the wall. Vagt, Fernholz (1979)

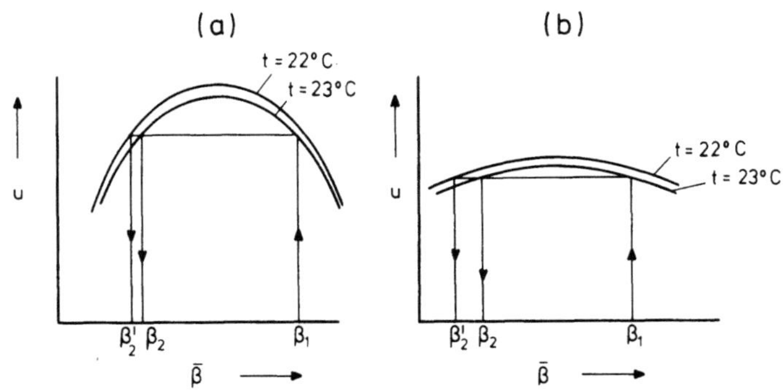

Fig. 7. Influence of temperature variation on flow angle characteristic in outer and near wall region. Vagt, Fernholz (1979)

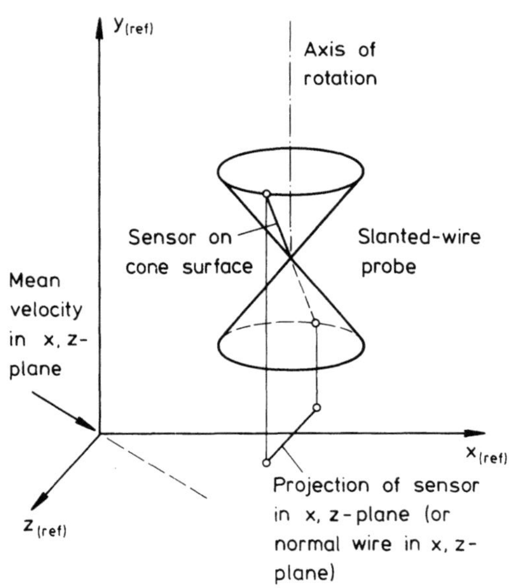

Fig. 8. Rotating hot-wire device. Bissonnette, Mellor (1974)

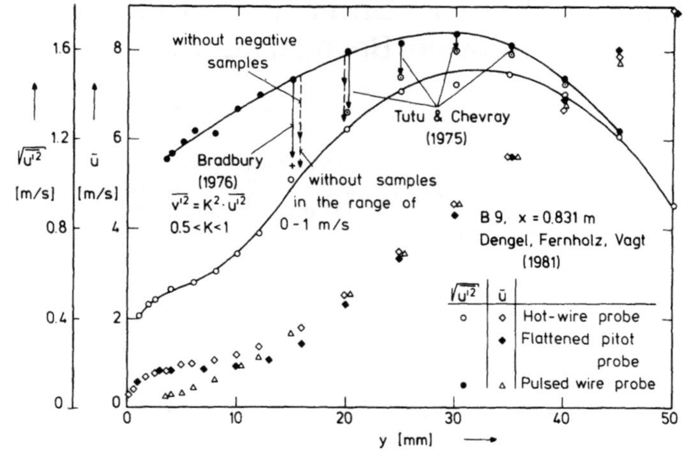

Fig. 9. Comparison between pulsed-wire and hot-wire. Dengel, Vagt

Measurements in a Pressure-Driven and a Shear-Driven Three-Dimensional Turbulent Boundary Layer

F. J. PIERCE

Mechanical Engineering
Virginia Polytechnic Institute and State University
Blacksburg, Virginia 24061

J. E. MCALLISTER

Savannah River Laboratory
E.I. duPont de Nemours & Co.
Aiken, South Carolina 29801

Summary

Results of mean velocity field, wall static pressure field and simultaneous, direct force measurements of the local wall shear stress magnitude and direction are reported for a pressure-driven and a shear-driven three-dimensional turbulent boundary layer. These data, particularly the direct force local wall shear data, were obtained to test the validity of several of the near-wall similarity models proposed in the literature for such flows.

Introduction

Many three-dimensional turbulent boundary layers* can be characterized as pressure-driven or shear-driven, according to the principal mechanism driving the secondary flow. Pressure-driven flows are typically characterized by freestream streamlines which curve with attendant transverse pressure gradients. The majority of the flows which are reported in the literature are in this category. Shear-driven flows occur when the shearing motion of adjacent fluid layers provide the principal driving force for the secondary boundary layer flow.

Direct force measurements of the local wall shear stress magnitude and direction in a 3DTBL coupled with corresponding velocity and pressure field data provide a unique means for the experimental validation of near-wall similarity models proposed for such flows. Since mean velocity and wall pressure measurements are common, emphasis will be placed on the unique measurements of wall shear stress in these three-dimensional flows.

*The short forms 3DTBL and 2DTBL will be used for convenience.

Only direct force measurements of wall shear stress can provide the means for verifying any three-dimensional near-wall similarity model (1,2,3) since essentially all indirect techniques (heated elements, Preston tubes, etc.) require that 1) near-wall similarity exist in a 3DTBL and that 2) a calibration of such an indirect device be made in a three-dimensional flow. To date, no three-dimensional near-wall similarity models have been experimentally verified, nor has any indirect diagnostic device been calibrated in a three-dimensional flow. The use of any indirect wall shear sensing device in a three-dimensional flow using a two-dimensional calibration assumes, a priori and without justification, that the near-wall flow is identical for the two- and three-dimensional cases. One large concern over such an assumption is that the two-dimensional similarity law is a scalar law, while the near-wall flow in the three-dimensional case can show a strong vector character down to the wall itself (4,5,6,7).

Flow Geometries and Instrumentation

The pressure-driven flow in a forward quadrant of a teardrop-shaped cylinder placed normal to the floor of a wind tunnel as shown in Fig. 1 was studied. The cylinder has a diameter of 12.7 cm, is 25.4 cm high and has a tapered trailing edge giving an overall length of 29.2 cm.

The body was placed in the test section of a large, open circuit, 0.61 x 0.91-m wind tunnel. The wind tunnel can provide air speeds up to about 26 m/sec with a freestream turbulence level of $< 0.6\%$ at the test section. Locations for data measurement are designated by two symbols. The first symbol is alphabetic and represents a line parallel to the tunnel axis, with the centerline of the body designated with the symbol A. The second symbol is numeric and represents the location along the tunnel axis, in inches, with the stagnation line of the body designated with the symbol 0. Measurement stations were placed on a 5.08 cm grid.

The shear-driven flow was created by a 3.8 cm high rectangular step, 19 cm wide oriented at a 30° yaw angle. The data stations were also located on a 5.08-cm grid. The station identification scheme is with two numbers XX YY where XX indicates the distance in inches off the tunnel centerline and YY is the distance in inches measured from the step's trailing edge on the tunnel centerline.

A floating element direct force-sensing wall shear meter was designed for the simultaneous measurement of the magnitude and direction of wall shear stress vectors. The 28.6-mm diameter circular sensing element is free to move in the plane of the tunnel flow. The uniform gap between the button and housing is 0.127 mm. Details of the meter's design, characteristics of operation, and calibration (both static and dynamic) are discussed in reference 8. Owing to its delicate nature, the meter was held stationary while the teardrop and step bodies were moved relative to the meter location.

The mean velocity field was measured with a standard three-hole claw probe. The probe, first calibrated in a uniform flow, was nulled for flow direction in the plane of the floor, and the center stagnation hole together with a wall pressure orifice were used to determine the local flow velocity magnitude in the plane of the floor.

The wall pressure field was measured using a tunnel floor plate containing 54 static pressure orifices. Pressures were read using a Scanivalve system with a Setra ultra low range pressure transducer. Pressure gradients were estimated using a regression method for successive order polynominal equations, subsequently differentiated to obtain estimates of the local pressure gradients.

<u>Uncertainties in Direct Wall Shear Measurements</u>

A summary of nine studies of various errors identified with floating element shear meters is presented in McAllister, Pierce, and Tennant (9). The largest errors appear to be as a result of element misalignment and possible pressure gradients imposed on the element. Two experimental studies, both carried out in two-dimensional supersonic flows, have been made concerning errors induced by floating element misalignment, and while some researchers have noted similar behavior in subsonic flows, little supporting data have been reported.

Three experimental studies of pressure gradient effects on floating elements of the type used here have been reported. In these three experiments the reference wall shear stress taken to be exact was variously inferred from the law of the wall, a momentum balance, or some indirect measurement technique. There is wide scatter among the results, with sometimes contradictory trends, and the likelihood of strong geometric dependence of each study makes any generalization of these results somewhat speculative and very difficult, if not impossible, to apply.

Application of misalignment or pressure gradient corrections obtained from two-dimensional devices to floating element measurements in three-dimensional flows is questionable. Direct application of pressure gradient corrections is complicated since the pressure gradient and wall shear stress vector directions are rarely coincident in three-dimensional flows. The obvious lack of consensus among the proposed corrections is also difficult to reconcile.

Pressure-Driven Flow Results

Local wall shear stresses, pressure gradients, and nearest wall velocity directions are shown in Fig. 1. These data were taken at a tunnel inlet unit Reynolds number, U/ν, of about 1.30 x 10^6/m. The uncertainty bands for the magnitude and direction of each of the shear stress vectors define a small curvilinear rectangle at the shear stress vector tip. All uncertainties were estimated after Kline and McClintock (10). The largest uncertainties occur near the separation horseshoe vortex where measurements are very difficult because of larger fluctuations in the detected wall shear.

As the leading edge of the body was approached, a reversal of the wall shear direction was noted near Station A3. A complete flow reversal was recorded at station A1 with a high local shear stress identified with the horseshoe vortex behind the upstream separation sheet. The separation line shown in Fig. 1 is based on wall oil streak patterns obtained with titanium dioxide parti-

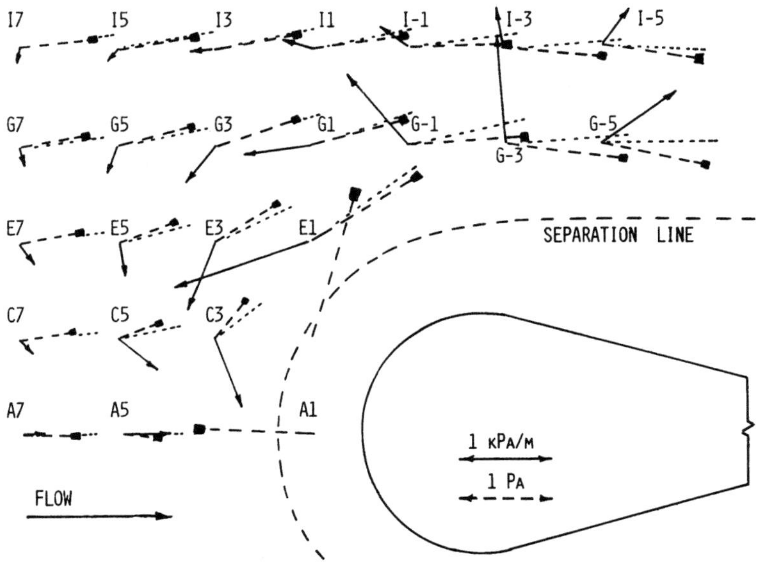

Fig. 1. Wall shears, pressure gradients, and nearest wall velocity directions for the pressure-driven flow.

cles suspended in a diesel fuel-mineral oil mix. It appears that station C1 is also within the separation region. The shear vector direction indicates a significant downstream wall flow component develops in a relatively short distance from the symmetry plane. This separated flow near the cylinder-flow corner appears to be very complex and the structure of this separated region is currently being investigated.

Omitting the singular zero pressure gradient case, the pressure gradient-wall shear vector orientations in two-dimensional flows are limited. These vectors may be directed in the same sense or in the opposite sense, but they are always collinear and may be treated in a scalar sense. The three-dimensional case is more complex. Figure 1 also shows a map of the pressure gradient field about the body. The relative directions of the pressure gradient and wall shear stress vectors vary from nearly collateral along the centerline (the A) stations to nearly orthogonal at Station G-1. These results support the anticipated difficulties in the application of any proposed pressure gradient corrections for two-dimensional flows to three-dimensional flow measurements.

Figure 1 also compares the velocity vector direction at 0.25 mm from the wall and the measured wall shear stress direction. The wall shear stress and the velocity vector angles can differ significantly. For example, at station C3 the angles differ by over 13°, and are well outside any reasonable uncertainties for both measurements. These results are consistent with recent experiments (4,5) and analyses (6,7) do not support the assumption of a collateral near-wall layer in pressure-driven 3DTBL flows as suggested by many polar representations of velocity profiles. Since the measurements of local wall shear stress direction can be significantly different than the typical nearest wall velocity direction, the common practice of inferring the limiting wall streamline angle from the latter, can yield significant errors for at least some flow conditions.

Limiting wall streamline directions were measured from the oil streak pattern and, with only a few exceptions, there is excellent agreement between the oil streak directions and the direct force wall shear directions, again supporting calculations in pressure-driven flows predicting turning of the velocity vector down to the wall itself.

The strong agreement between the oil streak limiting wall streamline directions and the direct force wall shear directions suggest that any pressure gradient effects on the direct force sensing shear meter affected the wall shear directions minimally, if at all. It is noted that when using a miniature, dual sensor, buried wire heat meters in a three-dimensional turbulent flow, Higuchi and Peake (11) show consistent and typical differences ranging from about 5° to as much as 15° when comparing indicated wall flow directions to surface flow visualizations. Those results suggest one can question the ability of miniature heat sensors to measure limiting wall streamline directions accurately.

Representative velocity profiles are shown in Figs. 2 and 3 as total velocity and flow angle relative to the local freestream direction. Station A5 is representative of the pseudo two-dimensional plane of symmetry flow, stations E7, G3, and E1

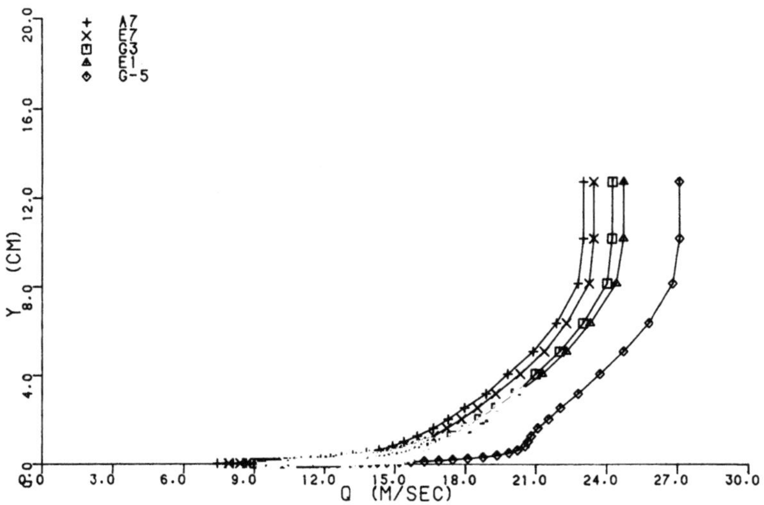

Fig. 2. Representative velocity magnitude profiles for the pressure-driven flow.

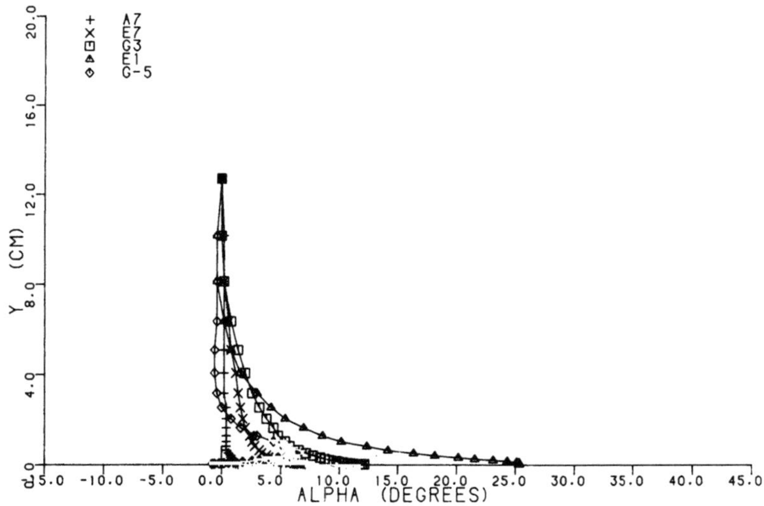

Fig. 3. Representative velocity angle profiles for the pressure-driven flow.

all for monotone increasing skew with a maximum skew angle of 7°, 12°, and 25° respectively. Station G-5 is the only station with bilateral skew, albeit very small.

Shear-Driven Results

Figure 4 shows a map of shears, pressure gradients, and nearest wall velocity directions for the step flow. Data were taken for 15 stations. Data stations are designated by two number groups XX YY, where XX is the transverse distance in inches from the tunnel centerline and YY is the distance in inches from the trailing edge along the tunnel centerline. All the velocity profiles are characterized as having a monotone increasing skew angle with the maximum measured skew angle ranging from a maximum of 67° just behind the step trailing edge to a minimum of 5° well downstream.

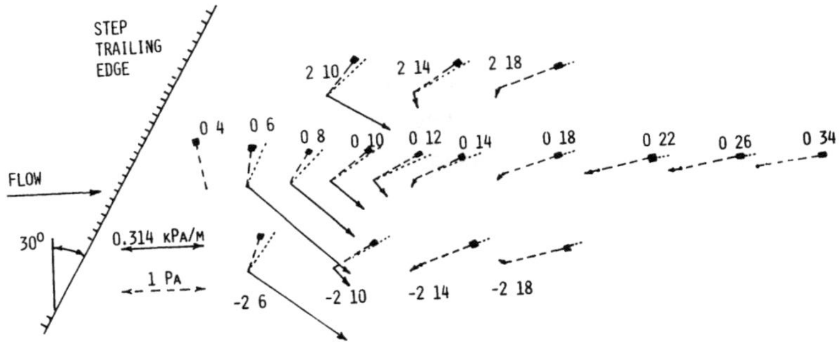

Fig. 4. Wall shears, pressure gradients, and nearest wall velocity directions for the shear-driven flow.

The wall shear stress gradually increases in magnitude as one initially moves downstream from the step. The direction relaxes gradually from its large initial skew relative to the tunnel centerline and approaches a two-dimensional flow situation with the wall shear stress vector and tunnel centerline nearly aligned. This skew relaxation is rapid near the step but becomes more gradual as one moves downstream. The estimated uncertainties are again shown as curvilinear rectangles.

The pressure gradient vector field downstream of the step in Fig. 3 shows that the vectors closest to the step have the largest magnitudes and are approximately normal to the step. The pressure gradient vectors quickly reduce to the nominal two-dimensional pressure gradient value and direction of -14 Pa/m at station 0 26. When compared to the pressure gradient magnitudes in Fig. 1, even the largest pressure gradient magnitudes for the step flow are relatively small. Any possible pressure gradient effects on the wall shear data in Fig. 3 should be essentially negligible for the step flow wall shear data.

Figure 3 also shows the direction of the velocity nearest to the wall. As in the pressure-driven flow results in Fig. 1, there is an appreciable difference in these two directions where pressure gradients exist. As the flow proceeds downstream, these differences decrease and tend to vanish as the pressure gradient field becomes small and aligns itself with the wall flow direction as the flow relaxes toward a two-dimensional condition.

Representative velocity profiles are shown as total velocity and skew angle with distance from the wall in Figs. 5 and 6. Profiles 0 18 and 0 26 have a maximum skew at the wall up to 15°. Profiles 0 12 and 0 14 have a maximum skew of between 15° and

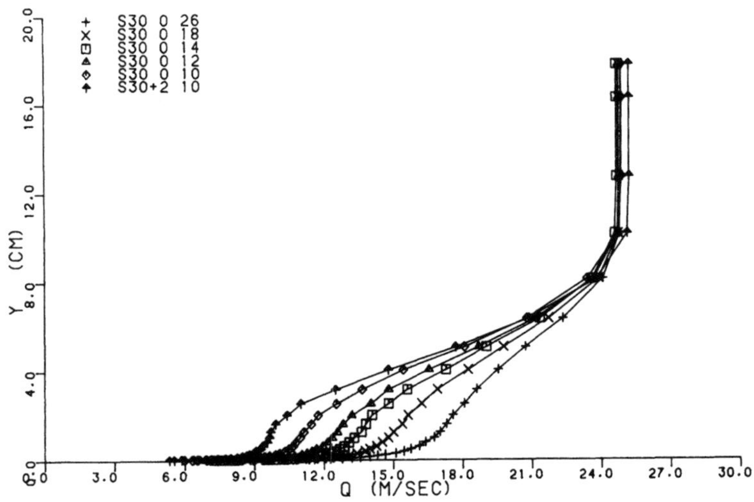

Fig. 5. Representative velocity magnitude profiles for the shear-driven flow.

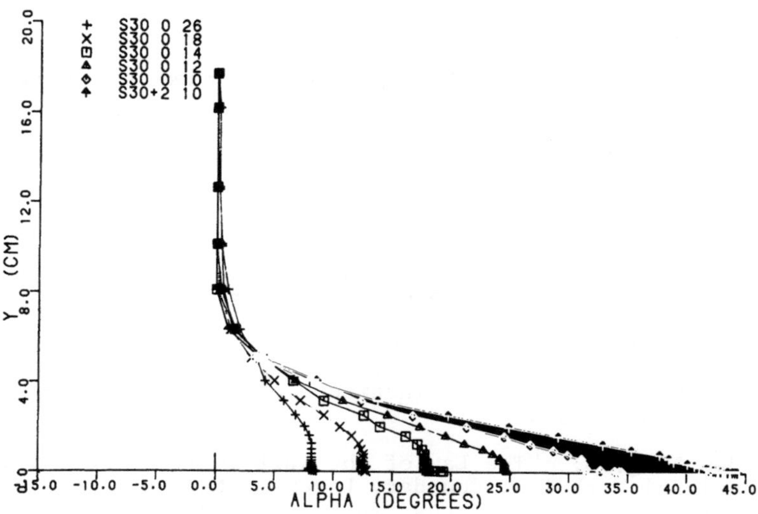

Fig. 6. Representative velocity angle profiles for the shear-driven flow.

30°, and profiles 0 10 and 2 10 have a maximum skew greater than 30°. As the flow relaxes in the downstream direction, the velocity magnitude profiles evolve toward a shape characteristic of equilibrium two-dimensional flows with the flow angle also decreasing. Similar experimental results were obtained for a step yawed at 45°.

<u>Summary</u>

Self-consistent data sets including direct force, simultaneous measurements of local wall shear stress magnitude and direction, mean velocity field, and wall pressure field data are presented for several data stations in both pressure-driven and shear driven 3DTBL flows. These data sets will be used to test the existence of and limits on ten of 11 near-wall similarity models proposed in the literature for such flows. These data may also be useful in turbulence modeling and/or in evaluating numerical techniques for flows as well. These data, indicate the existence of a velocity vector turning continuously to the wall in pressure-driven flows. These data also suggest that pressure gradient effects on the direct wall shear measurements are minimal.

Acknowledgements

This research was initiated with NSF support under Dr. George Lea and subsequently supported under NASA-Ames grant funds monitored by Mr. Joseph G. Marvin.

References

1. Nash, J. F., and V. C. Patel, Three-Dimensional Turbulent Boundary Layers, SBC Technical Books, 1972.

2. Chandrashekhar, N., "Some Studies on Three-Dimensional Turbulent Boundary Layers," Ph.D. Dissertation, Department of Applied Mechanics, Indian Institute of Technology, Madras, India, October 1974.

3. van den Berg, B., "The Law of the Wall in Two- and Three-Dimensional Turbulent Boundary Layers," National Aerospace Laboratory, The Netherlands, TR 72111 U, January 1973.

4. Rogers, B. K., and M. R. Head, "Measurement of Three-Dimensional Turbulent Boundary Layers," The Aero. Journal of the Royal Aero. Society, Vol. 73, September 1969, pp. 796-798.

5. Hebbar, K. S., and W. L. Melnik, "Wall Region of a Relaxing Three-Dimensional Incompressible Turbulent Boundary Layer," Journal of Fluid Mechanics, Vol. 52, Pt. 1, 1978, pp. 33-58.

6. Pierce, F. J., and J. L. East, "Near-Wall Collateral Flow in Three-Dimensional Turbulent Boundary Layers," AIAA Journal, Vol. 10, No. 3, March 1972, pp. 334-336.

7. Klinksiek, W. F., and F. J. Pierce, "A Finite Difference Solution of the Two- and Three-Dimensional Incompressible Turbulent Boundary Layer Equations," Journal of Fluids Engineering, Trans. ASME, Vol. 95, September 1973, pp. 445-458.

8. Tennant, M. H., F. J. Pierce, and J. E. McAllister, "An Omnidirectional Wall Shear Meter," Journal of Fluids Engineering, Trans. ASME, Vol. 102, March 1980, pp. 21-25.

9. McAllister, J. E., F. J. Pierce, and M. H. Tennant, "Direct Force Wall Shear Measurements in Pressure-Driven Three-Dimensional Turbulent Boundary Layers," to be published in the Journal of Fluids Engineering, Trans. ASME.

10. Kline, S. J., and F. A. McClintock, "Describing Uncertainties in Single-Sample Experiments," Mechanical Engineering, January 1953, p. 3.

11. Higuchi, H., and D. J. Peake, "Bi-Directional, Buried-Wire Skin-Friction Gage," NASA TM 78531, Ames Research Center, 1978.

Turbulence Measurements in Ship-Model Flow

T. OKUNO and Y. HIMENO

College of Engineering
University of Osaka Prefecture
Sakai, Osaka, Japan

Summary

The present paper describes an experimental investigation of the three-dimensional turbulence around a ship model in a circulating water channel. The measured results at the stern indicate a great three-dimensionality of the turbulent flow and a strong influence of the so-called longitudinal vortex. An improved method is presented for determining mean velocities and Reynolds stresses by use of a slanted-wedge hot-film technique. The applicability of the method is examined in a pipe flow and in a flat-plate boundary layer flow.

1. Introduction

Research findings related to the boundary layer around a ship model have provided significant advances in recent years using a three-dimensional boundary layer theory. However, further experimental investigations, for example, the measurement of Reynolds stress in the stern flow, are necessary in order to reveal the turbulent flow mechanism in more details.

A number of turbulence measurements have made by means of a hot wire. For the three-dimensional measurement, the following techniques are available. One is the so-called several-wire method, the triple-wire or the x-wire method, by which the time history of the turbulent flow-velocity can be obtained generally. The other is a single-wire method which is restricted to the stationary flows. Hayashi et al. [1] and Acrivlellis [2] have measured the velocity components and Reynolds stresses. Recently, Fukuda et al. [3] have also measured the three-dimensional turbulence around a ship-double-model in air flow in the same manner as Hayashi el al. have developed.

The study on three-dimensional turbulence measurements in water flow, however, has scarcely published because of the difficulties, i.e., the severe sensor contamination with dirt, the strong mechanical vibration of device caused by vortex shedding, the intermix of noise due to electric conduction of water, the influence of water-temperature change and so on.

In this paper, the usefulness of a slanted-wedge hot-film probe in water is shown. An improved method for determining the mean velocity and Reynolds stress is presented. Before the turbulence measurement around the ship model, the method is tested in a circular pipe flow and in a boundary layer flow over a flat plate.

2. Hot-film probe and models

Webster [4] and Champagne et al. [5] have shown that the directional sensitivity of a inclined-wire is represented as follows:

$$U_e{}^2 = \left\{\sin^2(\alpha+\beta) + k^2\cos^2(\alpha+\beta)\right\}U_\infty^2 \qquad (1)$$

where U_e is the so-called effective cooling velocity, α the angle between the flow and the probe axis and β the inclined angle For the slanted-wedge hot-film probe, it is found that the same relation as for the wire can be used and the values of β and the coefficient k are obtained by curve fitting as shown in Fig. 1.

The models for the experiments are a straight circular pipe, a flat plate and a ship model which are set in the circulating water channel. The experimental conditions and the principal dimensions of models are shown in Table 1.

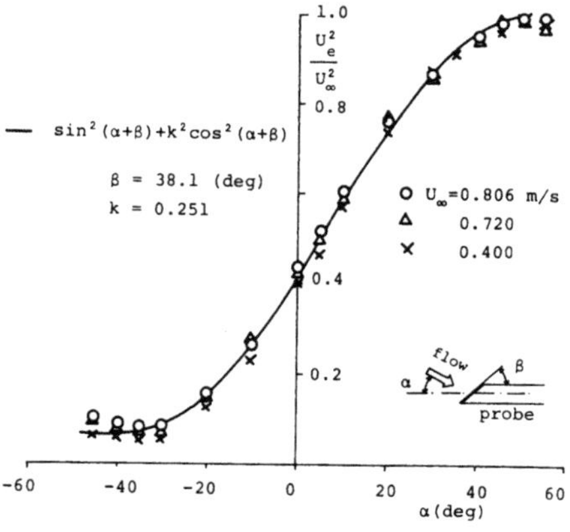

Fig.1. Directional sensitivity of slanted-wedge probe

3. Velocity and Reynolds stress analysis

Table 1 Experimental conditions and models

Model	Dimension (m)	Position	Velocity (m/s)
Circular pipe	L × d(inside dia) =3.5 × 0.05	3.45 m from inlet	0.590 $\left(\frac{u_* R}{\nu}=1.01\times10^3\right)$
Flat plate	L × d × t =2.0×0.4×0.0021 (d: draft)	1.85 m from leading edge	0.839 $\left(\frac{Ux}{\nu}=1.4\times10^6\right)$
Ship (Series 60)	L × B × d =1.8×0.257×0.1028 C_B =0.7	S.S.$\frac{1}{2}$, S.S.$\frac{1}{4}$	0.879 $\left(\frac{UL}{\nu}=1.21\times10^6\right)$

As described above, Hayashi et al. have measured the three-dimensional turbulence by use of the inclined-wire technique. The measurements are performed by rotating the probe on its axis. The equation system is needed for calculating the nine unknowns, the three components of mean velocity and the six Reynolds stresses. The conventional method for determining these quantities is described in the references [1] and [2].

3.1 Conventional method

For the inclined hot-wire, the effective cooling velocity U_e can be written in the three-dimensional form:

$$U_e{}^2 = \sum_{i=1}^{3} \sum_{j=1}^{3} C_{ij} \, U_i U_j \tag{2}$$

where U_i (i=1,2,3) are the velocity components and the coefficient C_{ij} represents the directional cosine of the wire. Taking time average of the eq.(2), we obtain:

$$\bar{U}_e{}^2 + \overline{u_e^2} = \sum_{i=1}^{3} \sum_{j=i}^{3} C_{ij} (\bar{U}_i \bar{U}_j + \overline{u_i u_j}) \tag{3}$$

where $\bar{U}_e$ and $\bar{U}_i$ represent the time-mean values, and u_e and u_i the fluctuating velocity respectively. Assuming the eq. (3) can be separated into the following equations.

$$\bar{U}_e{}^2 = \sum_{i=1}^{3} \sum_{j=1}^{3} C_{ij} \, \bar{U}_i \bar{U}_j \tag{4}$$

$$\overline{u_e^2} = \sum_{i=1}^{3} \sum_{j=1}^{3} C_{ij} \, \overline{u_i u_j} \tag{5}$$

Three components of velocity are determined from the eq. (4) and the six Reynolds stresses from the eq. (5). This is an outline of the conventional method.

However, we can understand that it is impossible to obtain the six Reynolds stresses from the eq. (5) in the case of single wire because the coefficients C_{ij} are the quadratic equations of $\sin\theta$ and $\cos\theta$, where θ is the rotation angle of the probe, in other words, the number of the independent variable is not enough to determine the nine unknowns. Fukuda et al. have pointed out that the determinant of the coefficients in the eq. (5) is zero for the single wire. They have used two or three types of probe to avoid the singularity.

The most doubtful point lies in that the derivation of the eqs. (4) and (5) itself is not represented in the references [1], [2] and [3]. Gaulier [6] wrote in his paper "Since the eq. (3) is identically true in all cases, the eq. (4) is also true for the flow of $u_e=u_i=0$. Thus we obtain the eq. (5)". This is not correct for the turbulent flow as Larsen et al. [7] and Bartenwerfer [8] have pointed out.

3.2 Improved method

The correct equation system should be derived as follows. The eq. (2) is employed as a starting point and is written in the alternative form:

$$U_e^2 = X + 2Y + Z \tag{6}$$

where $X=\Sigma\Sigma C_{ij}U_iU_j$, $Y=\Sigma\Sigma C_{ij}U_iu_j$ and $Z=\Sigma\Sigma C_{ij}u_iu_j$. The root of the eq. (6) is expanded as follows:

$$U_e = \sqrt{X}\ \{1 + \frac{Y}{X} + \frac{Z}{2X} - \frac{Y^2}{2X^2}\} + O(\varepsilon^3) \tag{7}$$

where we assume that the mean velocity $U_i=O(1)$ and the fluctuating velocity $u_e=O(\varepsilon)$. Taking time average of the eq. (7), we obtain $\bar{U}_e$. Then the fluctuating effective velocity u_e can be written:

$$u_e = U_e - \bar{U}_e$$

$$= \sqrt{X}\ \{\frac{Y}{X} + \frac{Z-\bar{Z}}{2X} - \frac{Y^2-\overline{Y^2}}{2X^2}\} + O(\varepsilon^3) \qquad (8)$$

After all, the variance $\overline{u_e{}^2}$ can be represented:

$$\overline{u_e^2} = \frac{\overline{Y^2}}{X} \qquad (9)$$

Now, if we neglect the second and higher order terms in the eq. (7), the eq. (4), i.e., the conventional method, is obtained. The use of the eq. (5) for determining Reynolds stress is not correct even if the eq. (4) can be used for the calculation of the first approximation of mean velocity.

We should employ the eqs. (3) and (9) which include the all second-order small terms. The coefficients in eq. (9) are the fourth-degree of sinθ and cosθ. This means that there is no singularity in the eq. (9) except for the case of uniform flow.

In the uniform flow, $U_1=1$ and $U_2=U_3=0$, it is impossible to solve the euations because the independent variables are reduced. The equations in the present method become similar to those in the conventional method. It can be note by examining the coefficients in the equations that the values of $\overline{u_1^2}$, $\overline{u_1 u_2}$ and $\overline{u_3 u_1}$ obtained by the conventional method are almost same as those by the present method and the sign of $\overline{u_2 u_3}$ value is opposite.

It is not so easy in practical view to solve the simultaneous equations (3) and (9) using only the data of the single probe rotating on its axis, since the equation system is still nearly singular. In this paper, the slanted-wedge probe is used by changing the setting angle or by combining with a conical probe which have often been used in water flow.

4. Experimental results and discussions

At first, the present method is tested in the pipe flow and in the boundary layer flow over the flat plate. Fig. 2 shows the

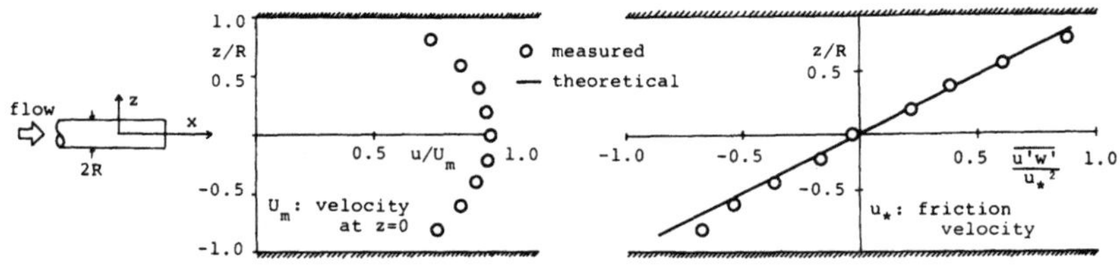

Fig.2. Velocity and turbulent shear stress distribution in the fully developed pipe flow

velocity and the shear stress distribution of the fully developed turbulent flow through the circular pipe. The measured Reynolds stress agrees well with the theoretical one. The friction velocity u_*, shown in the figure, is evaluated by using the measured streamwise pressure distribution.

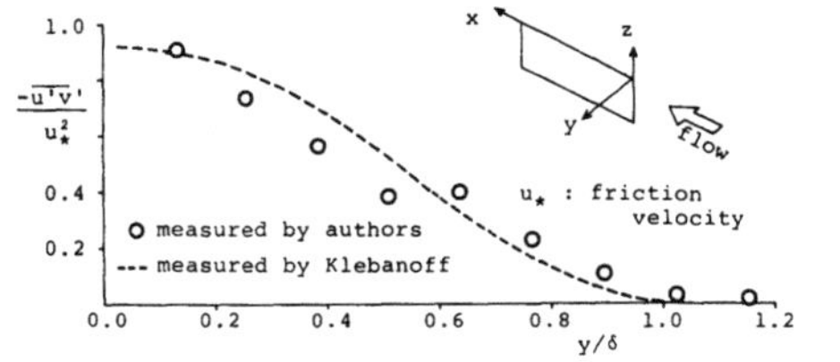

Fig.3. Shear stress distribution in flat-plate boundary layer flow

The shear stress and the normal stress profiles in the boundary layer over the flat plate are in approximate agreement with the results measured by Klebanoff [9] except for the result of $\overline{v'^2}$ as shown in Figs. 3 and 4. In the measurement, the conical probe was used together with the slanted-wedge probe. The friction velocity is obtained from the measured velocity profile.

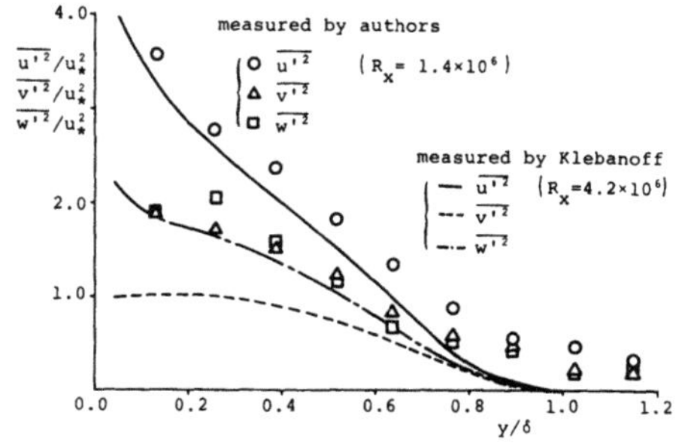

Fig.4. Normal stresses in flat-plate boundary layer flow

Now, we discuss the results

of the ship-model flow. Fig. 5 represents the Reynolds stress distribution along the water line, z/draft = -0.903, where the coordinate (x,y,z) is defined as shown in the figure. The results obtained by the conventional method are also shown. It is found that the values of $\overline{u'v'}$ and $\overline{w'u'}$ obtained by two different methods almost agree but the sign of $\overline{v'w'}$ is opposite. The normal stress distributions are not so monotonous as in the flat-plate boundary layer owing to the influence of the cross flow.

It is known that there is a longitudinal vortex near the stern bilge generated by the three-dimensional separation. The cross plane flow at S.S.1/4 shows clearly the existence of the longitudinal vortex as in Fig. 6. It is noticed that the wake contours (longitudinal velocity contours) near the bilge are affected by the vortex.

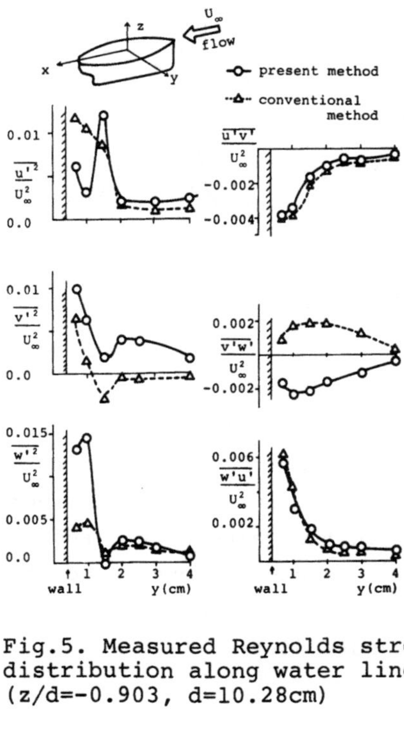

Fig.5. Measured Reynolds stress distribution along water line (z/d=-0.903, d=10.28cm)

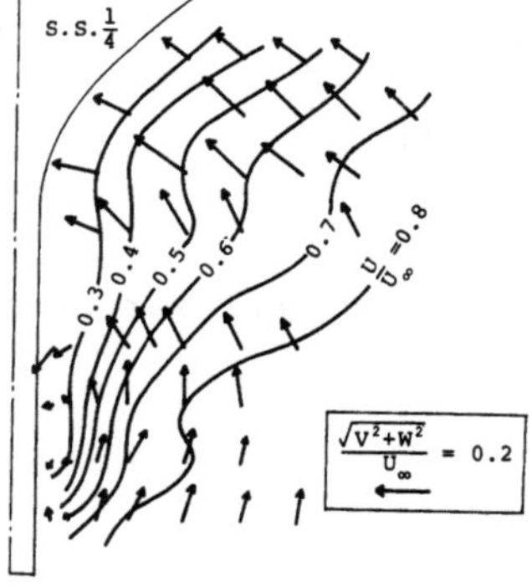

Fig.6. Mean velocity distribution at S.S.1/4 (Series 60, $0.7C_B$)

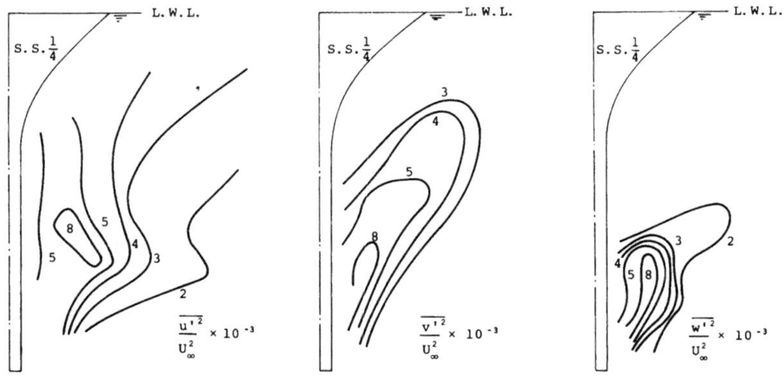

Fig.7. Normal stress distribution at S.S.1/4

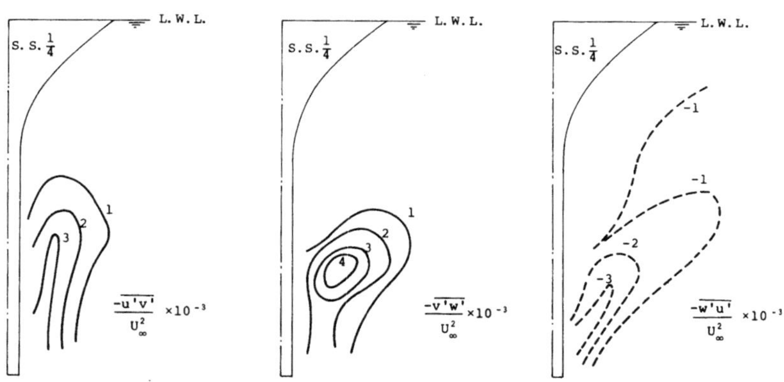

Fig.8. Shear stress distribution at S.S.1/4

Fig.7 shows the normal stress distribution at S.S.1/4. The distribution of $\overline{u'^2}$ is slightly different from the results of $\overline{v'^2}$ and $\overline{w'^2}$. The location of the peak value of $\overline{u'^2}$ does not coincide with the vortex core though the value of $\overline{v'^2}$ and the $\overline{w'^2}$ are high near the vortex core. It can be considered that the distribution of $\overline{u'^2}$ is influenced by the vortex and also by the longitudinal velocity and pressure distribution.

The shear stress in Fig. 8 are also affected by vortex. The value of $\overline{u'v'}$, $\overline{v'w'}$ and $\overline{w'u'}$ are almost of same order. The same order of magnitude of the turbulence as shown in Figs. 7 and 8 suggests the strong three-dimensionality of the turbulent flow near the stern.

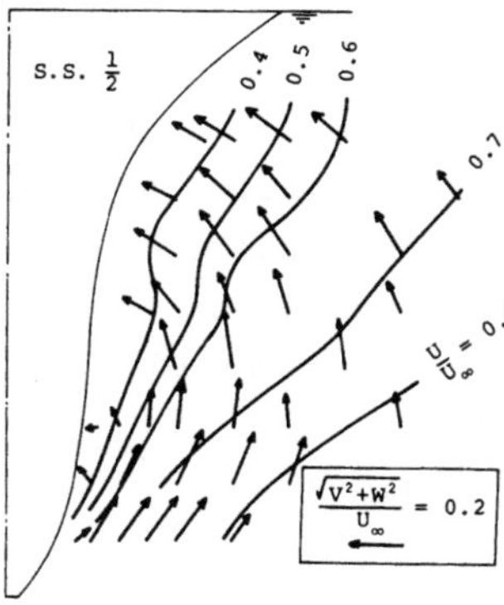

Fig.9. Mean velocity distribution at S.S.1/2

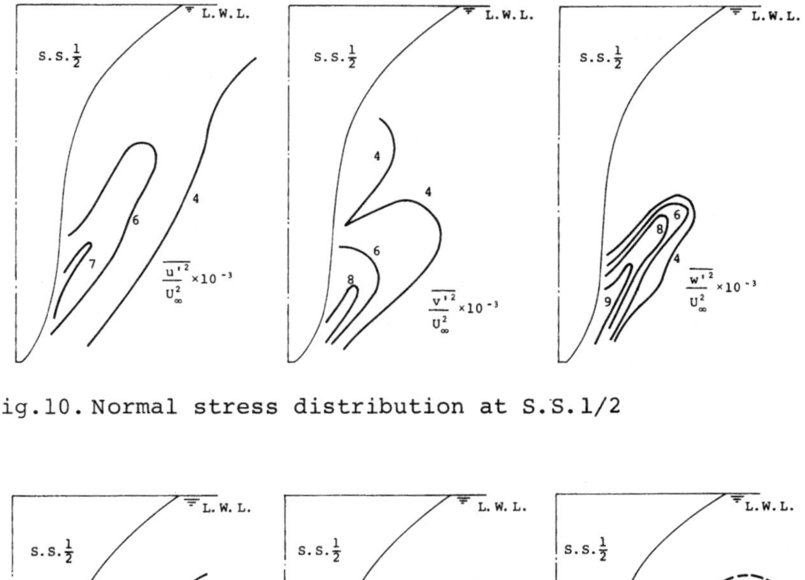

Fig.10. Normal stress distribution at S.S.1/2

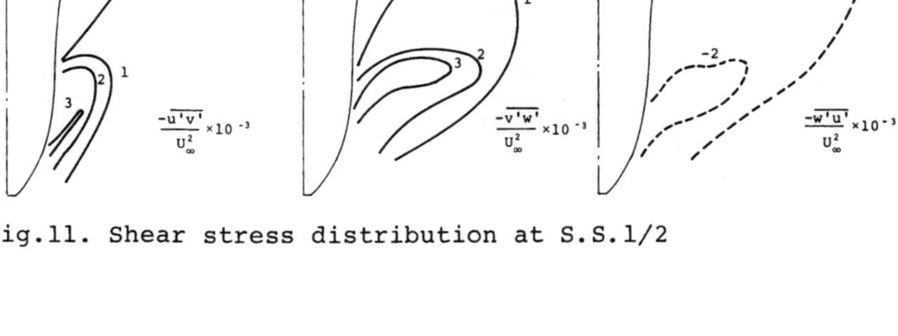

Fig.11. Shear stress distribution at S.S.1/2

Since the flow near S.S.1/2 also includes the vortex as shown in Fig. 9, the same results as that described for S.S.1/4 are obtained from Figs. 10 and 11 which show the measured Reynolds stress distribution at S.S.1/2 of Series 60, 0.7 C_B ship model. It seems that the ordinary thin boundary layer assumption, in which the normal stress and one of the shear stresses are neglected, is not valid in the stern flow.

5. Conclusions

The three-dimensional turbulence measurements near the stern flow of the ship model were carried out in the circulating water channel. The major results of the present investigation are summarized as follows:

1) In the stern flow, the three-dimensionality of the turbulent flow is considerably great.
2) The turbulence intensity is high near the longitudinal vortex core generated by the three-dimensional separation.
3) The slanted-wedge hot-film probe is useful in water flow.
4) The conventional method for determining the mean velocity and Reynolds stress is not correct. The improved method is presented.

The authors are grateful to Prof. N. Tanaka, University of Osaka Prefecture, for his encouragements and to the students, Mr. Tai, Mr. Kitano, Mr. Takeda and Mr. Ishii, for their assistance in the experiments.

References

1. Hayashi, Y. and Nakatani, T. :Measurement of Three-Dimensional Mean Velocity Vector and Reynolds Stress by a Single Rotatable Hot-wire, Technical Report of National Aerospace Laboratory, TR-242, 1971 (in Japanese).

2. Acrivlellis, M. :An Improved Method for Determining the Flow Field of Multidimensional Flows of Any Turbulence Intensity, DISA Information, No.23, 1978.

3. Fukuda, K. and Fujii, A.:Turbulence Measurements in Three Dimensional Boundary Layer and Wake around a Ship Model, Journal of the Society of Naval Architects of Japan, Vol.150, 1981 (in Japanese).

4. Webster, C. A. G. :A note on the sensitivity of yaw of a hot-wire anemometer, Journal of Fluid Mechanics, vol. 13, part 1, 1962.

5. Champagne, F. H., Sleicher, C. A. and Wehrmann, O. H. : Turbulence measurements with inclined hot-wires, Journal of Fluid Mechanics, vol. 28, part 1, 1967.

6. Gaulier, C. :Measurement of Air Velocity by Means of a Triple Hot-wire Probe, DISA Information, No. 21, 1977.

7. Larsen, S. E. and Rasmussen, K. R. :Comment on :Measurement of Air Velocity by Means of Triple Hot-wire Probe, DISA Information, No.23, 1978.

8. Bartenwerfer, M. :Remarks on Hot-wire Anemometry Using "Squared Signals", DISA Information, No. 24, 1979.

9. Klebanoff, P. S. :Characteristics of Turbulence in a Boundary Layer with Zero Pressure Gradient, NACA Report 1247, 1955.

Measurements of Reynolds-Stress Profiles in the Stern Region of a Ship Model

L. LÖFDAHL and L. LARSSON

The Swedish Maritime Research Centre SSPA
P O Box 24001
400 22 GÖTEBORG
Sweden

Abstract

In the present investigation Reynolds stress profiles have been measured at different positions in the stern region of a ship model. The measurements were carried out using a hot-wire technique in which the mean velocity component perpendicular to the surface was considered. From the turbulence profiles it may be concluded that streamline convergence (divergence) as well as wall curvature strongly influence the Reynolds stresses.

Introduction

It is evident from recent comparative calculations, Larsson [1], that the thin boundary layer approximations cannot be applied for the flow near the stern of a ship. A more accurate computational method is required, and to develop such a method it is necessary to have information on all components of the Reynolds stress tensor. Unfortunately, for ship model flows, such data are practically non-existent. Only very limited information on the longitudinal normal stress has been reported. In the present investigation profiles of all Reynolds stresses have been measured at 19 positions in the stern region of a ship model. Using these data it should be possible to develop a suitable turbulence model, and the results of a complete calculation method can be evaluated. In this paper, only a summary of the investigation can be given. For full account of the work, see Löfdahl [2].

Experimental equipment

All measurements were carried out using hot-wire anemometry in a low-speed wind tunnel, at a constant Reynolds number, $5 \cdot 10^6$. The hull tested was the SSPA model 720, earlier used by Larsson [3] to investigate the mean flow in the boundary layer. In fig. 1 a lines plan of the afterbody is shown, and in fig. 2 the distributions of some curvature parameters along the potential flow streamlines 3 and 5 are given.

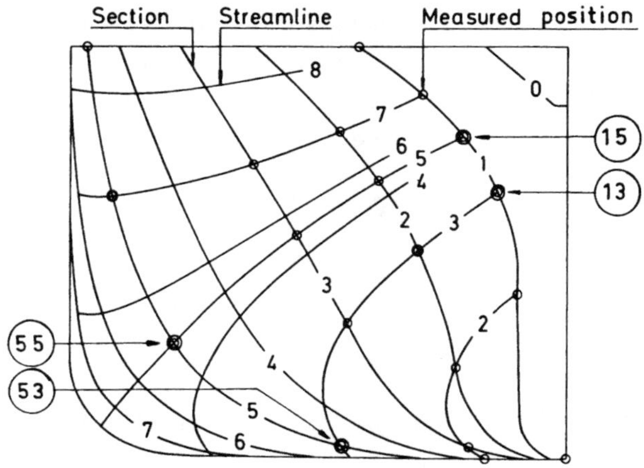

Fig.1. Lines plan of the afterbody.

K_{13} represents the geodesic curvature, and K_{31} the convergence (divergence) of the streamlines. K_{12} is the normal curvature of the surface in the streamline direction.

The measurements of the Reynolds stress profiles were made using a DISA 55M10 constant temperature anemometer connected via an AD-converter to a mini-computer. Two probes were employed, a DISA 55P04 (90° bent) and a DISA 55P02 (45° slanted).

Hot-wire technique

In the derivation of the equations for the Reynolds stresses, the effective cooling velocity of the sensor is expressed as the product of the instantaneous velocity and a function of the angle between the sensor and the velocity vector. This vector is expressed as the sum of the mean and fluctuating velocity components, while the angular dependence is assumed to follow Hinze's law. By time averaging of the effective cooling velocity and subtraction from the corresponding instantaneous expression, the fluctuating part of this velocity can be obtained. Finally, squaring and averaging the fluctuation yields an expression containing all the Reynolds stresses. The independent variable is the angle of rotation, so from RMS-measurements at several angles, a system of equations for the Reynolds stresses is obtained. Unfortunately, to get enough sensitivity to all six stresses, two different probes have to be traversed separately through the boundary layer. This principle has been used in earlier investigations, where, however, the normal component (upwash) of the mean velocity vector has been neglected.

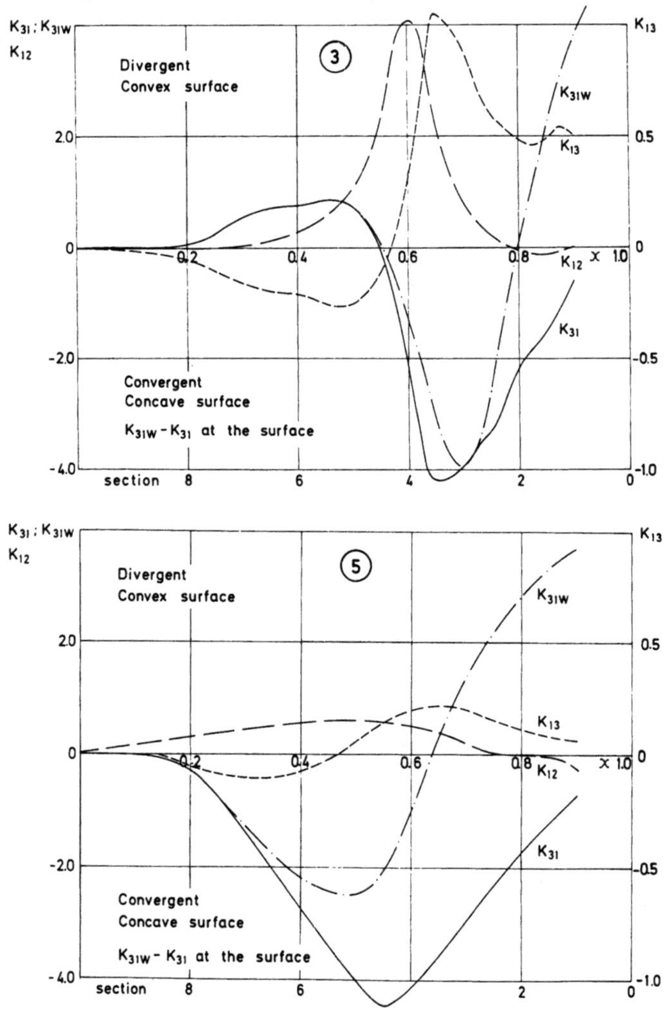

Fig.2. Distributions of the curvature parameters K_{12}, K_{13} and K_{31} along streamlines 3 and 5 .

Accuracy

The best way of estimating the accuracy of the measuring technique is to test it on a known case. Therefore a flat plate was manufactured and placed in the wind tunnel. The normal stress components $(\overline{u^2})$, $(\overline{v^2})$ and $(\overline{w^2})$[1] and the shear stress $(\overline{uv})$ were measured using standard cross-wire technique, and the results could be used together with Klebanoff's as reference data.

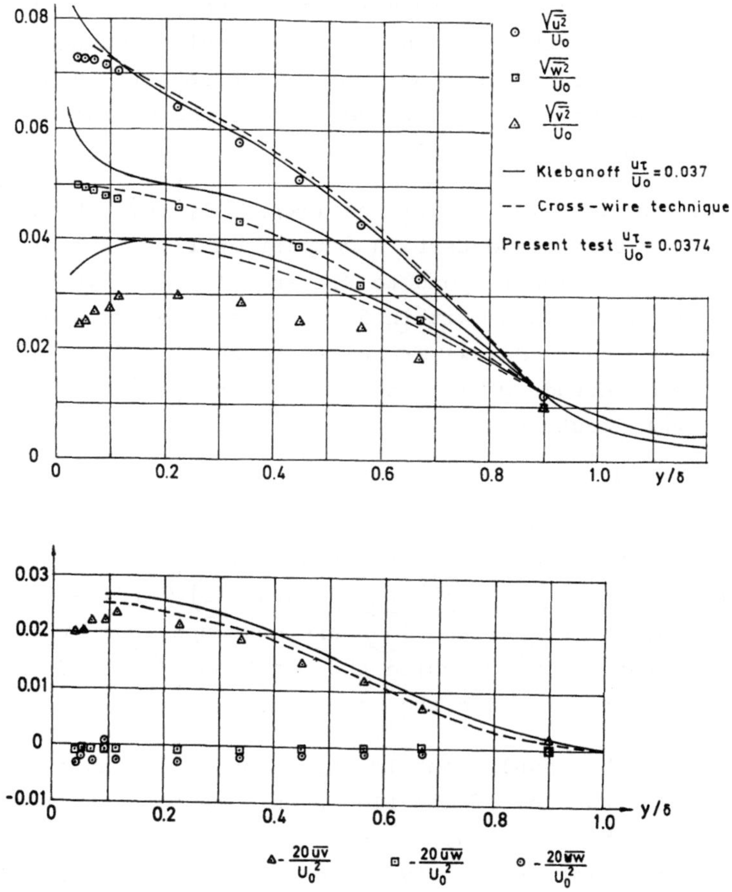

Fig.3. Reynolds stresses determined on the flat plate by cross-wire and the present hot-wire technique respectively

---

[1] ρ is dropped when discussing the Reynolds stresses. u, v and w are the fluctuating velocities in X, y and Z directions respectively. X is along the streamlines of fig.1, and y is along the normal to the surface.

In fig. 3 comparisons with the results of the present technique are shown. The $(\overline{u^2})$ component corresponds with the cross-wire results and Klebanoff's data within 3 % in the outer 90 % of the boundary layer. The peak value close to the wall is, however, not reached, so deviations in this region up to 10 % may occur. The $(\overline{w^2})$ component is approximately 10 % lower than Klebanoff's but very close, 1-2 %, to the cross-wire measurements. The deviation in the normal stress perpendicular to the wall, $(\overline{v^2})$, is larger than for the other components, since this component is most difficult to determine. Compared with Klebanoff, approximately 25 % lower values are obtained, while the deviation from the cross-wire data is about 20 %.

Of the shear stresses, $(\overline{uv})$, $(\overline{vw})$ and $(\overline{uw})$, only $(\overline{uv})$ should be non-zero in the flat plate test. A comparison reveals that the present measurements correspond to Klebanoff's data within 10 % and to the cross-wire data within 5 % in the outer 90 % of the boundary layer. $(\overline{uw})$ and $(\overline{vw})$ are close to zero in fig. 3. Referred to the magnitude of $(\overline{uv})$ the deviation in $(\overline{uw})$ is about 3 % and in $(\overline{vw})$ about 10 %.

Results and discussion

In the following, the discussion has to be restricted to four of the measured profiles, considered typical, namely the ones located at positions (53), (55), (13), and (15)[1)] indicated in fig. 1. The positions (53) and (55) are most upstream at streamlines 3 and 5 respectively, while the positions (13) and (15) are located downstream along the same lines.

When discussing the results one important observations as to the convergence (divergence) of the flow should be borne in mind. The streamlines shown in fig. 1 represent tracings based on potential flow calculation and may approximately be considered as projections on the surface of the real streamlines at the boundary layer edge. The convergence (divergence) computed from these lines should thus be representative for the outer portion of the boundary layer. However, in the inner part the cross flow

[1)] Position number ij, at section i on streamline j.

may cause an entirely different situation. This is particularly obvious near position (15), where the flow is convergent in the outer part and divergent close to the surface. In fig. 2 the convergence (divergence), K_{31}, computed from the potential flow tracing is given together with an estimation of the corresponding quantity, K_{31W}, close to the surface.

The upwash angle, ε, depends strongly on the convergence (divergence) of the flow. If the streamlines converge, normally a more positive upwash angle is found. Conversely, a streamline divergence generates more negative upwash. In fig. 4, ε-profiles at the positions (53), (55), (13) and (15) are shown. If the outer part of the boundary layer is considered first, it appears that the convergence of the streamlines near the edge causes an increasing upwash in the downstream direction. In the inner part of the boundary layer, where the flow changes from convergent to divergent, the upwash becomes negative. As expected this effect is most accentuated at position (15).

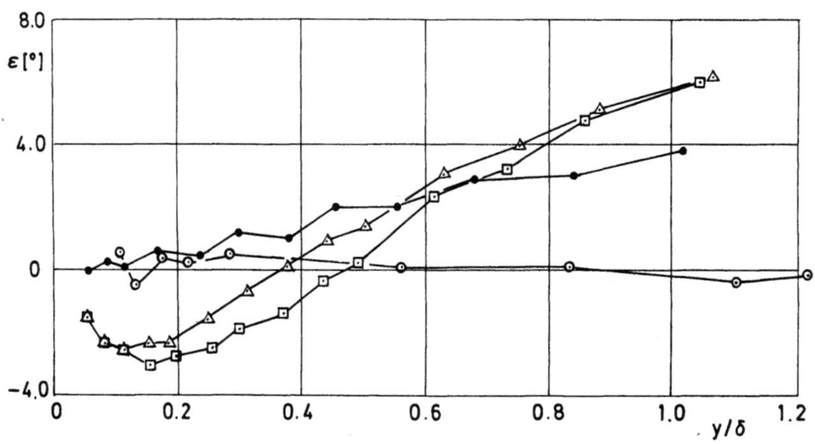

Fig.4 Profiles of the upwash angle ε at positions ⊙ - (53), ● - (55), ▲ - (13) and ⊡ - (15)

The Reynolds stresses are also sensitive to the flow convergence (divergence). If the flow is stretched laterally, i e if the

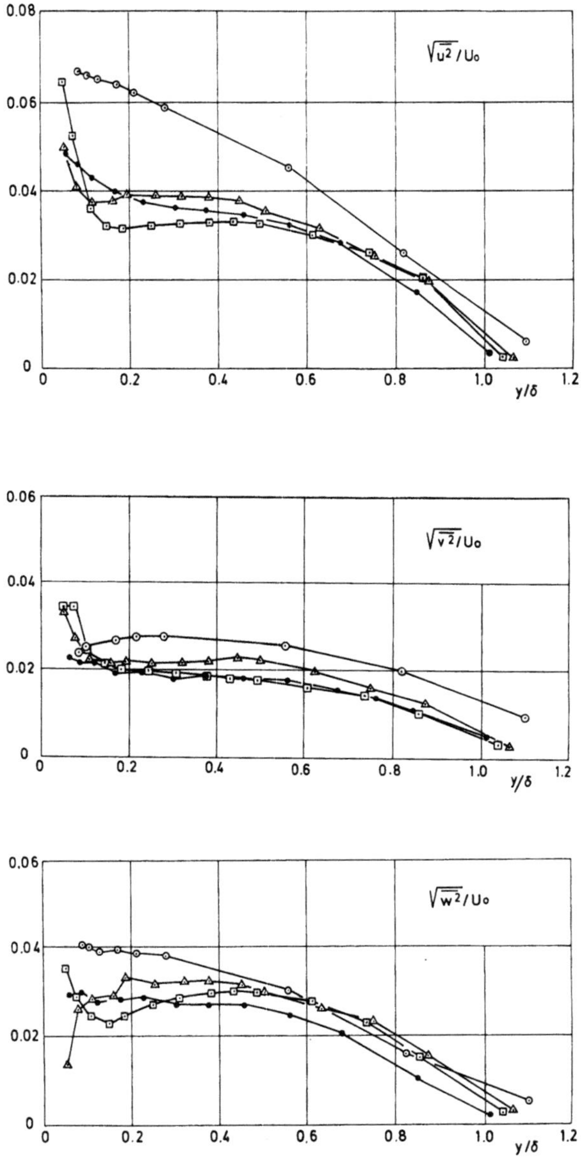

Fig.5. Profiles of the normal stresses at positions (53), (55), (13) and (15), symbols as in fig. 4

streamlines diverge, the turbulence intensity generally increases, and vice versa. Another quantity of importance is the normal curvature of the surface. If the surface is convex the flow is stabilized and the opposite is true if the surface (in the flow direction) is concave.

Profiles of the Reynolds stresses are shown in figs. 5 and 6. At position (53), where the flow is slightly divergent and the surface somewhat convex, the flat plate levels of the stresses are obtained. However, further downstream along streamline 3 the flow turns convergent, which in co-operation with a fairly large curvature of the surface causes a large drop in the normal stresses and the $(\overline{uv})$ shear stress. This is most obvious at position (13), but as a matter of fact, this co-operation appears in the major part of the investigated region, so in general, the levels of the stresses are reduced as compared to the ones found on the plate.

An interesting feature of the stress distribution at position (13) is the sharp increase close to the wall. This is due to the fact that the flow near the surface turns divergent at section 2, as appears from fig. 2.

The situation at streamline 5 is very different from the one at streamline 3. Judging from fig. 2 the largest convergence found in the flow appears very close to position (55), while the normal curvature of the surface is fairly small. Thus at this position a considerable stress reduction may be found as compared with the flat plate data. In the outer and middle parts of the boundary layer this low level is maintained all the way to the stern, but since the inner part turns divergent near section 4, very sharp increases are found near the surface.

A question of considerable interest in boundary layer calculations is the magnitude and the variation of the $(\overline{uw})$ shear stress. It is mainly the gradient of the $(\overline{uw})$ shear stress in the transverse direction, which is of interest. An estimation of this gradient between, for example, position (13) and (15) reveals that it is at least of the same magnitude as the gradient of

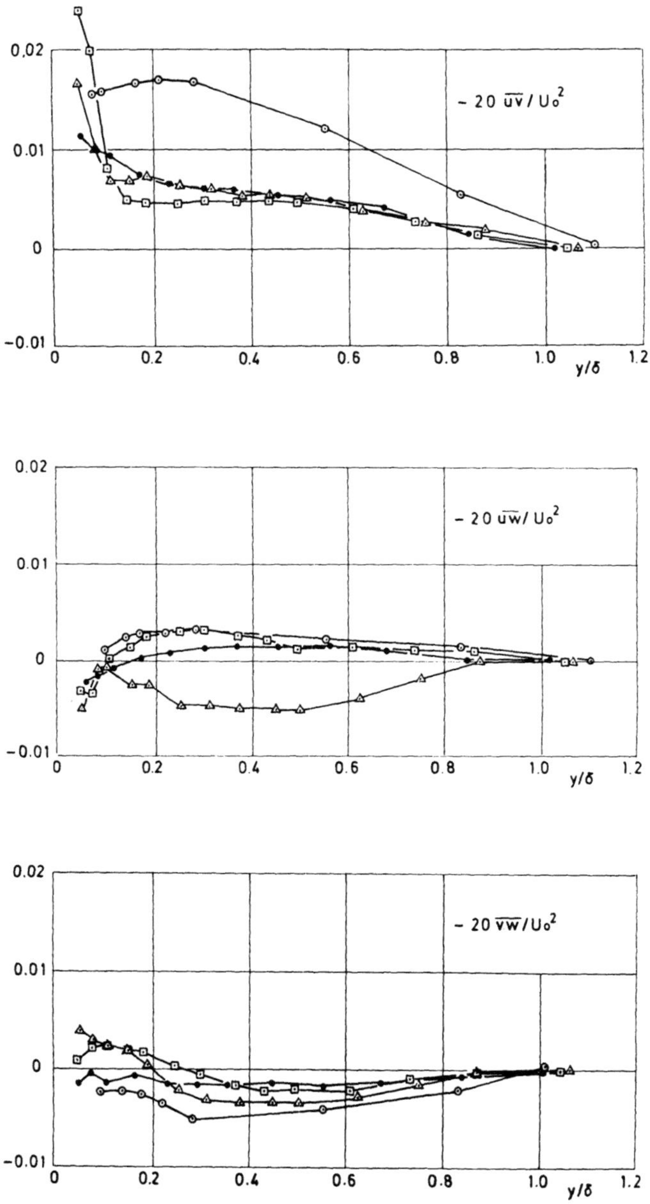

Fig.6. Profiles of the shear stresses at positions (53), (55), (13) and (15), symbols as in fig. 4

the ($\overline{uv}$) shear stress perpendicular to the surface, in the major part of the boundary layer. Thus it can be concluded that the ($\overline{uw}$) shear stress should be taken into consideration in future turbulence models.

Profiles of the skew angle, $\gamma^{1)}$, for the shear stress vector, $\zeta^{1)}$, for the mean velocity gradient vector, and $\beta^{1)}$ for the cross flow, are given in fig. 7. There appears to be no strong correlation between γ and ζ, as us usually assumed in boundary layer calculations, but the levels and trends of the two angles are generally the same. As expected the angles are much smaller at section 5 (positions (53) and (55)) than at section 1 (positions (13) and (15)).

In fig. 8 the structural coefficient, a_1, and the mixing length, ℓ_{mix}, of positions (53), (55), (13) and (15) are shown. The structural coefficient is defined as the ratio between the shear stress component in the plane of the surface and the turbulent kinetic energy. In calculation methods the assumption is usually made that the structural coefficient is constant across the boundary layer, and equal to 0.15.

If all measured profiles are considered the coefficient is generally reduced from section 5 to section 1, starting from a level close to, or somewhat higher than, 0.15 and dropping to 0.10-0.12. Except for the aftermost sections the levels are fairly constant across the boundary layer. Both at position (13) and (15) there is a sharp increase in a_1 close to the surface. One possible explanation for this is that the strong divergence encountered here increases the shear stress more than the normal stress. Another effect is the large negative upwash, which in fact increases the shear stress, evaluated in a co-ordinate system parallel to the surface, see Löfdahl [2].

According to the definition of the mixing length, it relates the shear stress and the local mean velocity gradient. In the wall region of a two-dimensional boundary layer, the mixing length can be empirically expressed as the product of a constant, the von Kármán constant, and the wall distance. Further out ℓ_{mix}

1) Referred to the X-direction

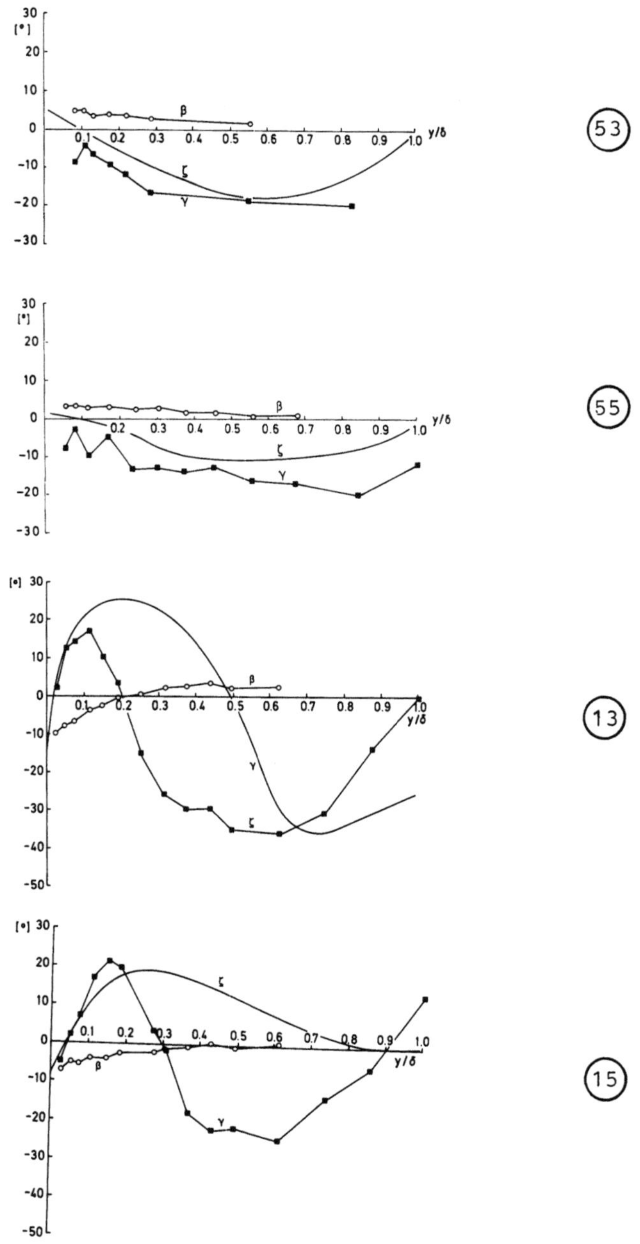

Fig.7. Profiles of the skew angles γ and ζ, and the yaw angle β at positions (53), (55), (13) and (15), symbols as in fig.4

is usually assumed to be constant. As appears from fig. 8 generally a reduction in the mixing length compared with the two-dimensional values occurs. This reduction may indicate that the dissipation length will also be reduced. If this is true the dissipation would be larger than in a thin boundary layer, and since the shear stresses are small, the production of turbulent kinetic energy is reduced. Accordingly the balance between production and dissipation is no longer maintained for thick boundary layers. Thus the convection and diffusion may be of higher importance in the energy balance. This interpretation of the reduction in mixing length has originally been suggested by V.C. Patel.

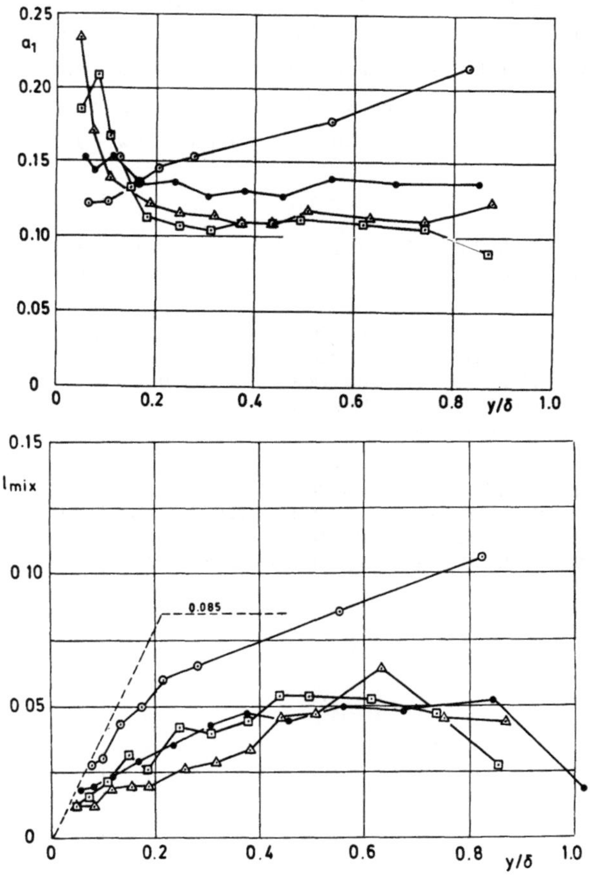

Fig.8. Profiles of a_1 and ℓ_{mix} at positions (53), (55), (13) and (15), symbols as in fig.4

Conclusions

An experimental investigation, in which Reynolds stress profiles were measured in the thick three-dimensional turbulent boundary layer at the stern of a ship model has been carried out. The measurements were performed using a specially developed hot-wire technique, in which the mean velocity component perpendicular to the surface was considered. A strong influence of streamline convergence(divergence) on the Reynolds stresses have been revealed in the measured profiles. Also the effects of wall curvature are of importance, and since in most parts of the investigated region, streamline convergence cooperates with convex curvature, the average level of the stresses is reduced. It may also be concluded that the ($\overline{uw}$)-shear stress should be taken into account in future turbulence models, and that standard values of the structural coefficient as well as of the mixing length may not be used in calculation of this complex flow case.

References

1. Larsson, L.: SSPA-ITTC Workshop on Ship Boundary Layers 1980. Publications of the Swedish Maritime Research Centre SSPA, No. 90, (1981).
2. Löfdahl, L.: Measurement of the Reynolds Stress Tensor in the Thick Three-Dimensional Boundary Layer Near the Stern of a Ship Model. Ph.D. thesis Chalmers University of Technology, (1982). Also available as Report No. 61 of the Swedish Maritime Research Centre SSPA.
3. Larsson, L.: Boundary Layers of Ships (Three-dimensional effects), Ph.D. thesis, Chalmers University of Technology, (1975). Also available as Reports Nos. 44, 45, 46, 47 and 49 of the Swedish State Shipbuilding Experimental Tank (SSPA).

A Method of Flow Field Determination for Moderately Skewed, Three-Dimensional Flows

F. B. Gessner
S. H. Arterberry

Department of Mechanical Engineering
University of Washington
Seattle, Washington 98195

Abstract

A hot-wire response model is described which can be used to determine the local mean velocity vector and Reynolds stress components in moderately skewed, three-dimensional flows. Initial discussion centers on the development of the model, with particular emphasis on its unique features. Reynolds stress distributions measured in fully-developed turbulent pipe flow under simulated skewed flow conditions are presented. Comparisons are made which demonstrate the validity of the proposed technique and its relative advantages over conventional methods of measurement.

Introduction

This paper presents a novel method for determining the magnitude and direction of the local mean velocity vector and all six components of the Reynolds stress tensor in a low-to-moderate intensity, three-dimensional flow field. This type of flow exists, for example, within curved ducts under wholly attached flow conditions, and within uncurved ducts with swirling inlet flow or with swirl induced by duct rotation. In these and other similar flow situations, it is often not possible to align hot-wire probes in a known manner relative to the local mean velocity vector at each point in the flow before data are taken. The present technique can accommodate a skewed mean flow condition across the wire. By selectively locating a normal wire probe and a rotatable slant wire probe in so-called "optimum planes", response sensitivity can be limited to no more than two mutually perpendicular fluctuations at any one time. This, in effect, uncouples the response equations to the extent that a 3 x 3 linear system can be solved for three of the six Reynolds stress components, and the remaining stress components can be determined by solving a 2 x 2 system and an explicit algebraic equation. The relative merits of this approach will be demonstrated by means of comparisons with data obtained in fully-developed turbulent pipe flow under simulated skewed flow conditions.

The Present Method

The response model is based on the conventional, non-linear form of the cooling law for a hot wire with the effective cooling velocity defined as shown in Fig. 1 for a wire in the xy plane. The slant wire is exposed to a skewed mean-flow condition (V/U, W/U arbitrary) and is considered to be mounted on a probe that can be rotated about its axial centerline, which is coincident with the x axis. If the wire is positioned sequentially in the four positions shown in Fig. 2, the magnitude and direction of the mean velocity vector can be determined from mean wire voltage measurements in the four positions. This was done by applying a bionomial series expansion to the instantaneous form of the cooling law, assuming low intensity fluctuations, and retaining terms up to the fourth order in V/U and W/U. By time-averaging the results and combining expressions for the four positions shown in Fig. 2, two coupled, fourth-order, non-linear algebraic equations for V/U and W/U were developed. These equations were solved by means of the Newton-Raphson algorithm, either directly when the yaw sensitivity factor (k) was assumed constant or by iteration when the yaw dependency of k was modeled by means of the correlation proposed by Friehe and Schwarz [1]. In performing these calculations, the reference value for k (0.15) was selected on the basis of pipe flow calibration data, and the pitch sensitivity factor (h) was specified as a constant (0.9) on the basis of calibration data obtained in the potential core of a free jet. Solving for V/U and W/U enables one to specify the magnitude of V_r/U or, equivalently, ϕ, the resultant flow angle shown in Fig. 3. In order to complete specification of the mean velocity vector, its magnitude (U_0) was evaluated by averaging the U_0 values calculated with the wire in each of the four positions shown in Fig. 2.

If a rotatable slant wire were to be located in any of the positions shown in Fig. 2 for the purpose of making Reynolds stress measurements, the wire would be sensitive to simultaneous fluctuations in three mutually perpendicular directions (u, v, and w, refer to Fig. 3) whenever V and W are both nonzero. If, however, the wire is located in certain symmetric positions relative to the plane of the resultant mean flow (the xy' plane in Fig. 3), fluctuation sensitivity is reduced, and sensitivity patterns are similar to those which exist for non-skewed flow conditions. Figure 4a shows the eight planar positions (A through H) of a rotatable slant wire which are commonly employed in order to evaluate the six Reynolds stress components for non-skewed flow conditions (ϕ = 0 deg). These positions correspond to the condition that $S_{w'} = 0$ (positions A and B), $S_{v'} = 0$ (positions C and D), and $S_{v'} = \pm S_{w'}$ (positions E, F, G, and H), where $S_{v'}$ and $S_{w'}$ are v' and w' fluctuation sensitivities referred to the xy'z' coordinate system shown in Fig. 4a. Under these conditions, a slant wire in the A and B positions (in conjunction with, say, a straight (normal) wire in the xy' plane) can be used to determine the $\overline{u^2}$, $\overline{v'^2}$, and $\overline{uv'}$ stress components. By positioning the slant wire in the C and D positions, $\overline{w'^2}$ and $\overline{uw'}$ can also be determined (since $\overline{u^2}$ is known), and $\overline{v'w'}$ can be evaluated from measurements in the E, F, G, and H positions following the method originally proposed by Rodet [2].

For a skewed flow condition (say, $\phi = 30$ deg) the above conditions on $S_{v'}$ and $S_{w'}$ also apply exactly for the eight planar positions (A through H) shown in Fig. 4b. These positions (and intermediate positions for $0 < \phi < 30$ deg) were determined by analyzing an expression for the instantaneous cooling velocity written in terms of the u, v', and w' fluctuations, the resultant flow angle ϕ, and the rotation angle θ, defined as shown in Fig. 3. Figure 4 shows that as ϕ changes from zero to some nominal value, the A and B positions remain unchanged, which implies that hot-wire measurements in the 0 and 180 degree planes can still be used to determine the $\overline{u^2}$, $\overline{v'^2}$ and $\overline{uv'}$ stress components when ϕ is non-zero. This figure also indicates, however, that when ϕ is non-zero, positions C and D do not lie in a common plane and positions E, F, G, and H no longer form an orthogonal configuration. These results imply that the remaining stress components ($\overline{w'^2}$, $\overline{uw'}$, and $\overline{v'w'}$) should be evaluated by rotating the wire to positions which are non-co-planar and non-orthogonal (e.g., positions C, D, E, F, G, and H for $\phi = 30$ deg, as shown in Fig. 4b).

In order to develop a response equation compatible with the above results, the effective cooling velocity was first written in terms of the u, v', and w' fluctuations, the resultant flow angle ϕ (actually V_r/U) and the rotation angle θ, and then expanded in terms of a binomial series while retaining terms up to fourth order in V_r/U. This expression was subsequently incorporated into the instantaneous form of the cooling law, squared, and time-averaged to yield a general response equation for the Reynolds stresses in terms of sensitivity coefficients: $F(\overline{u^2})$, $F(\overline{v'^2})$, $F(\overline{w'^2})$, $F(\overline{uv'})$, $F(\overline{uw'})$ and $F(\overline{v'w'})$. Figures 5-7 show the variation of these coefficients with rotation angle for a slant wire when $n = 0.45$ (cooling law exponent), $\alpha = 45$ deg, $k = 0.15$ and $h = 0.9$. The results in Fig. 5 indicate that a rotatable slant wire in the A and B positions has relatively high sensitivity to the $\overline{u^2}$, $\overline{v'^2}$, and $\overline{uv'}$ stress components, regardless of the skewness level. Figure 6 shows that the wire retains a high degree of sensitivity to the $\overline{u^2}$, $\overline{w'^2}$, and $\overline{uw'}$ stress components as ϕ changes from 0 to 30 deg when the wire is located in the C and D positions. Similar behavior can be seen in Fig. 7, where the values for $F(\overline{v'w'})$ at the E, F, G, and H positions do not vary markedly as ϕ ranges from 0 to 30 deg. For the wire positions shown in Fig. 4 (A through H), a 3 x 3 linear system was solved for the $\overline{u^2}$, $\overline{v'^2}$, and $\overline{uv'}$ stress components by combining slant wire data referred to the A and B positions with normal wire data. The $\overline{w'^2}$ and $\overline{uw'}$ stress components were determined by solving a 2 x 2 system from slant wire data referred to the C and D positions and $\overline{v'w'}$ was determined by means of an algebraic expression utilizing slant wire data referred to the E, F, G, and H positions. In the present study Gaussian elimination was used to solve the 3 x 3 and 2 x 2 systems, and orthogonal transformations were used to convert the results to alternate coordinate systems to facilitate interpretation of the results.

Experimental Results

1. Method of Simulation In order to examine the feasibility of the proposed technique, normal and slant wire measurements were made near the exit plane of fully-developed turbulent pipe flow for a fixed Reynolds number ($Re_b = 10^5$) with the wires positioned as shown in Fig. 8. Details of the instrumentation and data reduction procedures are given by Arterberry [3]. Initial data were taken under non-skewed flow conditions (ϕ = 0 deg) in order to generate a reference set of Reynolds stress distributions which could serve as a basis for comparison. Figure 9 shows that normal stress distributions measured under non-skewed flow conditions in the present study are compatible with previous results obtained by Lawn [4] and Perry and Abell [5]. In order to examine the influence of a skewed mean-flow condition, slant wire data were taken by first positioning the probe at a fixed angle of inclination relative to the pipe axis (ϕ = 10, 20, or 30 deg). The probe was then rotated about its axis in fixed angular increments (10 to 20 positions depending on the imposed skewness level) and mean and mean-square fluctuating bridge voltage data were recorded at each position.

In the present study, both pure yaw (W/V = 0, Fig. 8b) and combined pitch and yaw (W/V = 1, Fig. 8a) operating conditions were simulated for a given inclination angle ϕ. In order to simulate pure yaw, mean-bridge voltage data were interpolated initially at the θ = 0, 90, 180, and 270 deg positions shown in Fig. 8b by means of a local fourth-order polynomial fit, and the mean flow analysis procedure described earlier was used to determine the magnitude of the mean velocity vector, U_0, the apparent flow angle, ϕ, and values for W/V and V_r/U. The calculated value for W/V was used to determine the orientation of V_r/U and the corresponding positions of the optimum planes, which differed slightly from the nominal positions shown in Fig. 4 whenever V_r/U did not coincide exactly with the y' axis. Mean and mean-square fluctuating bridge voltage values, as well as k values, were then evaluated at the optimum planar locations by interpolation, and the calculation procedure described earlier was used to determine the six Reynolds stress components. For the case of simulated combined pitch and yaw (W/V = 1) the same basic procedure was followed, except that mean-bridge voltage data were initially interpolated at the θ = 45, 135, 225, and 315 deg positions shown in Fig. 8a. For both cases (W/V = 0 and W/V = 1) the solution procedure yielded Reynolds stresses referred to the xy'z' coordinate system shown in Fig. 8c. Orthogonal transformations were then applied in order to convert these results to results referred either to pipe flow coordinates (rθz coordinates) or to "laboratory" coordinates (the xyz coordinate systems shown in Figs. 8a and 8b).

2. Case study: W/V = 0 The pipe flow Reynolds stress distributions measured by the present technique for the case of simulated pure yaw (W/V = 0) and k variable are shown in Fig. 10. In general, the distributions determined for the different set angles of inclination (ϕ = 10, 20, and 30 deg) agree very well with the distributions measured under non-skewed flow conditions (ϕ = 0 deg). The slight deviation of the distribution for u'_θ/u_τ when ϕ = 30 deg from the distributions measured at ϕ = 0, 10, and 20 deg (refer to Fig. 10) is attributable to series truncation effects, rather than a basic flaw in the data or model. The overall level of agreement is generally better than the agreement noted by Moffat and his co-workers [6,7] when they compared their triple-wire results with distributions measured in two-dimensional channel flow under non-skewed flow conditions. The present method, for example, yields $\overline{u_r u_z}/u_\tau^2$ values which generally agree to within 5 percent with the theoretical (linear) distribution for skewness levels as great as 30 deg (Fig. 10), whereas shear stress values based on the triple-wire technique of Moffat, et al., deviate by as much as 15 percent from the theoretical distribution at a skewness level of only 15 deg.

In order to examine the influence of a constant or variable yaw sensitivity factor on the results, pipe flow Reynolds stress distributions were calculated with k specified either as a constant (0.15) or as a variable using pure yaw simulation. In order to isolate effects, the calculations were performed using the set values for ϕ (i.e., 10, 20, or 30 deg), rather than calculated values based on solution of the coupled equations for V/U and W/U. Data were also interpolated at exact optimum planar locations (i.e., with θ referenced to V_r aligned with the y' axis, as shown in Fig. 8c), rather than interpolating data at optimum planar locations determined by referencing θ to the calculated orientation of V_r, as determined from the mean flow analysis. The results are shown in Fig. 11 and indicate that there is essentially no difference between corresponding distributions based on constant or variable k when $\phi \leq 20$ deg. When $\phi = 30$ deg, however, the $\overline{u_r u_z}/u_\tau^2$ and u_r'/u_τ distributions are elevated by approximately 10 percent above their counterparts referred to a variable k model. It should be noted that the magnitude of the deviation would be even greater if shorter wires with larger k values ($k \geq 0.2$) had been used in the present study.

3. Case Study: W/V = 1 In order to investigate the applicability of the present method when combined pitch and yaw are present (W/V = 1), flow angle and Reynolds stress distributions were evaluated by starting the calculations with data initially interpolated at the 45, 135, 225, and 315 deg positions shown in Fig. 8a. The Reynolds stress components determined with respect to the xy'z' coordinate system shown in Fig. 8c were then transformed to components referred to the xyz system shown in Fig. 8a to facilitate interpretation of the results relative to a laboratory coordinate reference frame. For purposes of comparison, the pipe flow Reynolds stress distributions measured for non-skewed flow conditions ($\phi = 0$ deg) were transformed to distributions referred to the xyz coordinate system shown in Fig. 8a. This was done by assuming a linear variation for the primary ($\overline{u_r u_z}$) shear stress distribution in the pipe and letting $\overline{u_r u_\theta} = \overline{u_\theta u_z} = 0$ in order to generate an idealized set of transformed distributions which could serve as a standard for comparison.

The Reynolds distributions measured under combined pitch and yaw simulation with k variable are shown in Fig. 12. For the imposed condition (W/V = 1), $\overline{v^2}$ and $\overline{w^2}$, as well as $\overline{uv}$ and $\overline{uw}$, should be identically equal for all inclination angles, inasmuch as $\overline{u_r u_\theta}$ and $\overline{u_\theta u_z}$ were specified as zero before transforming the pipe flow Reynolds stresses. Figure 12 shows that all six Reynolds stress components referred to the xyz system are non-zero, and that calculated stress distributions based on the present method are in good agreement with transformed pipe flow results when $\phi \leq 20$ deg. When ϕ is 30 deg, the level of agreement between corresponding distributions is not as good, especially between the $\overline{uw}$ distribution and its pipe flow counterpart, because of series truncation effects.

Comparison with Other Methods

The discussion thus far has centered on comparisons based on data interpolated at positions which correspond to the so-called "optimum planes". In order to justify the use of this terminology, calculations were also performed by interpolating slant wire data at the 2 - 8, 3 - 7, and 4 - 6 positions shown in Fig. 13. These six positions correspond to planar orientations where the slant wire is simultaneously sensitive to all three fluctuation components referred to the xy'z' system (u, v', and w') whenever the skewness angle, ϕ, is non-zero. In contrast, the wire is sensitive only to the u and w' fluctuations when it lies in either of the two optimum planes corresponding to a prescribed skewness angle (i.e., the planes denoted by the dashed, dashed-dot, and solid lines in the lowest inset figure in Fig. 13). The consequences of this behavior were examined by evaluating the binormal pipe flow intensity component (u_θ'/u_τ) from interpolated data at each fixed position (i.e., positions 2 through 8) and positions corresponding to the optimum planes. In order to isolate effects, the same procedure as that outlined earlier for calculating Reynolds stress distributions based on constant or variable k was followed. For non-skewed flow conditions, (ϕ = 0 deg), Fig. 13 shows that there are only slight differences, at most, between distributions based on the four different methods of measurement. Calculated u_θ'/u_τ values based on interpolated data at the 3 - 7 positions and at positions corresponding to the optimum planes are, in fact, identical because these positions coincide when ϕ = 0 deg. Values based on interpolated data at the 2 - 8 and 4 - 6 positions exhibit slightly greater scatter about the reference (solid line) distribution when ϕ = 0 deg than results referred to data interpolated at the 3 - 7 (optimum plane) positions. This occurs primarily as a result of cumulative errors associated with slant wire sensitivity to all three fluctuation components (u, v', and w') when the wire is in the 2 - 8 or 4 - 6 positions.

The consequences of maintaining a slant-wire in fixed positions when the wire is exposed to increased skewness levels can be seen by comparing the four calculated distributions corresponding to ϕ = 0, 10, 20, and 30 deg and a given method of measurement. When data are interpolated at the 2 - 8 positions, for example, calculated u_θ'/u_τ values show increased scatter as ϕ increases, until $\overline{u_\theta^2}/u_\tau^2$ values, in fact, all become negative when ϕ = 30 deg. For data interpolated at the 3 - 7 positions, calculated u_θ'/u_τ values lie above the reference distribution when $\phi \geq 20$ deg, and overestimate actual values by 50 percent or more when ϕ = 30 deg. These results have important implications, because they imply that conventional wire rotation techniques which rely on analyzing data obtained with a wire in fixed or arbitrary positions may lead to significant errors in calculated Reynolds stress values when the wire is exposed to a moderately skewed mean flow ($\phi \geq 20$ deg). This supposition applies to previously proposed single and X-wire rotation methods, which are all based on obtaining data with the wire or X array in positions that are either fixed (Mojola [8], Dukler and Chen [9], Acrivlellis [10], Müller and Krause [11]) or arbitrarily specified (Hoffmeister [12], Bissonnette and Mellor [13], DeGrande and Kool [14]).

If data are referenced to the optimum planes, Fig. 13 shows that calculated u_θ'/u_τ distributions agree well with the reference (solid line) distribution, provided $0 \leq \phi \leq 20$ deg. When ϕ = 30 deg, calculated u_θ'/u_τ values lie slightly below the reference distribution as a result of the series truncation effects noted earlier. If calculated values of u_θ'/u_τ are based on data interpolated at the 4 - 6 positions, Fig. 13 shows that the level of agreement is comparable to that observed when calculations are based on data referenced to the optimum planes. As noted earlier, however, values based on data referenced to the 4 - 6 positions exhibit somewhat greater scatter about the reference distribution when ϕ = 0 deg. Furthermore, when ϕ = 20 or 30 deg, calculated values lie below the reference distribution by a slightly greater amount than the optimum planar values. On the basis of these results, it is reasonable to conclude that in moderately skewed flows, Reynolds stress distributions based on single-wire measurements in the optimum planes will be inherently more accurate than distributions based on measurements with single wire probes or X-wire arrays located in fixed or arbitrarily specified positions.

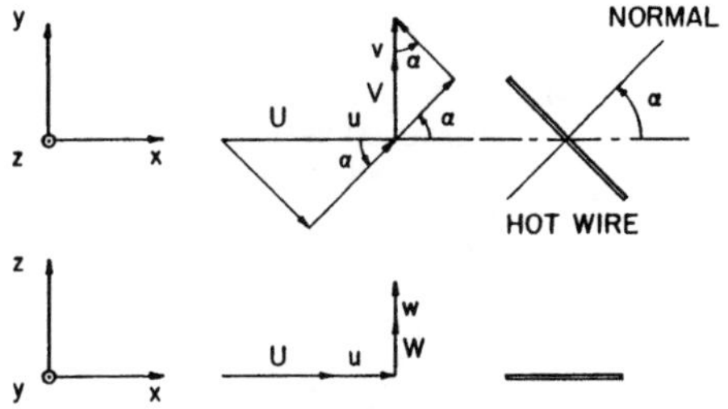

$$q_n^2 = \left[(U+u)\cos\alpha + (V+v)\sin\alpha\right]^2$$

$$q_p^2 = \left[(U+u)\sin\alpha - (V+v)\cos\alpha\right]^2$$

$$q_{bn}^2 = (W+w)^2$$

AND

$$q_{eff}^2 = q_n^2 + k^2 q_p^2 + h^2 q_{bn}^2$$

Fig. 1 Effective cooling velocity for an inclined wire exposed to an arbitrarily skewed mean flow (wire in xy plane)

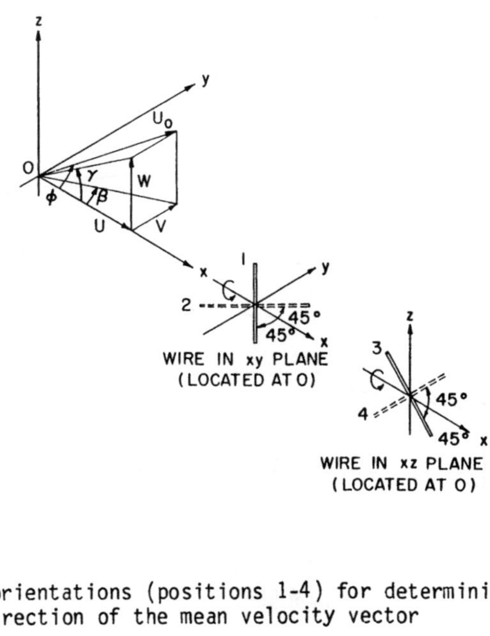

Fig. 2 Wire orientations (positions 1-4) for determining the magnitude and direction of the mean velocity vector

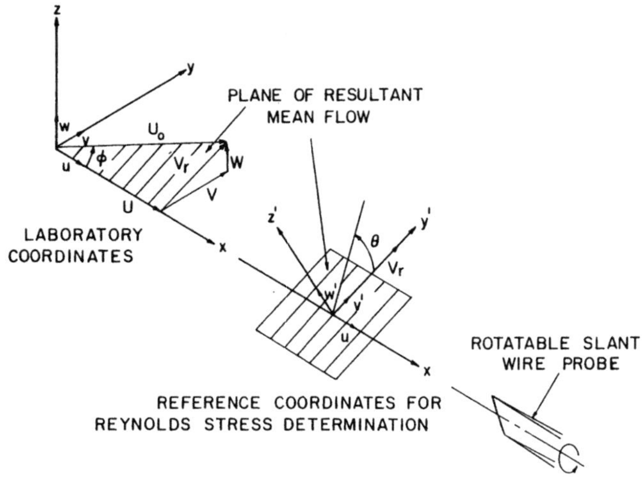

Fig. 3 Reference coordinates for Reynolds stress determination (xy'z' system) relative to laboratory coordinates (xyz system)

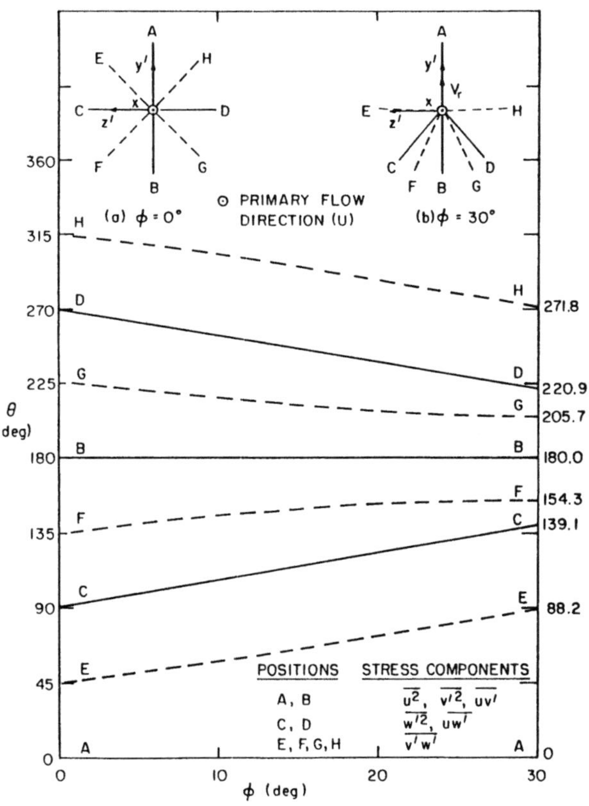

Fig. 4 Optimum planar locations (positions A through H) as a function of the skewness level (wire orientations are shown for (a) ϕ = 0 deg and (b) ϕ = 30 deg)

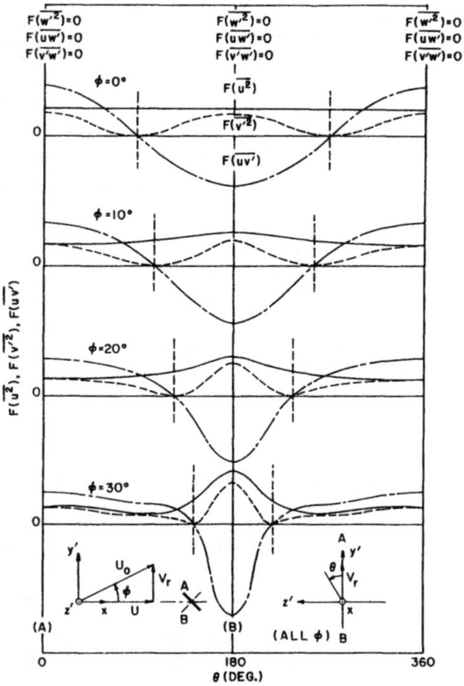

Fig. 5 $F(\overline{u^2})$, $F(\overline{v'^2})$, and $F(\overline{uv'})$ dependency on angle θ

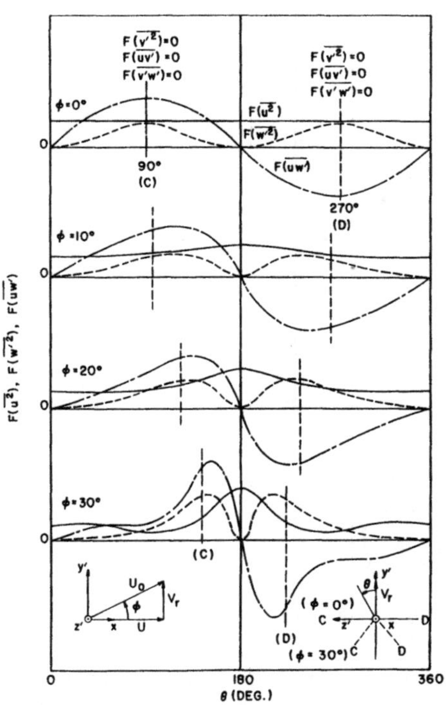

Fig. 6 $F(\overline{u^2})$, $F(\overline{w'^2})$, and $F(\overline{uw'})$ dependency on angle θ

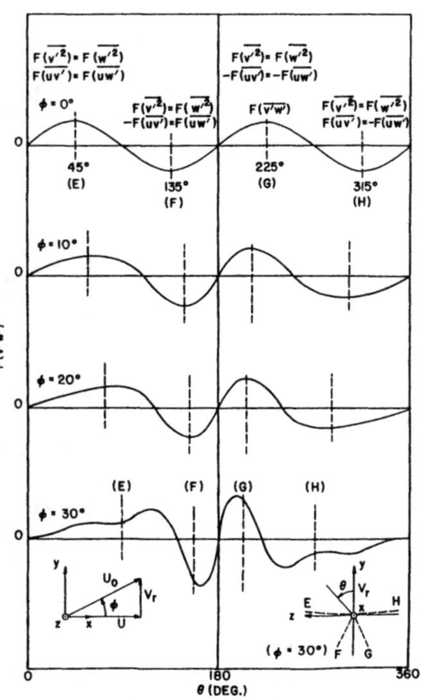

Fig. 7 $F(\overline{v'w'})$ dependency on angle θ

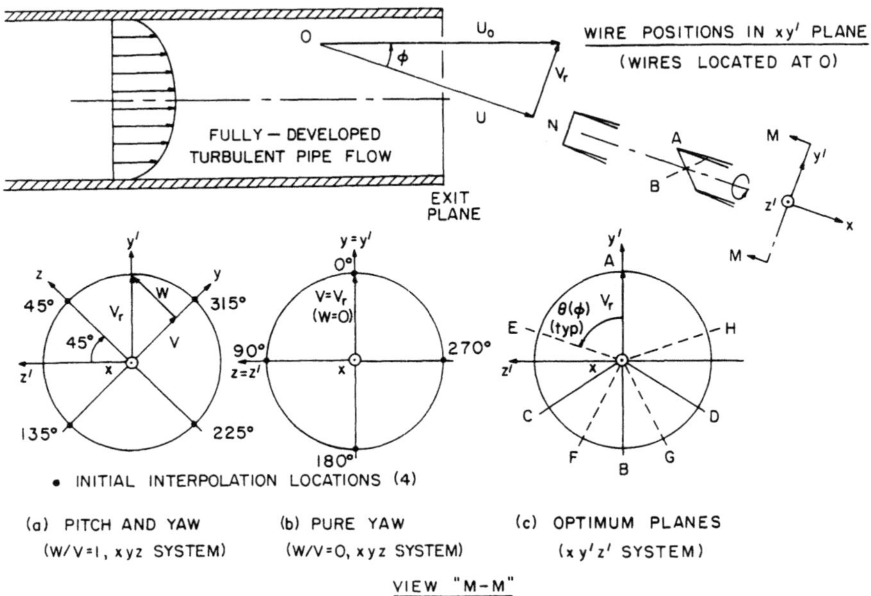

Fig. 8 Method of imposing various skewness levels, as interpreted in terms of (a) combined pitch and yaw and (b) pure yaw simulation

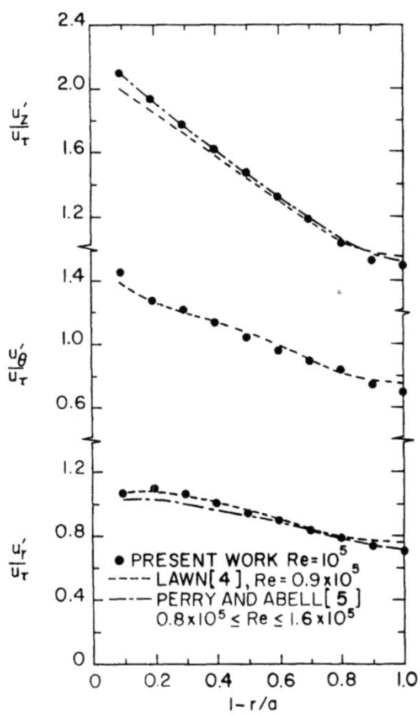

Fig. 9 Normal stress (rms) distributions measured in fully-developed turbulent pipe flow under non-skewed flow conditions (ϕ = 0 deg)

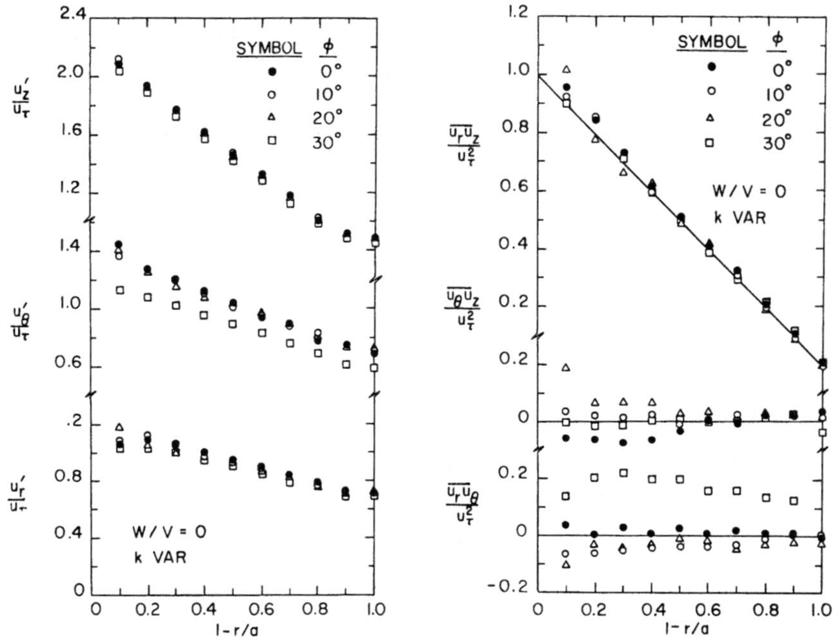

Fig. 10 Pipe flow Reynolds stress distributions, as evaluated for pure yaw simulation (W/V = 0, k var, $Re_b = 10^5$)

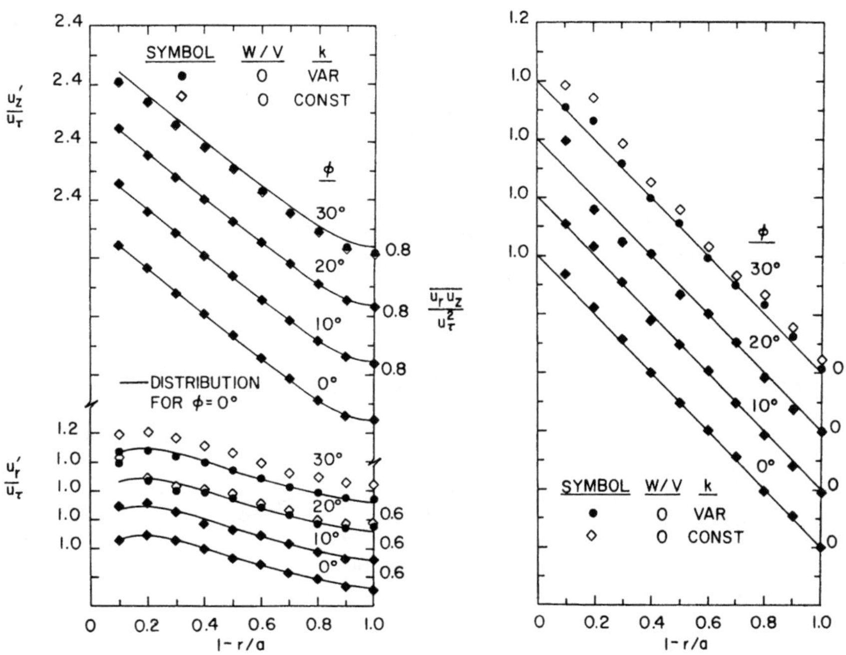

Fig. 11 Pipe flow Reynolds stress distributions, as evaluated for pure yaw simulation with k constant or k variable (W/V = 0, $Re_b = 10^5$)

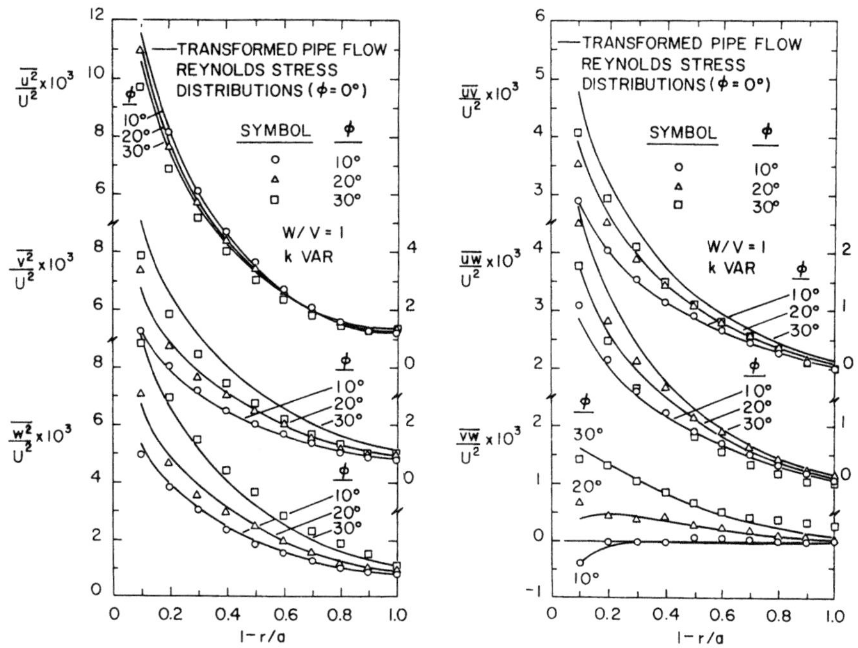

Fig. 12 Reynolds stress distributions referred to laboratory coordinates for combined pitch and yaw simulation (W/V = 1, k var, $Re_b = 10^5$)

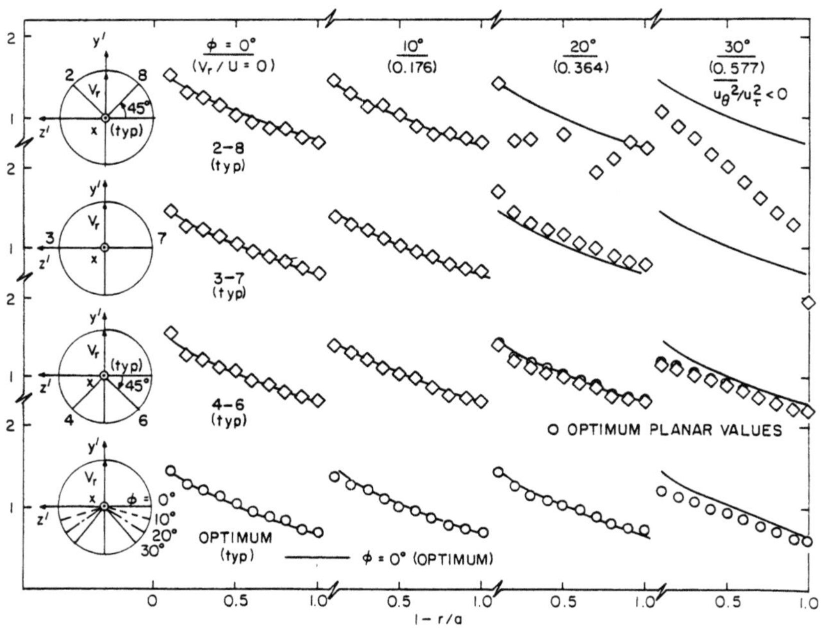

Fig. 13 Binormal pipe flow intensity distributions, as evaluated for pure yaw simulation by four different methods of measurement (W/V = 0, var, $Re_b = 10^5$)

References

1. Friehe, C. A., and Schwarz, W. H.: Deviations from the cosine law for yawed cylindrical anemometer sensors. J. Appl. Mech. 35 (1968) 655-662.

2. Rodet, E.: Étude de l'écoulement d'un fluide dans un tunnel prisimatique de section trapézoidale. Publ. Scient. du Ministère de L'Air, Paris, No. 369 1960.

3. Arterberry, S. H.: An improved method of flow field determination in three-dimensional flows of low intensity and moderate skewness. MS Thesis, Dept. of Mech. Engrg., Univ. of Washington 1982.

4. Lawn, C. J.: The determination of the rate of dissipation in turbulent pipe flow. JFM 48 (1971) 477-505.

5. Perry, A. E., and Abell, C. J.: Scaling laws for pipe-flow turbulence. JFM 67 (1975) 257-271.

6. Moffat, R. J., Yavuzkurt, S., and Crawford, M.E.: Real-time measurements of turbulence quantities with a triple hot-wire system. Proc. Dyn. Flow Conf., Johns Hopkins University 1978.

7. Yavuzkurt, S., Crawford, M. E., and Moffat, R. J.: Real time hot-wire measurements in three dimensional flows. Proc. 5th Bienn. Symp. Turb., Univ. of Missouri-Rolla 1977.

8. Mojola, O. O.: A hot-wire method for three-dimensional shear flows. DISA Info. 16 (1974) 11-14.

9. Dukler, A. E., and Chen, W. N.: Measurement of mean velocities and turbulence structure using hot wires or films in two-dimensional flow. Air flow over a wavy surface. Ind. Engrg. Chem. Fundam. 14 (1975) 359-362.

10. Acrivlellis, M.: An improved method of determining the flow field of multidimensional flows of any turbulence intensity. DISA Info. 23 (1978) 11-16.

11. Müller, U., and Krause, E.: Measurements of mean velocities and Reynolds stresses in an incompressible three-dimensional turbulent boundary layer. Proc. 2nd Symp. Turb. Shear Flows, Imperial College (London) 1979.

12. Hoffmeister, M.: Using a single hot-wire probe in three-dimensional turbulent flow fields. DISA Info. 13 (1972) 26-28.

13. Bissonnette, L. R., and Mellor, G. L.: Experiments on the behavior of an axisymmetric turbulent boundary layer with sudden circumferential strain. JFM 63 (1974) 369-413.

14. DeGrande, G., and Kool, P.: An improved experimental method to determine the complete Reynolds stress tensor using a single rotating, slanting hot-wire. J. Phys. E. Sci. Instrum. 14 (1981) 196-201.

New Developments in Open Separation

K. C. WANG

Department of Aerospace Engineering
and Engineering Mechanics
San Diego State University
San Diego, CA 92182

Summary

New developments in open separation are discussed. First a number of related experiments and calculations reported since 1976 are briefly reviewed. Then the idea that an open separation line starts from a regular point in the middle of surface flow field is demonstrated with new supporting evidence. Finally the open-vs-closed separations developed originally for three dimensional, steady boundary layers are extended by analogy to unsteady cases.

1. Introduction

The open-vs-closed separation concept for three-dimensional flow was introduced by this author [1] about a decade ago. A closed separation is consistent with the usual concept of separation carried over from two-dimensional studies, an open-separation is a new idea. This new idea was evolved at first from the solutions of a symmetry-plane boundary layer [2] over an inclined body of revolution and from judicious interpretation of surface-flow experiments such as those of Werle [3] and Stetson [4]. Further supports prior to 1976 were reported later [5]. In the intervening years since then, a number of new developments [6-20] appeared in the literature. All except Ref. 12 lend additional supports to the open-separation idea, even though some (11,13) raised certain questions. In the present paper, we intend (1) to review briefly these new developments, (2) to clarify the questions about where an open-separation starts, and (3) to extend the same idea to unsteady boundary layer flows. A separate paper is being prepared to answer critics [11,13].

To begin with, we recapitulate some definitions of terminology and notations. By an "open" separation, it is meant that the separation line (figure 2a) on the body surface is not closed in the front leeside surface and does not originate from a singular point where both skin-friction components vanish, the

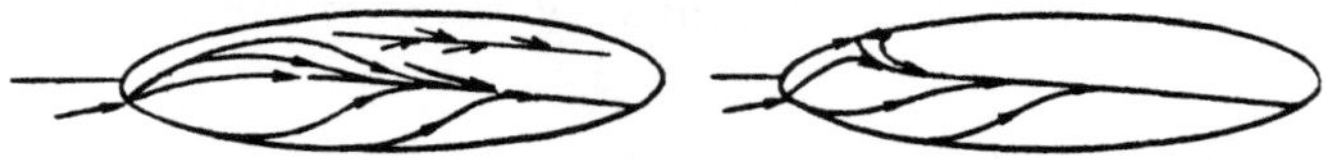

(a) Open (b) Closed

Fig. 1. Open and closed separations

limiting streamlines on both sides of the separation line originate from the same front attachment (or stagnation) point. In contrast, for a closed separation, the separation line is closed in the front, it passes through the singular points of the limiting streamlines and the limiting streamlines on two sides of the separation line originate respectively from the front and the rear attachment points (or lines).

The differences between Maskell's [18] free-vortex-layer separation and our open-separation warrant brief comments, although this question was already discussed in Ref. 5. Referring to Maskell's classical sketch (Fig. 1c) it is understood that the

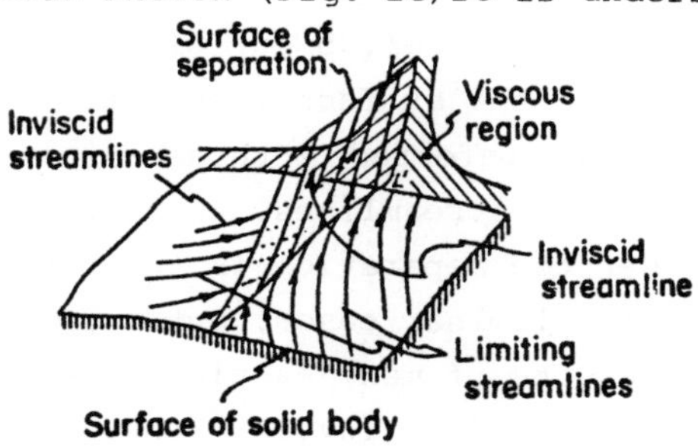

Fig. 1c. Maskell's free vortex layer separation

separation line, LL', contains only ordinary points, the same is emphasized for an open-separation line. In this respect, there is no difference between these two. However, Fig. 1c gives no idea to the following questions: (1) whether the line LL' is open or closed on the front portion of separation (see Figs. 1a, b)?, (2) where the limiting streamlines on two sides of the line LL' may come from?, and (3) where the line LL' begins in an actual flow? In contrast, these questions were clearly defined in the open separation concept. The idea of "openness" is especially crucial, because it is the "openness" which contrasts strikingly to the long-held characterization of separation by "closeness" or "inaccessibility". As a result, the whole concept of separation has fundamentally been changed.

The terms "open" and "closed" were used by Maskell to describe the streamlines. It is evident that whether a separation line is open or closed is a different matter as to whether a streamline is open or closed. Corresponding to our conceptual depiction of Figs. 1a,b, Maskell's sketches for an inclined body were presented in Fig. 9 of his original report. The reader may verify that his sketches there do not contain the open separation idea as we defined.

In short, both free-vortex-layer separation and open separation stress to contain regular points only, but there are subtle differences. A number of features which the open separation idea brought to light were not spelled out in the free-vortex-layer case.

In this paper as well as in Ref. 5 our discussion of separation will be based largely on the surface flow patterns. "Regular" or "singular" points are referred to those associated with the equation of the limiting streamlines. Also no effort is made to distinguish whether the separation is laminar or turbulent. Calculations cited are laminar cases, but experiments involved turbulent flows. Inspite of many subtle differences between laminar and turbulent boundary layers, general separation characteristics are nevertheless very similar in both cases. They differ in extent, but not in character. It is in fact due to this connection which justifies the author's participation to this symposium. Otherwise this author's work has been exclusively confined to laminar cases.

2. Recent Contributions

Most of those recent experiments and calculations after 1976 were made for an ellipsoid of revolution, the same geometry which was used in the author's original calculations. This fact makes direct comparison much easier. In some cases, the major-minor axis ratio is 1/4 which is the same as ours. In others, the value 1/6 was used.

Patel and Han [6] used colored dyes to display the surface flow pattern over an ellipsoid of revolution ($b/a \simeq 1/4$) in water tunnel. Although the limiting streamlines were not sharp enough due to the diffusion of dye, several features were clearly demonstrated in agreement with our predictions. Unfortunately the original color pictures are too expensive to reproduce. Fig. 2a for a low incidence ($\alpha = 5^{o}$) corresponds closely to our

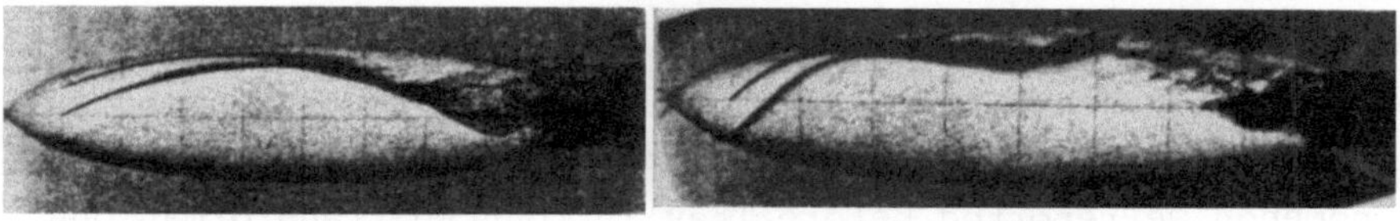

(a) $\alpha=5°$ (b) $\alpha=30°$

Fig. 2. Patel and Han's experiment

calculated case of $\alpha = 6°$ (17). Two features are noteworthy: (1) the streamlines over the front curve downward, confirming significant area of reversed circumferential-flow ahead of separation, and (2) the separation line assumes an unusual tongued shape as revealed by calculation. Fig. 2b for a high incidence (30°) demonstrated the separation being open. More recently, Ramaprian, Patel and Choi [7] measured turbulent boundary layer profiles and skin frictions. Open separation pattern was again reported.

Kreplin, Vollmers and Meir [8] studied the transitions and separation on 1/6 ellipsoid of revolution by means of surface hot film probes and oil flow visualizations. They measured the skin frictions and derived therefrom the limiting flow patterns. Fig. 3 shows their results for the case of $\alpha=30°$ and $U_{\infty}= 45m/s$. An open separation is clearly shown.

Bippes [9] used the hydrogen bubble technique to visualize the boundary layer flow over a 1/6 ellipsoid of revolution. With

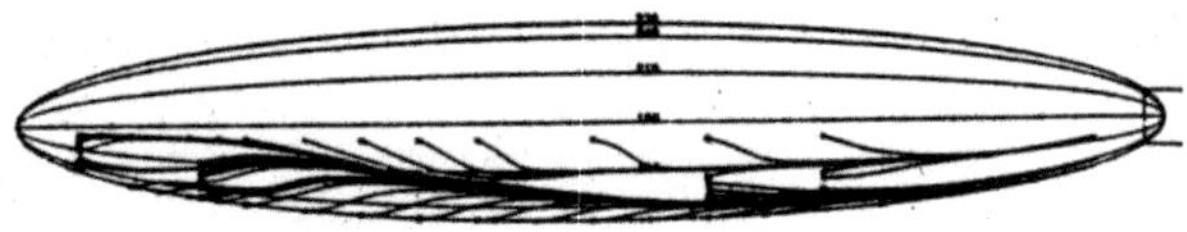

Fig. 3. Limiting flow pattern by Kreplin etc.(Top view)

this method, the symmetry plane flow was beautifully displayed. Open separation was reported to move upstream with increasing incidence which agrees with out description [5]. Discrepancy was reported over the question of separation jump. This discrepancy may be largely due to the influence of a long support stint in his experiment. During a recent visit to Gottingen, Dr. Bippes was kind to show the author his unpublished results which are extensive and extremely stimulating.

Cebeci, Khattab and Stewartson [10] reconsidered our symmetry-plane boundary layer problem and confirmed the separation jump phenomenon we reported before. They introduced transformations which facilitated calculation near the front vertex of the body and demonstrated the same separation jump phenomenon to hold for different values of the major-minor axis ratio.

The same authors [10] also repeated our three-dimensional boundary layer calculations. Our results were again nowhere contradicted by theirs. They acknowledged that open-separation occurs in real flows, but they considered it inappropriate to speak of open-separation within the context of boundary layer theory whereas the solution on the leeside of the separation line could not be calculated. We shall answer their criticism in the separate paper mentioned above.

Peake and Tobak attempted first [12] to contradict our open-separation idea by adhering strictly to Lighthill's concept of separation. But recently they [13] shifted their stand so as to be essentially the same as ours, although they preferred to speak in terms of a different set of terminology. The details will also be discussed in the separate paper mentioned.

An application of open-vs-closed separation to a hatch-back cars is found interesting. Morel [21] reported that the drag coefficient C_D increases as the slanting angle β increases (Fig. 4a), but at $\beta=30°$, C_D suddenly drops and remains almost unchanged thereafter. Such sudden drop of C_D is attributed [22] to a change-over of separation pattern from open to closed (Fig. 4b,c).

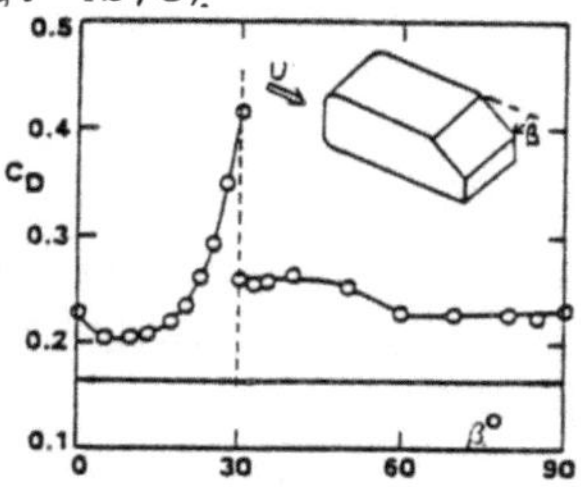

(a) C_D vs β

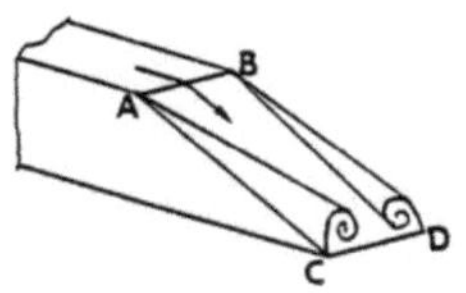

(b) open separation

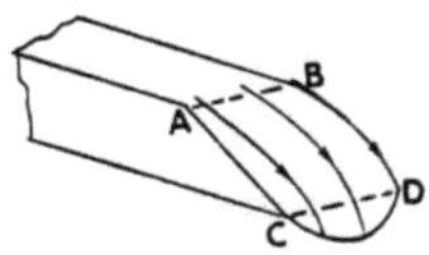

(c) closed separation

Fig. 4. Separation on a hatch-back car

Other recent contributions will be discussed in Sect. 3

3. Origin of Open Separation

Previously documented calculations and experiments [5] did confirm some basic features of an open separation, but the evidence to ascertain where an open separation starts was not sufficient. For example, Stetson's [4] for a blunt cone (Fig. 5a) shows very clearly the separation line being open in the front, but the picture does not indicate the beginning point of the separation line. Likewise, the calculated limiting flow [20] for an ellipsoid of

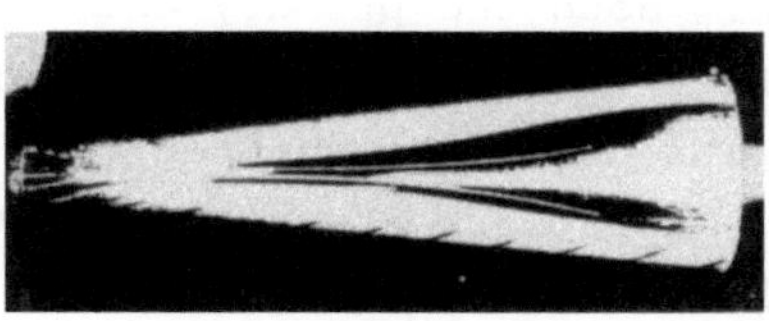

(a) Stetson's experiment (top view) $\alpha=10°$, $M=14.2$

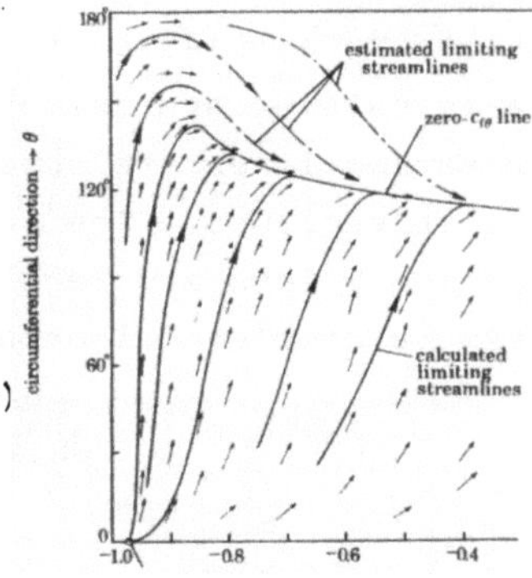

(b) Wang's calculation

Fig. 5. Previous evidence of open separation

revolution at 30° incidence (Fig. 5b) shows again that the separation line is not connected to the leeside symmetry-plane (i.e. $\theta = 180°$) and hence it is open. However, the leeside calculation was very limited, so the flow structure in the vicinity of the separation starting point again was not known for certain. This situation has cast doubts to the whole idea of open separation. We shall now present evidence to demonstrate that an open-separation starts indeed from a regular point in the middle of surface flow field inspite of the fact that this contradicts the long-held notion that a separation line must start from a singularity.

In 1968, extensive surface flow visualizations on an ellipsoid were reported by Atraghji [23], some of his results were widely quoted in the literature [5], but the specific aspect which is of particular interest to our present problem has received little attention. Fig. 6 shows a cut-out of his picture, viewed at $\theta=135°$, over an ellipsoid of revolution ($b/a=1/6$, $M=0.74$, $\alpha=25°$). The primary and secondary separation lines are clearly visible. They start respectively at points A and B where there is no sign of any singular behavior.

Fig. 7 due to Hsieh [24] was quoted in Ref. 5 to represent an example of open separation and nose vortices. It is included here to demonstrate specifically the separation line starting at a regular point.

Figure 8 shows the calculated limiting flow pattern by Risz, Chausse and McRae [14] over a hypersonic inclined sphere-cone using a thin-layer approximation. Both the primary and secondary separation lines are revealed by the running-together of limiting streamlines. The separations are of an open type and the starting points are regular. This work presented for the first time a complete rigorous leeside solution.

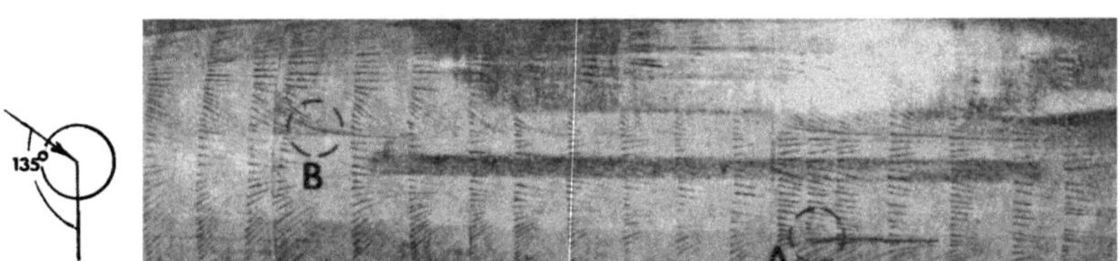

Fig. 6. Atraghji's experiment

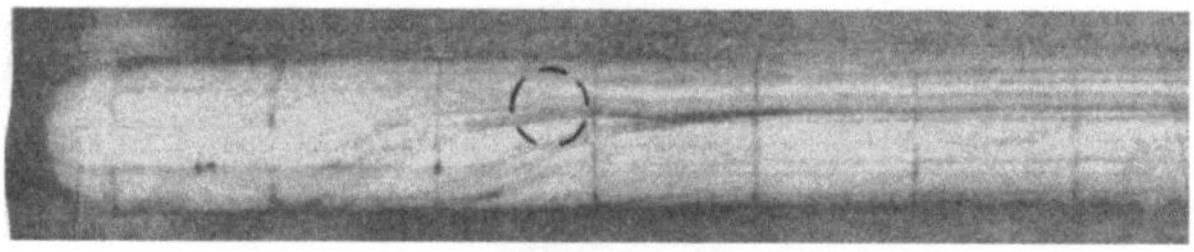

Fig. 7. Sphere-cylinder experiment (Hsieh), side view

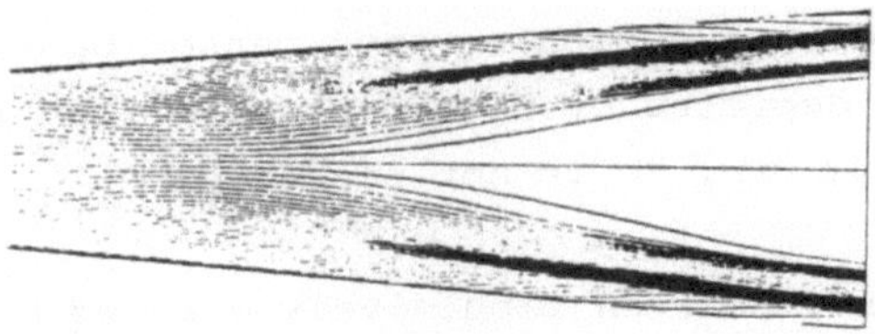

Fig. 8. Calculated pattern on a sphere-cone by Rizk, etc. (Top view)

4. Extension to Unsteady Flow

Due to the similarities between the equations of three-dimensional, steady boundary layers and those of two-dimensional, unsteady boundary layers, several ideas developed for the former may be analogously carried over to the latter. The rule of the zones of dependence and influence [25] was an example in this light, here we give another example related to separation.

The open and closed separations developed for three-dimensional, steady boundary layers are determined by the running together of limiting streamlines on a body surface (see Fig. 1). To carry this idea to two-dimensional, unsteady cases, one can define an analogous set of limiting lines in a x,t-plane [15] where x is measured along a two-dimensional body, and t the time. Similarly unsteady separation may also be classified into open and closed types. An open unsteady separation occurs when there is no separation initially, but separation develops at later times (Fig. 9a). On the other hand, a closed unsteady separation occurs when separation starts at the very beginning of an unsteady motion (Fig. 9b). Like the steady counterpart, unsteady

(a) open

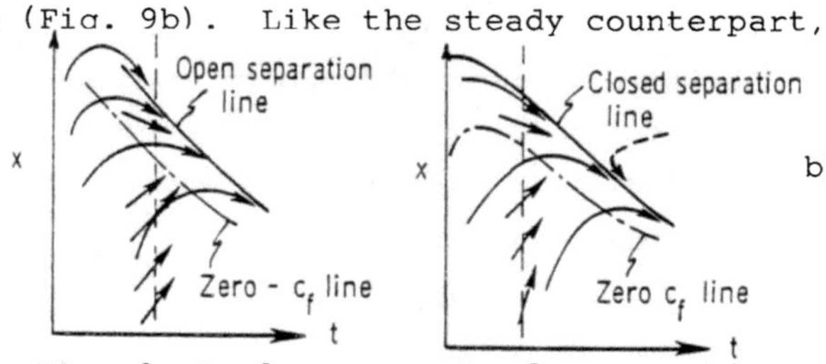

b) closed

Fig. 9. Analogous unsteady separations

open separation means that the limiting streamlines may reach both sides of the separation line, while in a closed separation, the limiting lines will be confined to the same side of the separation line, i.e., the lower side in Fig. 9b. Unlike in the three-dimensional steady case where reversal along both surface directions is common, events cannot be reversed in time. Hence the trend depicted by the dotted limiting line above the separation line in Fig. 9b is not possible.

The classical problem of an impulsively-started circular cylinder was calculated to provide an example. The limiting streamlines in the θ, t-plane (Fig. 10) are shown to run together at

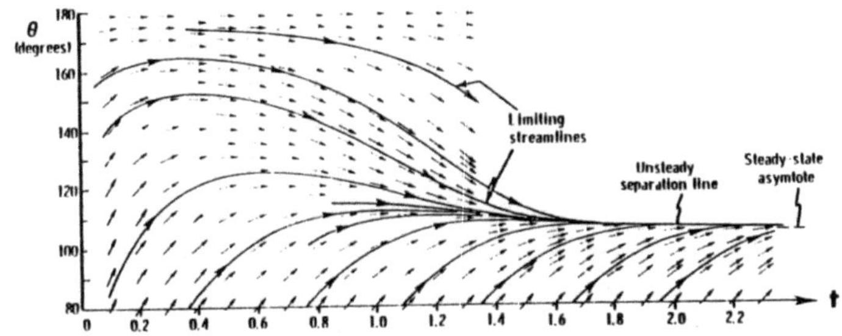

Fig. 10. Calculated unsteady separation

later times, so the separation is of an open type. This conclusion was once involved in a controversy, but the dispute has since been resolved [14]. The case of a closed separation was also calculated in Ref. 15.

Another unsteady separation problem was recently considered [17]. For a steady flow over an ellipsoid of recolution (b/a= 1/4) at high α(=45°), the separation was known [5] to be of a closed type. The question here is how such a steady separation is developed in time if the same body is impulsively started from rest? Intuitively one would expect an unsteady growth sequence as depicted in Fig. 11 a,b. In other words, the separated region gradually extends forward and a closed separation

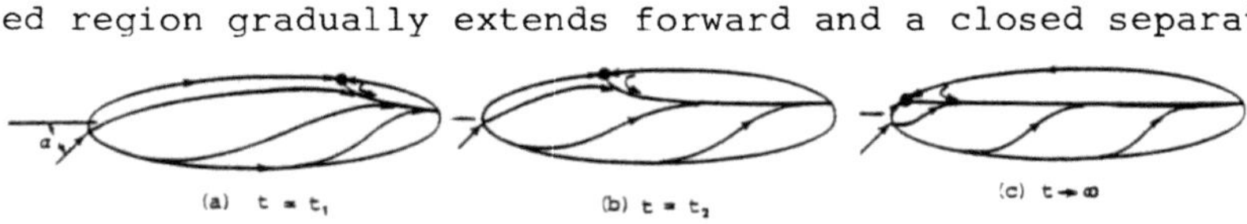

Fig. 11. Conventionally expected unsteady sequence

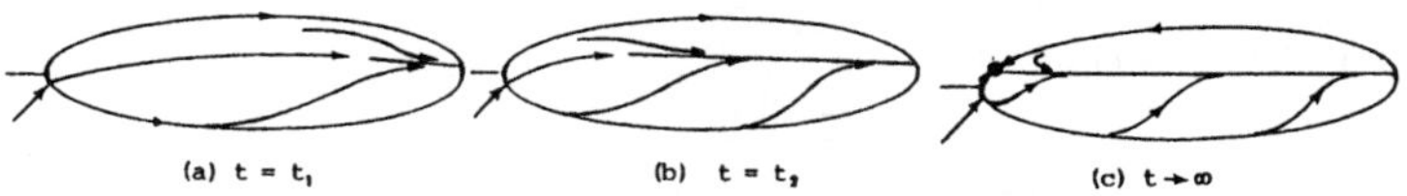

Fig. 12. Proposed sequence of unsteady separation

prevails at every intermediate instant.

However our calculation of the unsteady symmetry-plane boundary layer [17] suggested that the likely sequence is as depicted in Fig. 12, i.e. an open separation prevails throughout the entire growth process, while the closed type of separation would only be reached at the final steady condition. This trend holds only for the high incidence case.

5. Concluding Remarks

After a decade since its inception, the open separation concept has gained wide acceptance and additional confirmations from both experiments and calculations continue being reported in the literature. Its application has been extended from steady to unsteady cases, from aerospace area to automobile design and others.

This work was supported by the Air Force Office of Scientific Research, Air Force Systems Command, U. S. Air Force, under AFOSR Grant -81-0109.

6. References

1. Wang, K. C.: Separation patterns of boundary layer over an inclined body of revolution. AIAA Journal, 10,8, (1972) 1044-1050.

2. Wang, K. C. : Three-dimensional boundary layer near the plane of a spheroid at incidence. J. Fluid Mech. 43, 1 (1970) 187-209.

3. Werle, H.: Separation on axisymmetrical bodies at low speed. La Recherche Aeronautique, 90, Sept.-Oct.(1962)

4. Stetson, K. F.: Boundary layer separation over slender cones at angle of attack. AIAA Journal, 10,5, (1972) 642-648.

5. Wang, K. C.: Separation of three-dimensional flow. Proc. of Viscous Flow Symposium (by invitation),

Lockheed Georgia Company, LG 77ER-0044, Atlanta, GA (1976) 341-414.

6. Han, T. and Patel, V. C.: Flow separation on a spheroid at incidence. J. Fluid Mech. 92,4, (1979) 643-657.

7. Ramaprian, B. R.; Patel, V. C. and Choi, D. H.: Mean flow measurements in the three-dimensional boundary layerover a body of revolution at incidence. J. Fluid Mech. 103 (1981) 479-504.

8. Kreplin, H.-P.; Vollmers, H. and Meir, H. U.: Experimental determination of wall shear stress vectors on an inclined prolate spheroid. DFVLR IB 251 80 A 08 (1980).

9. Bippes, H.: Visualization of the flow along a prolate spheroid at high incidence in the water towing tank of DFVLR. DFVLR Rep. IB 251-79. A 18 (1979).

10. Cebeci, T.; Khattab, A. K. and Stewartson, K.: On nose separation. J. Fluid Mech. 97, 3 (1980) 435-454.

11. Cebeci, T.; Khattab, A. K. and Stewartson, K.: Three-dimensional boundary layer and the OK of accessibility. J. Fluid Mech., 107 (1981) 57-87.

12. Tobak, M. and Peake, D. J.: Topology of two-dimensional and three-dimensional separated flows. Paper AIAA-79-1481, (1979).

13. Peake, D. and Tobak, M. : Topological structure of three-dimensional separated flow. AIAA paper-81-1260, (1981).

14. Rizk, Y. M.; Chaussee, D. S. and McRae, D. S.: Computation of hypersonic viscous flow around three-dimensional bodies at high angles of attack. Paper AIAA-81-1261 (1981).

15. Wang, K. C.: Unsteady boundary layer separation. Martin Marietta Labs. TR-79-16c (1979).

16. Wang, K. C.: On current controversy of unsteady separation. Proc. of the symposium on numerical and physical aspects of aerodynamic flow, Long Beach 1981.

17. Wang, K. C. and Fan, Z. Q.:Unsteady symmetry-plane boundary layer and 3-D unsteady separation, Part I. high incidence. San Diego State University. AE&EM TR 82-01 (1982).

18. Maskell, E. C.: Flow separation in three dimensions. Royal Aircraft Establishment, RAE Rep. Aero 2565 (1955)

19. Wang, K. C.: Boundary layer over a blunt body at low

incidence with circumferential reversed flow. J. Fluid Mech. 72, 1 (1975), 49-65.

20. Wang, K. C. : Boundary layer over a blunt body at high incidence with an open-type separation. Proc. Royal Soc. London A 340 (1974) 33-55.

21. Morel, T.: Aerodynamic drag of bluff body shapes characteristic of hatch-back cars. Research publication GMR-2581, General Motors Labs. (1977).

22. Sedney, Ray: Private communication (1981)

23. Atraghji, E. G.: Surface flow visualization, surface pressure and surface Preston tube pitot pressure measurement over a 6:1 ellipsoid at incidence at M= 0.3 and 0.74. NRC Rep. 5x5/0032,Ottawa, Canada (1968).

24. Hsieh, T.: An investigation of flow about a hemisphere-cylinder at 0- to 19-deg. incidence in the Mach number range from 0.6 to 1.5. AEDC-TR-76-112, (1976).

25. Wang, K. C.: Aspect of "multitime initial-value problem" originating from boundary layer equations. Phys. of Fluids, 18, 8, (1975) 951-955.

Unsteady Structures of a Separated Three-Dimensional Turbulent Boundary Layer

J. ISHII and S. HONAMI

Department of Mechanical Engineering
Science University of Tokyo
Shinjuku, Tokyo 162, Japan

Summary

A discription is given of the unsteady flow behaviors at a plane of symmetry in a confined jet impinging on a back wall including separation of three-dimensional turbulent boundary layers. Flow direction intermittency by a thermal tuft and wall pressure fluctuations are obtained in the separated region. A combined simultaneous flow visualization with the thermal tuft is also made. Two kinds of vortices are found to occur. The primary separation vortex just ahead of the back wall has a low frequency in the wall pressure fluctuation and a long period of the flow intermittency. The secondary vortex near the separation point with smaller pressure fluctuations has more periodic feature than the primary one. Little propagation of the unsteady disturbance due to vortex motions towards the upstream direction is observed.

Introduction

The investigations of a three-dimensional turbulent boundary layer along with separation have become a particularly important subject of concern. A comprehemsive survey of the literature on three-dimensional turbulent boundary layers, prior to 1976, was given by Johnston [1]. He showed the various categories of experiments which were fully three-dimensional. One of them is a free jet or a confined jet over a flat wall impinging on a back wall or an obstacle. In this type of flow, boundary layer separation associated with a horse-shoe vortex is a common phenomenon. There exist two kinds of separation in the three-dimensional boundary layer both on and off a plane of symmetry. Separation points occur at singular points or ordinary points as mentioned in [2]. The singular separation points is defined as a point where the wall shear stress goes to zero like two-dimensional separation. At the ordinary separation point, a streamline leaves the wall, but the wall shear stress need not vanish.

The early study [2] revealed that a simple relationship might exist between two mean velocity components of the skewed boundary layer. Dechow [3] demonstrated recently that the skew angle of two turbulent shear stress did not coincide with that of the mean velocity gradient across the boundary layer. All measurements of these studies were limited in a region upstream of the above separation points, since flow unsteadiness might appear behind separation.

Our understanding of the characteristics on the stall or separation in two-dimensional flows has been improved by the developments of instrumentation technology. The separated flow structure of a two-dimensional turbulent boundary layer was investigated by use of a directionally sensitive laser anemometer [4]. Unsteady detached flows in two-dimensional diffusers were examined by using a thermal tuft probe [5]. Furhtermore, flow unsteadiness in a two-dimensional flow downstream of a backward-facing step was explored [6]. The unsteady flow behaviors of a three-dimensional boundary layer including separation, however, are not understood in detail, although knowledge of this type of flow is needed.

The objective of the present paper is to present an experimental study on the unsteady flow behavior at the singular separation point and the dynamic behavior of the large scale vortex ahead of the back wall. Measurements are made on the plane of symmetry immersed in fully three-dimensional turbulent boundary layers.

Experimental Apparatus and Techniques

Fig.1 shows an experimental test rig. It consists of a straight duct with a rectangular cross section and a test section made of a 10 mm thick lucid plate. The boundary layer at a reference point 100 mm upstream of the test section inlet was confirmed to be turbulent by means of a tripping plate at a nozzle exit. The free stream velocity at the reference point was 20 m/s (±1%) with the turbulence intensity of 0.5 %. The momentum thickness Reynolds number for the tests was 1530.

The air issued from the straight duct proceeds as a confined jet between the upper and lower test walls and impinges on a back wall. The back wall causes the jet to diverge symmetrically with respect to the plane of symmetry which corresponds to a hatched portion in Fig.1. Mean velocity profiles on the plane of symmetry are collateral, though the flow field is three-dimensional. The flow at the plane of symmetry, however, is not two-dimensional since it is completely immersed in a three-dimensional flow. Therefore, there exists a divergent effect, i.e., $\partial W/\partial z$. With the three-dimensional boundary layer development on the upper and lower walls, the three-dimensional separation mentioned before occurs upstream of the back wall.

The smoke wire technique for flow visualization was used in the present experiment. The improved strobo light source with a shorter duration time

(10 μsec) was employed to obtain a frozen picture of an instantaneous flow. A combined simultaneous measurement of a thermal tuft with the flow visualization was also made.

The thermal tuft or flow direction probe is an instrument which indicates the instantaneous flow direction. Fig.2 shows a sketch of the probe used. The principle and the electronic circuit of the thermal tuft probe were discribed in detail [7]. The fraction of the time, γ_p, that the flow is in the downstream direction was obtained by use of both the wall and traversing probes. The time average γ_p reported here was obtained by averaging fifty sets of data measured during 1 sec with a sampling rate of 0.01 sec. A comparison of γ_p was also made by time-average for 256 sec of a volt meter with a Volt-Frequency converter; quantitative disagreement between them was negligible. The thermal tuft data was confined to measurements on the upper wall.

Wall static pressure fluctuations were measured by use of a diffusion type of a semiconductor pressure transducer flush-mounted on the upper wall. The resolution in pressure is 1.0 Pa. The power spectrum density of pressure fluctuations was reduced by the one-dimensional autoregressive program.

Results and Discussions

Fig.3 shows mean wall-static pressure coefficient C_{pw} and the pressure gradient dC_{pw}/dx differentiated by using a cubic spline data-smoothing program. The present flow situation is governed strongly by an adverse pressure gradient. The free stream does not decelerate monotonously but accelerates locally behind a singular separation point. The separation point was decided by thermal tuft measurements and flow visualizations which will be discussed later. The pressure gradient exhibits a maximum value and decays abruptly afterthere. The singular separation point locates after the peak pressure gradient, while separation occurs after decay of the pressure gradient in the two-dimensional cases [4,5].

Fig.4 shows mean velocity profiles in U^+-Y^+ coordinates. These profiles were measured with a cobra total head probe (O.D.=1mm) and wall static taps. The measured velocities were corrected for the effect of velocity gradient. Mean velocity profile shows the law of the wall at the reference point. The logarithmic law region becomes narrow with increasing x because of the adverse pressure gradient. The law of the wall is still valid near the separation point, even though the logarithmic law region is quite narrow.

Typical flow visualizations in the separated region are illustrated in

Fig.5. Also, the results of a combined simultaneous measurement of flow visualization with the thermal tuft are shown in Fig.6, where locations of the thermal tuft probe installed are specified by an arrow. The delay time t_s between the strobo flush and the camera's shutter trip which activates electrical heating of the smoke wire is adjusted in each photo. The flow visualization photos illustrate two kinds of vortices in the separated region. After examining a large number of photos, we define a vortex of a large scale (diameter) near x=450 mm just ahead of the back wall as a primary vortex and a small one near x=400 mm as a secondary vortex. The primary vortex is characteristic of impinging jets against the obstacle on the wall and different from a separation bubble as in two-dimensional flows. On the other hand, the secondary vortex might not be generated by the primary vortex motion but by an entrainment of the separation shear layer downstream of a separation point. The above two vortices rotate about a spanwise z-axis and also in the same direction; namely the direction of the vorticity vectors is the same each other. The primary vortex changes a size and location remarkably and sometimes grows up to a half height of the test section. When the vortex center moves around near x=450 mm, the size becomes large, whereas it is smaller off this location. However, the changes in size of the secondary vortex is not so remarkable. Since a separated region upstream of the primary vortex is not affected by the primary vortex motions, it is divided into two portions; the primary vortex region and the separating region.

Fig.7 shows the distribution of the percent downstream flow time, γ_p, measured at y=0.4 mm from the upper wall where the symbol I indicates the standard deviation. y was measured from the upper wall in Fig.7 and 9. The time mean separation point of two-dimensional flows was assumed to be a location where γ_p was 50 %, since the mean wall shear was zero at the location of 50 percent downstream flow [8]. Even in the case of three-dimensional turbulent boundary layer separation, the singular separation point may be decided by extension of the two-dimensional flow definition. Accordingly, the present singular separation point is located at x_s=338 mm. A range from an incipient detachment (ID) point (γ_p=99 %) to the singular separation point covers $3.2\delta_{ID}$, while it amounts to $4.7\delta_{ID}$ in the two-dimensional separation of Simpson [4] and 4.0-4.8δ_{ID} in the 10° two-dimensional diffuser by Ashjaee [5] (Fig.8). A divergent effect with respect to the plane of symmetry, i.e., $\partial W/\partial z$, retards separation in contrast to the two-dimensional boundary layer.

Fig.9 shows an isoplot of γ_p in the separated region. The existence of two kinds of separation vortices illustrated by flow visualization is reconfirmed, since two 20 percent lines are found near the wall in the separated region. The time average locations of the primary and secondary vortex centers are located at 450 mm and 400 mm, respectively. A 50 percent line corresponds to a dividing streamline and passes through two separation vortex centers. The local increase of γ_p at x=425 mm in Fig.7 and the dip of γ_p isoplot near x=425 mm are due to the interference of two kinds of vortices. Therefore, the isoplot of γ_p can show the time average flow situation in the separated region.

Samples of the thermal tuft probe output with a simultaneous flow visualization photo are shown in Fig.6, where an upstream flow corresponds to a reverse one. A trigger pulse activates a strobo flush. The flow reversal is certain to occur intermittently in the up-and-downstream directions. In Fig.6(a), a reversed flow near the thermal tuft probe is observed from the combined measurement; the flow situation from flow visualization corresponds to the probe output. Fig.6(b) shows the other combined measurement in the primary vortex region. The flow situation is not common but particular, since the intermittent downstream flow occurs even in this region. A free stream jet appears to inject toward the upper wall and cause the primary vortex to drift into the corner of the back wall. Nevertheless, the secondary vortex still exists near x=400 mm. Then, those measurements provide us with more useful informations on the vortex behaviors. A defect of flow visualization which indicates an instantaneous flow feature can be improved by a combination of a thermal tuft technique. Therefore, combined simultaneous measurements of flow visualization with the thermal tuft is a powerful and effective tool to understand the flow behavior and history in the unsteady flow field with detachment and reattachment.

Short-time averages of thremal tuft data were employed to clarify the unsteadiness of the separation point. Shift of a short-time-averaged separation point were obtained by use of three thermal tuft probes. The experimental procedure is similar to the previous measurements on the reattaching point in a two-dimensional backward-facing step flow [6], except a sample rate of 1 msec and a short-time average of 0.1 sec interval. A newly devised circuit of the thermal tuft probe was used to improve drift and noise characteristics [9]. One of the records is shown in Fig.10. The short-time-averaged separation point varies from about 310 mm to 370 mm around the time mean separation point at x=338 mm. Separation point motions in up-and-downstream directions are introduced by a complicated mechanism. The record or the other records show that the separation point seems to move around the time mean separation point for a relatively long period and shift intermittently by a large scale motion.

Fig.11 shows distributions of the RMS values and the power spectrum density of wall static pressure fluctuations. A natural frequency of the test rig was 84.8 Hz from free oscillation measurements using a strain gauge. The RMS value near the separation point has almost same level as that in a attached boundary layer, even if the separation point moves in up-and-downstream directions. The RMS value exhibits a noticeable increase after separation and a peak in the primary vortex having the dominant frequency of about 5 Hz in the power spectrum density. While the power spectrum has an obvious peak near 7 Hz in the secondary vortex, no discriminative peak is observed near the separation point, even though having a wide frequency range of the power spectrum less than 20 Hz. The dominant frequency of the power spectrum density at x=400 mm and 450 mm is 6.8 Hz and 5.3 Hz, respectively. In consequence, propagation of the large unsteady disturbances corresponding to pressure fluctuations toward the upstream direction is diminished, even if the primary vortex motion causes large pressure fluctuations.

Fig.12 shows the probability distributions of the flow reversal period measured at 0.5 mm from the wall. The period was reduced from the thermal tuft records using a minicomputer with a A-D converter. Most of flow reversals near the separation point have a short period less than 0.2 sec, while the result in the primary vortex shows a longer period more than 0.2 sec and low probability at a wide range. Accordingly, the flow reversals in the separated region including the separation point and the secondary vortex are of more periodic features than the primary vortex. Therefore, the period of the flow reversal in the primary vortex has a longer time compared with that in the vicinity of the singular separation point as in the case of the wall static pressure fluctuations.

Conclusions

According to the measurements of flow visualization, flow directional intermittency and wall pressure fluctuations in the separated region of a three-dimensional turbulent boundary layer, there exist two kinds of vortices. The primary vortex just ahead of the back wall has a low frequency in large wall pressure fluctuations and a long period of flow reversals. The flow reversals on the separation point and in the secondary vortex have more periodic features than those in the primary vortex. The flow in the separated region is governed by the unsteadiness with a low frequency less than 20 Hz. Propagation of the large unsteady disturbances toward the upstream direction is diminished, although the primary vortex motion causes large pressure fluctuations.

Acknowledgment

The present project was supported by the Japanese Ministry of Education under the Grant-in-Aid for Scientific Research. The authors are indebted to Prof. F.Hara and Mr. T.Yamashita of Science University of Tokyo for their advices. The authors wish to thank Messrs K.Takenaka and M.Okudaira who assisted in data acquisition.

References

1. Johnston, J.P.: Experimental studies in three-dimensional turbulent boundary layers. Stanford University Report MD-34 (1976).

2. Johnston, J.P.: Three-dimensional turbulent boundary layer. MIT Gas Turbine Laboratory, Report 39 (1957).

3. Dechow, R.: Mittlere Geschwindigkeit und Reynoldsscher Spannungstensor in der dreidimensionalen turbulenten Wandgrenzschicht vor einem stehenden Zylinder. Diss. University of Karlsruhe (1977).

4. Simpson, R.L.; Strickland, J.H.; Barr, P.W.: Features of a separating turbulent boundary layer in the vicinity of separation. Journal of Fluid Mechanics, Vol.79 (1977) 553-594.

5. Ashjaee, J.; Johnston, J.P.; Straight-walled, two-dimensional diffusers— transitory stall and peak pressure recovery. ASME Journal of Fluids engineering, Vol.102 (1980) 275-282.

6. Eaton, J.K.; Johnston, J.P.: Low frequency unsteadiness of a reattaching turbulent shear layer. Proceedings of Third Symposium on Turbulent Shear Flow, Davis, CA (1981).

7. Eaton, J.K. et al.: A wall-flow-direction probe for use in separating and reattaching flows. ASME Journal of Fluids Engineering, Vol.101 (1979) 364-366.

8. Westphal, R.V.; Eaton, J.K.; Johnston, J.P.: A new probe for measurement of velocity and wall shear stress in unsteady, reversing flow. ASME Journal of Fluids Engineering, Vol.103 (1981) 478-482.

9. Saitoh, S.; Honami, S.: Backward facing step flows in a curved channel. (1982) Submitted to Journal of Turbomachinery (in Japanese).

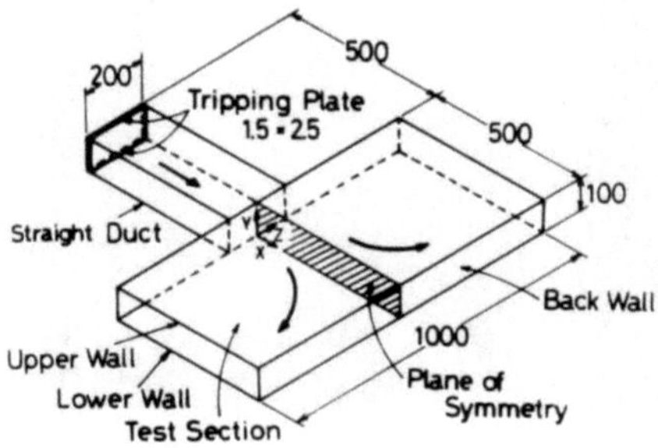

Fig. 1 Experimental apparatus - all dimensions in mm

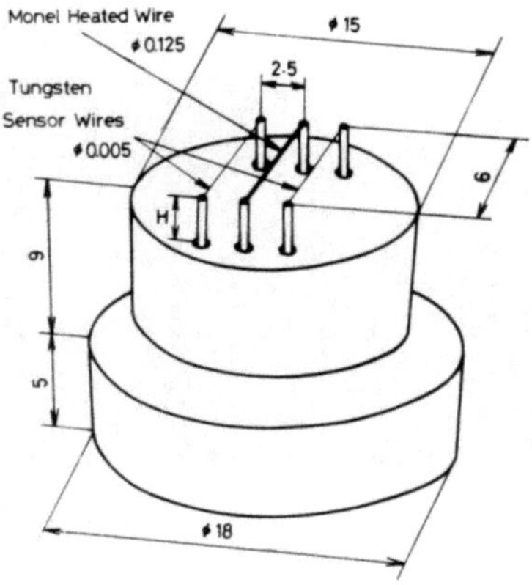

Fig. 2 Thermal tuft probe (dimensions in mm)

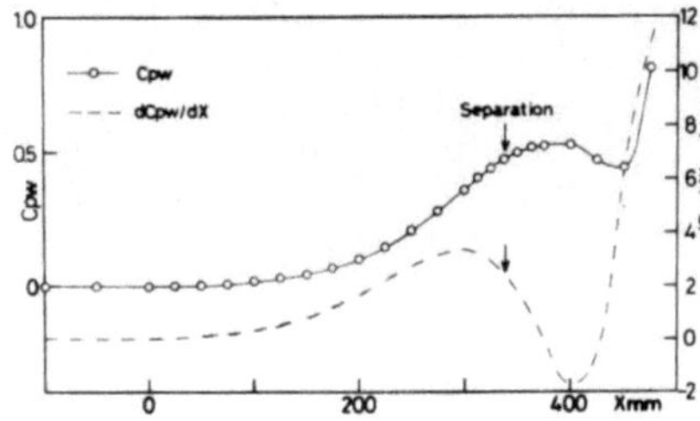

Fig. 3 Wall static pressure parameters: pressure coefficients and pressure gradients

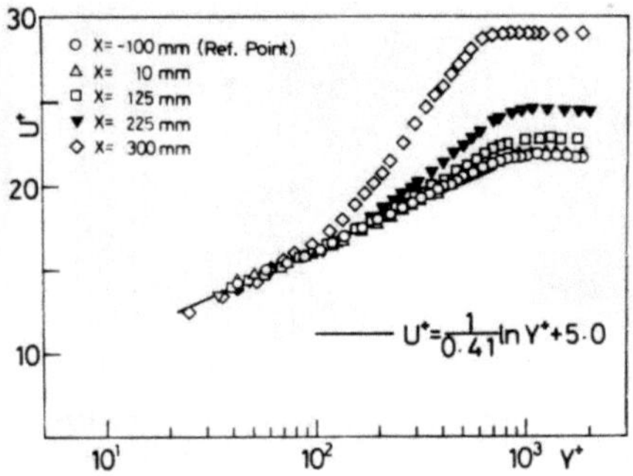

Fig. 4 Mean velocity profiles

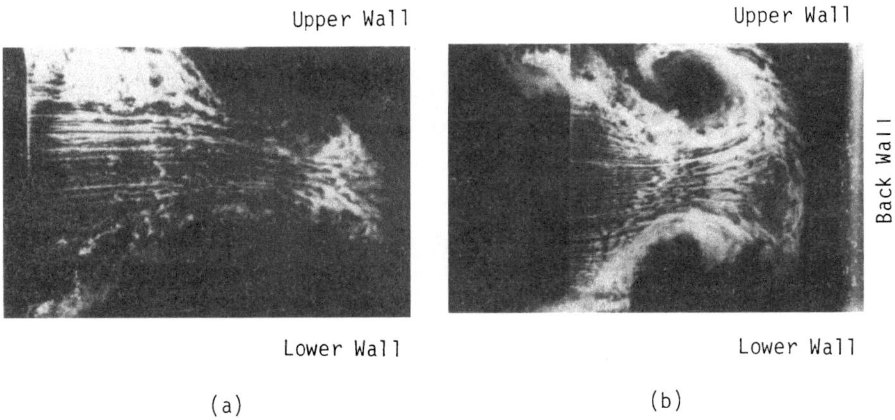

Fig. 5 Flow visualization in the separated region
(a) Side view near the singular separation point: smoke wire location x_W=300 mm; delay time t_S=13.2 msec
(b) Side view ahead of the back wall: smoke wire location x_W=400 mm; delay time t_S=13.4 msec

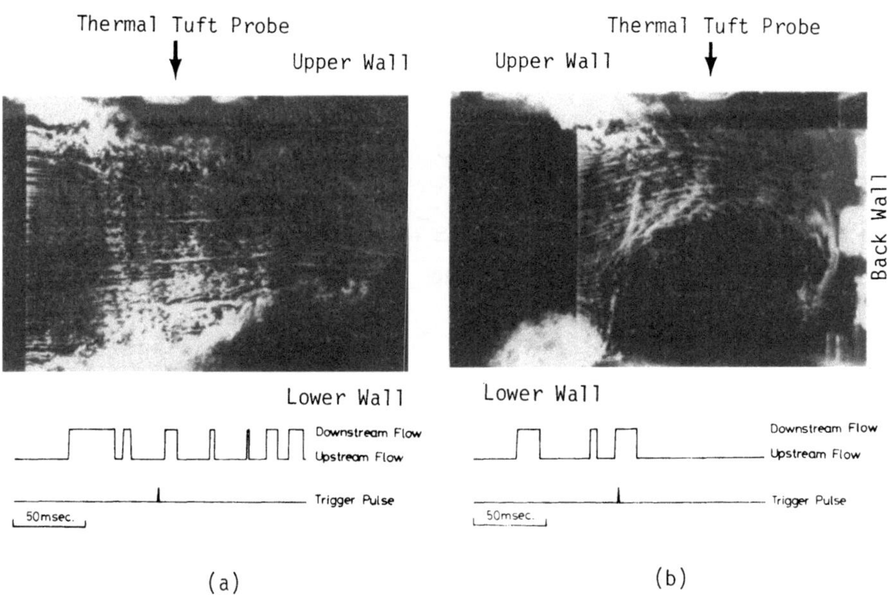

Fig. 6 Simultaneous visual and thermal tuft measurement
(a) Typical flow situation: smoke wire location x_W=300 mm; thermal tuft location x=350 mm; delay time t_S=15.0 msec
(b) Particular flow situation: smoke wire location x_W=400 mm; thermal tuft location x=450 mm; delay time t_S=13.2 msec

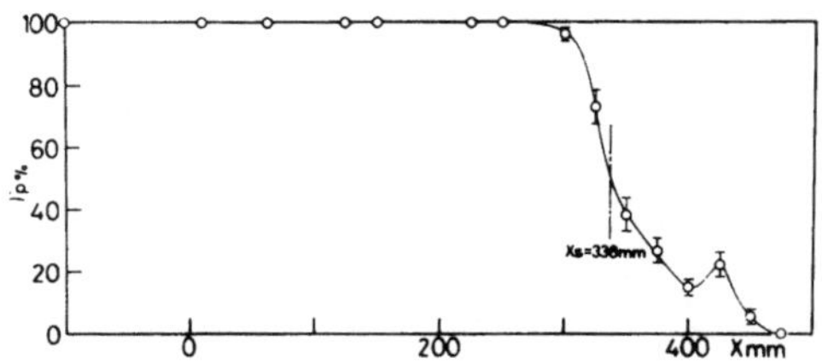

Fig. 7 Percent downstream flow time γ_p on the plane of symmetry

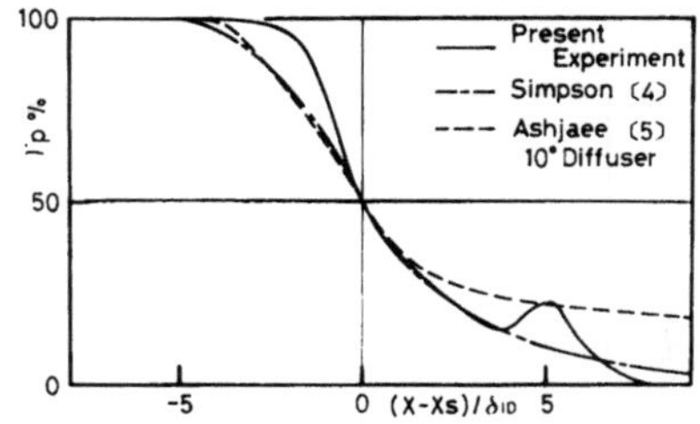

Fig. 8 Comparison of γ_p with two-dimensional boundary layer separation [4,5]

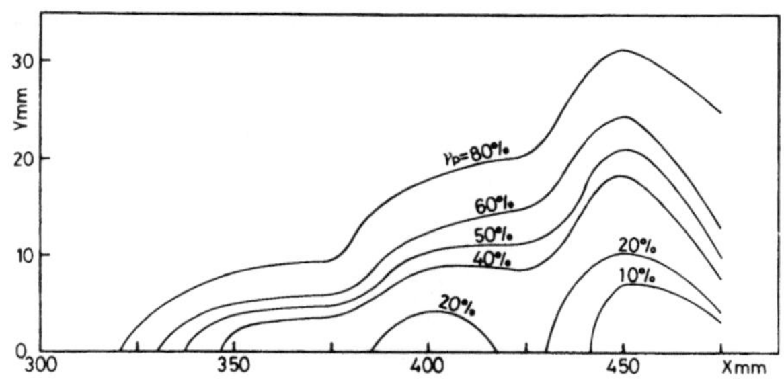

Fig. 9 Isoplot of the percent downstream flow time on the plane of symmetry

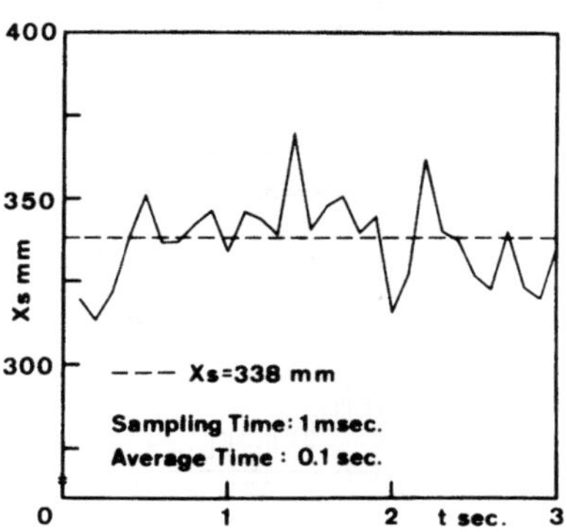

Fig. 10 Shift of the short-time-average (0.1 sec) separation point

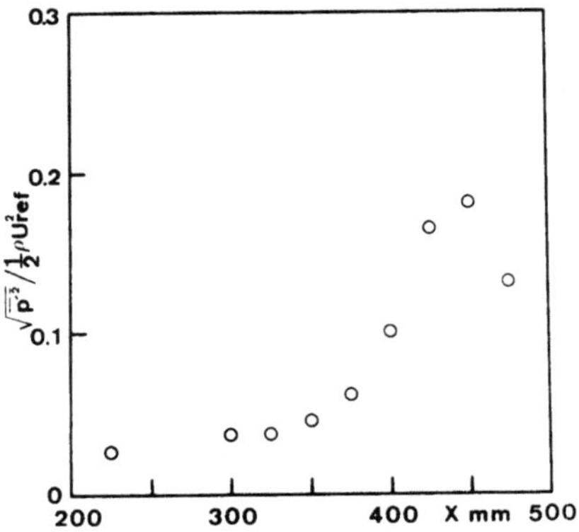

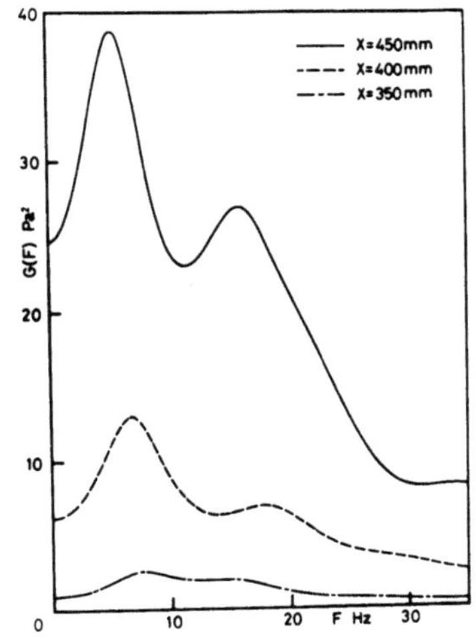

Fig. 11(a) RMS values of wall static pressure fluctuations

Fig. 11(b) Power spectrum density of wall static pressure fluctuations

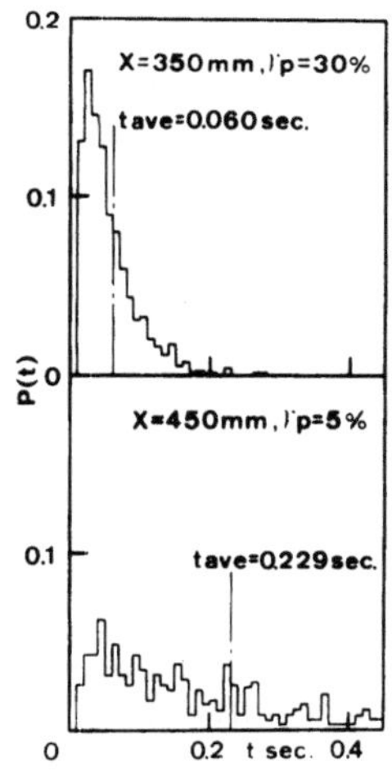

Fig. 12 Histograms of the flow reversal period at y=0.5 mm

Effect of Crossflow on the Vortex-Layer-Type Three-Dimensional Flow Separation

TSZE C. TAI

Aviation and Surface Effects Department
David Taylor Naval Ship Research and Development Center
Bethesda, Maryland, USA

Summary

A newly derived streamline equation indicates that the crossflow properties are solely responsible for the vortex-layer-type three-dimensional flow separation. A crossflow shear stress model based on modified Mager's crossflow velocity profile is proposed. Numerical results of three-dimensional separation lines on a prolate spheroid at angles of attack ranging from 10-30 degrees show favorable comparison with experimental measurements.

Introduction

A streamline method for determining the vortex-layer-type three-dimensional flow separation was developed recently by the author [1]. This approach is based on the Maskell postulation about separation patterns in three dimensions [2]. It was assumed that the envelope of converging viscous streamlines is very close to that of the converging wall streamlines. Therefore, the line of separation can be determined by the envelope of converging streamlines inside the viscous layer. To simulate the viscous effect, a friction model consisting of the inertia forces multiplied by empirical parameters was proposed.

The purpose of the present paper is to further explore this simple, economic method by modeling the shear stress with the aid of velocity profiles. As in the previous paper [1], a prolate spheroid in an incompressible flow at moderate angles of attack is used for illustration.

Viscous Streamline Equations

The streamline method for the determination of the vortex-layer-type separation line consists of three first-order ordinary-differential equations for calculating streamlines inside a boundary layer (see Fig. 1).

This research was sponsored by U.S. Naval Air Systems Command (AIR-320D, AIRTASK 9R023-02) under the cognizance of D. Kirkpatrick.

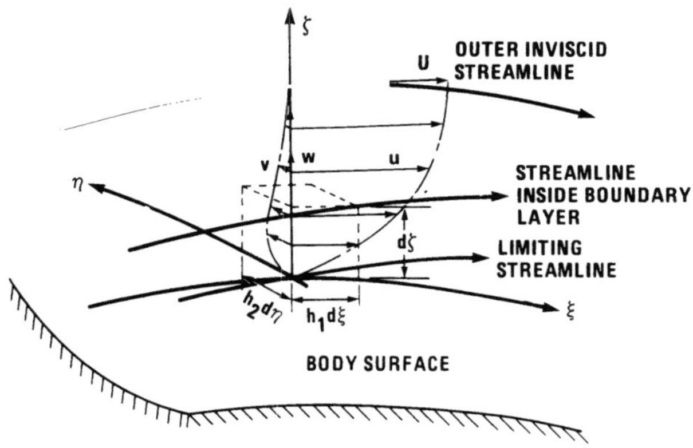

Fig. 1. Streamlines in a boundary layer

In particular, for the case of a prolate spheroid at an angle of attack, these equations take the simple axisymmetric form in orthogonal coordinates x and ϕ [1]:

$$\frac{D\Theta}{DS} = \frac{1}{\gamma M^2 P}\left[\left(\frac{\partial P}{\partial x} - \frac{\partial \tau_x}{\partial z}\right)\sin\Theta - \left(\frac{1}{f}\frac{\partial P}{\partial \phi} - \frac{\partial \tau_\phi}{\partial z}\right)\cos\Theta\right] - \frac{df}{dx}\frac{\sin\Theta}{f} \tag{1}$$

$$\frac{D\phi}{DS} = \frac{\sin\Theta}{f} \tag{2}$$

$$\frac{Dx}{DS} = \cos\Theta \tag{3}$$

where, with reference to Fig. 2,

f = radial distance measured from the centerline

S = distance along a viscous streamline inside the boundary layer

x, ϕ, z = body orthogonal coordinates

Θ = local streamline angle measured with respect to x coordinate

τ = shear stress

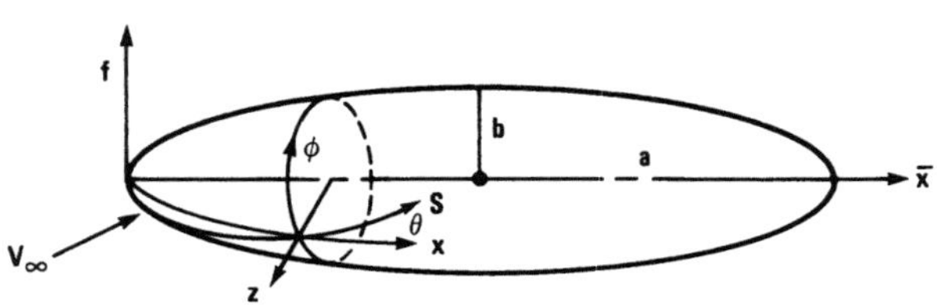

Fig. 2. Body orthogonal coordinate system

For incompressible flows, the potential flow solution [3] can be used for the pressure gradients, $\partial P/\partial x$ and $\partial P/\partial \phi$. The problem is, therefore, to provide proper formulas for the shear stress derivatives $\partial \tau_x/\partial z$ and $\partial \tau_\phi/\partial z$.

Crossflow Shear Stress

To facilitate modeling of the viscous terms, the $\partial \tau_x/\partial z$ and $\partial \tau_\phi/\partial z$ were evaluated in the streamwise coordinates ξ and η. The advantage of doing so is that existing crossflow prediction techniques, which almost all were developed using streamwise coordinates, may be employed. With the aid of the sketch below, the shear stress derivatives can be written in steamwise coordinates ξ and η as follows:

$$\frac{\partial \tau_x}{\partial z} = \frac{\partial \tau_\xi}{\partial z} \cos \Theta - \frac{\partial \tau_\eta}{\partial z} \sin \Theta \quad (4)$$

$$\frac{\partial \tau_\phi}{\partial z} = \frac{\partial \tau_\xi}{\partial z} \sin \Theta + \frac{\partial \tau_\eta}{\partial z} \cos \Theta \quad (5)$$

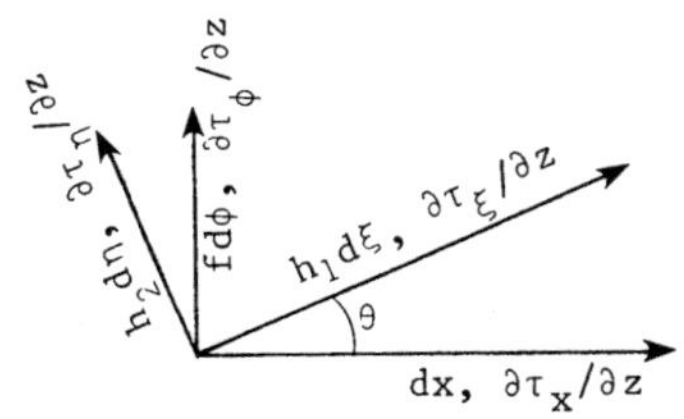

Upon substitution into Eq. (1), there results:

$$\frac{D\theta}{DS} = \frac{1}{\gamma M^2 P} \left(\frac{\partial P}{\partial x} \sin \Theta - \frac{1}{f} \frac{\partial P}{\partial \phi} \cos \Theta + \frac{\partial \tau_\eta}{\partial z} \right) - \frac{df}{dx} \frac{\sin \Theta}{f} \quad (6)$$

The new equation for the streamline contains only the shear stress derivative in the crossflow direction. (The same procedure is applicable to the pressure gradients as well. For convenience in computation, however, the pressure gradients are still written in x and ϕ coordinates.) It implies that theoretically the vortex-layer-type separation, or open separation [3], depends only on the crossflow properties. On the other hand, the streamwise resistance and pressure gradient seem to be closely related to the bubble-type separation, or closed separation [3], in three dimensions. It is consistent with Cooke and Hall's suggestion that for three-dimensional boundary layers, the flow in the streamline direction corresponds to a two-dimensional or axisymmetric type of boundary layer flow [4].

The problem is simplified in the sense that only the crossflow friction needs to be modeled. This is carried out by assuming a crossflow velocity profile suggested by Mager [5] and later used by others [6,7] for three-dimensional turbulent boundary layers. The quadratic term in Mager's form is modified to biquadratic so that its first- to third-order derivatives become zero at the edge of the boundary layer.

$$\frac{v}{U} = A\ (1 - \frac{z}{\delta})^4\ \frac{u}{U} \tag{7}$$

or

$$\bar{v} = A\ (1 - \zeta)^4\ \bar{u}$$

where $\bar{v} = v/U$, $\bar{u} = u/U$, and $\zeta = z/\delta$.

Its first and second derivatives are:

$$\partial\bar{v}/\partial\zeta = A\ (1 - \zeta)^3\ [(1 - \zeta)\ \partial\bar{u}/\partial\zeta - 4\ \bar{u}] \tag{8}$$

$$\partial^2\ \bar{v}/\partial\zeta^2 = A\ (1 - \zeta)^3\ [(1 - \zeta)\ \partial^2\ \bar{u}/\partial\zeta^2 - 8\ \partial\bar{u}/\partial\zeta] - 12\ A\ (1 - \zeta)^2\ \bar{u} \tag{9}$$

and the shear stress gradient is:

$$\frac{\partial\tau_\eta}{\partial z} = \frac{\rho U}{\delta^2}\ \frac{\partial}{\partial\zeta}\ (\varepsilon\ \frac{\partial\bar{v}}{\partial\zeta}) \tag{10}$$

It indicates that the resulting crossflow shear stress gradient has to be evaluated in terms of streamwise values. The separation line so calculated is, therefore, still influenced by the streamwise shear stress, although it does not explicitly appear in Eq. (10).

The streamwise velocity, $\bar{u}$, is assumed to be of the 1/7th-power-law type

$$\bar{u} = \zeta^{1/7} \tag{11}$$

This function gives good approximation of certain types of turbulent boundary layers. It is not differentiable at the wall, however. To obtain the derivatives near the wall, we proceed by correlating the streamwise velocity derivative with the shear stress:

$$\rho\ \varepsilon\ (\partial u/\partial z) = \tau_w\ F(z) \tag{12}$$

Since the purpose of the present paper is to demonstrate the effect of the crossflow friction model on flow separation rather than to elaborate such a model, simple, analytical forms are taken for the required functions in Eq. (12):

$$\varepsilon = 0.16\ z^2 |\partial u/\partial z| \qquad (13)$$

$$\tau_w = 0.0296\ \rho U^2/Re_s^{\ 0.2} \qquad (14)$$

$$F(z) = 1 - z/\delta \qquad (15)$$

$$\delta = 0.37\ S/Re_s^{\ 0.2} \qquad (16)$$

where Re_s is the Reynolds number in the streamwise direction, $Re_s = \rho US/\mu$.

Upon substitution of Eqs. (13) through (15) into Eq. (12) and rearranging, there yields:

$$\partial\bar{u}/\partial\zeta = 0.43\ (1 - \zeta)^{1/2}/(Re_s^{\ 0.1}\ \zeta) \qquad (17)$$

$$\partial^2\ \bar{u}/\partial\zeta^2 = -0.215\ (1 + 2\zeta - 2\zeta^2)/[Re_s\ \zeta\ (1 - \zeta)^{1/2}] \qquad (18)$$

The crossflow velocity and its derivatives, and thus the crossflow shear stress gradient, can now be evaluated. Figure 3 shows the modified Mager crossflow velocity profile, its first- and second-order derivatives, and the resulting shear stress gradient.

The coefficient A in the modified Mager profile is approximated empirically by:

$$A = -\ 0.5\ \beta\ (\frac{S}{a})^2\ \phi^{\alpha/10} \qquad (19)$$

where β is the streamwise pressure gradient parameter that serves as an indicator of flow acceleration or retardation of the mainstream. The value of β ranges between 2 and -0.199; positive values are for accelerated flows, and negative values for retarded flows. The physical aspects of the flow, the magnitude of the crossflow grows in the downstream region and/or at high angle of attack, are implemented in terms of S/a and ϕ. The constant 0.5 is arbitrary to be determined by correlation with experiments.

Numerical Results

The values of the crossflow shear stress gradient, along with potential flow pressure distribution, were then input to Eqs. (2), (3), and (6) for determining the streamline pattern and thus the separation line. The

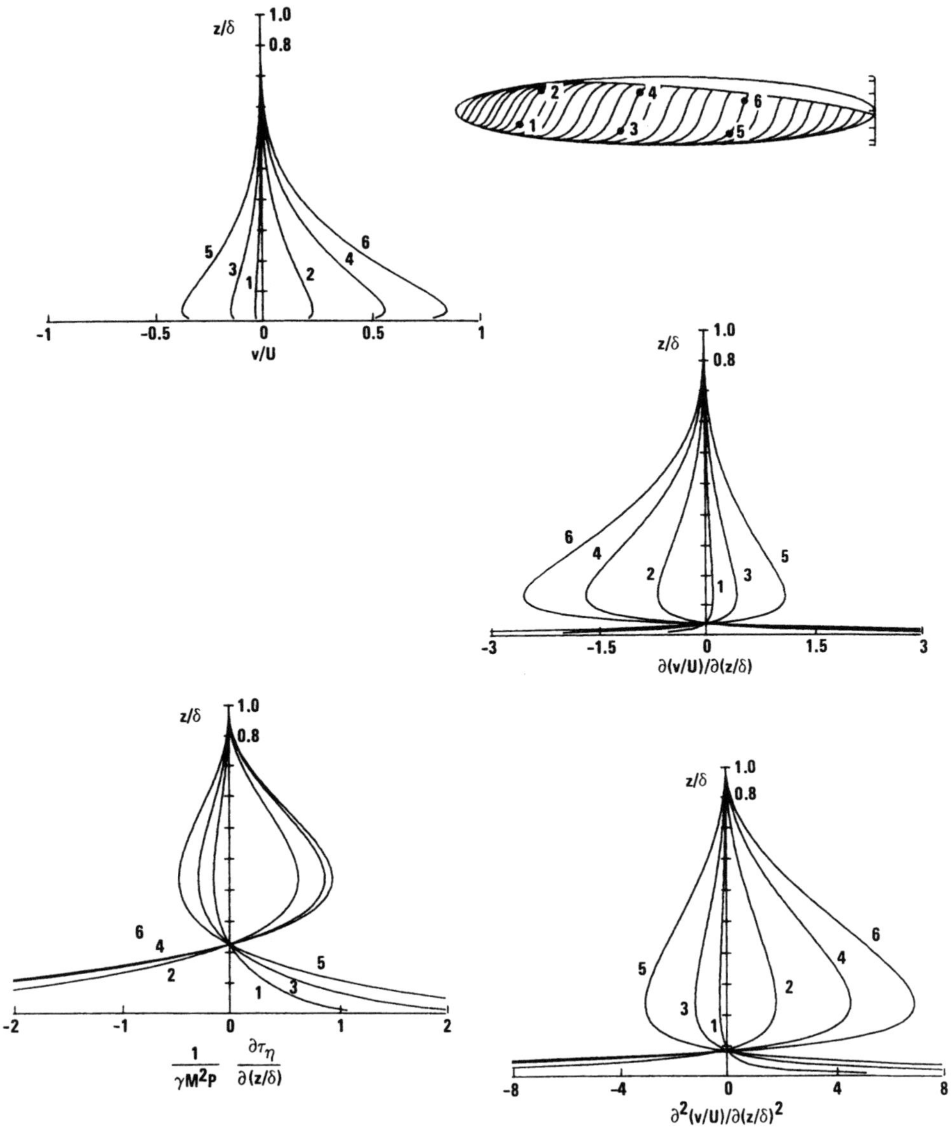

Fig. 3. Modified Mager crossflow velocity profile, derivatives, and resulting shear stress gradient on a 6:1 prolate spheroid at α = 30 degrees

complete computer program was written for a desk-top Tektronix 4052 computer with instant graphical display. The viscous layer in computational space is chosen so that the square of the resultant velocity is one-half of that at the edge of the boundary layer, i.e., $M = 0.707\ M_e$. According to the 1/7th-power law, this layer would lie somewhere between the displacement thickness and the momentum thickness of the overall boundary layer. Numerical results were computed for the case of a prolate spheroid of a/b = 6 in a turbulent boundary-layer flow at moderate angles of attack. Figures 4, 5, and 6 show the calculated streamline patterns and the separation lines at angles of attack of 10, 20, and 30 degrees, respectively. A comparison of the present result for the α = 10-degree case with the available numerical results of Cebeci [8] and the experimental data of Meier et al. [9] was made. Very good agreement is observed; see Fig. 4. For α = 30 degrees, a comparison of the present result with the measurement also is favorable; see Fig. 6. No comparison was made for the case of α = 20 degrees for which neither experimental nor numerical data were available.

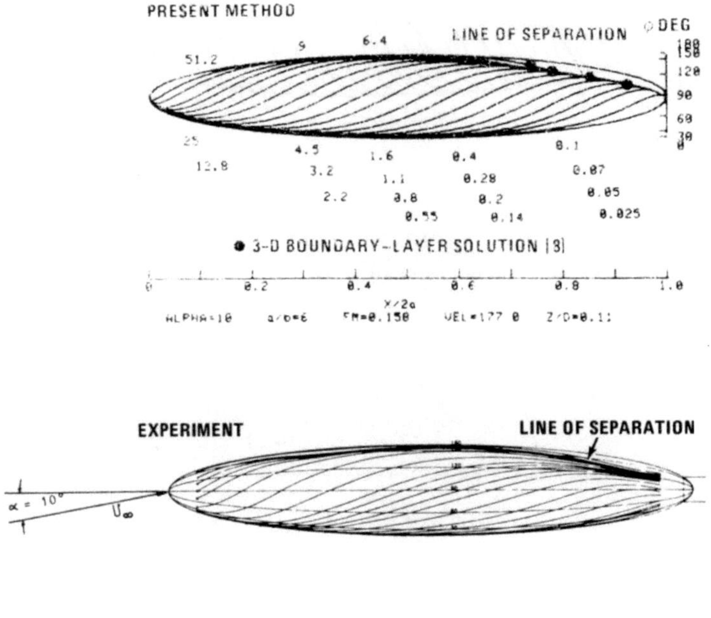

Fig. 4. Flow separation over a prolate spheroid at α = 10 degrees

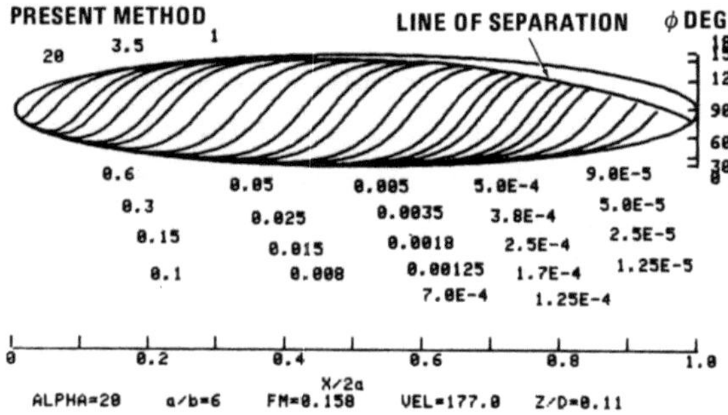

Fig. 5. Flow separation over a prolate spheroid at α = 20 degrees

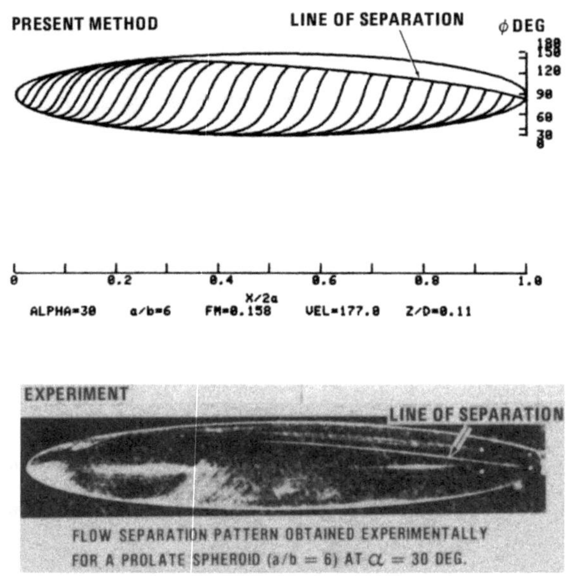

Fig. 6. Flow separation over a prolate spheroid at α = 30 degrees

Conclusions

In conclusion, the present work indicates that the effect of the cross flow is the main driving force of the vortex-layer-type (open) separation. An empirically correlated crossflow friction model leads to reasonably good results in using the streamline method for the determination of three-dimensional flow separation. Further improvement of the friction model is required for application to more general three-dimensional configurations.

References

1. Tai, T.C., "Determination of Three-Dimensional Flow Separation by a Streamline Method," AIAA Journal, Vol. 19, No. 10, Oct 1981, pp. 1264-1271.

2. Maskell, E.C., "Flow Separation in Three Dimensions," Royal Aircraft Establishment, Bedford, England, RAE Report Aero 2565, Nov 1955.

3. Wang, K.C., "Boundary Layer Over A Blunt Body at High Incidence with an Open Type of Separation," Proceedings of the Royal Society of London, A 340, pp. 35-55, 1974.

4. Cooke, J.C. and Hall, M.G., "Boundary Layers in Three Dimensions," Progress in Aeronautical Sciences, Vol. 2, Pergamon, 1962.

5. Mager, A., "Generalization of Boundary Layer Momentum Integral Equations to Three-Dimensional Flows Including Those of Rotating Systems," NACA Report 1067, 1952.

6. Johnston, J.P., "On the Three-Dimensional Turbulent Boundary Layer Generated by Secondary Flow," Journal of Basic Engineering, Mar 1960, pp, 233-250.

7. Stock, H.W., "Integral Method for the Calculation of Three-Dimensional Laminar and Turbulent Boundary Layers," NASA TM 75320, Jul 1978.

8. Cebeci, T., "Private Communications," Jul-Sep 1980.

9. Meier, H.U., Kreplin, H.P., and Vollmers, H., "Velocity Distributions in 3-D Boundary Layers and Vortex Flows Developing on an Inclined Polate Spheroid," Proceedings of the 6th U.S.-West Germany Data Exchange Agreement Meeting on Viscous and Interacting Flow Field Effects, Göttingen, Germany, Apr 1981.

Wind Tunnel Investigations of Some Three-Dimensional Separated Turbulent Boundary Layers

S. SAROHIA[+] & A.D. YOUNG

Queen Mary College, Mile End Road, London, E1 4NS, England.

[+] Now at University College, London.

Summary

Two kinds of separated flow were investigated, namely, that to be expected on a swept wing of large aspect ratio and that in the region of a wing-body junction. In both cases, the reverse flow vortices developed large velocity components in their axial direction near the surface. Reynolds stresses do not closely correlate with mean velocity components. Where appropriate comparisons are made with related 2-dimensional results.

1. Introduction and Some Experimental Details

There is a dearth of experimental information on turbulent boundary layers in all but some basic cases of two-dimensional attached flow. An investigation has therefore been in progress at Queen Mary College of some typical cases of three-dimensional separated flows. In view of the complexity of such flows the investigations have been kept relatively modest in scope. They are still in progress but it is hoped the results to date will be of interest. Only a small part of the data obtained so far can be presented here so the first named author will gladly supply on request more comprehensive results.

The investigation was concerned with two different kinds of separated flow. The first was the quasi two-dimensional flow over a swept wing of large aspect ratio. The second was that around a wing leading edge-fuselage junction where the fuselage boundary layer separates. Simple forms of models were chosen to provide the essential features of such flows.

Thus for the first kind, the arrangements used by Chu and Young (Ref. 1) for the investigation of two-dimensional separated flows were adapted to the corresponding flows over a swept plate, i.e. the turbulent boundary layer on a flat plate swept at 45 degrees was subjected to the pressure distribution produced by a nearby equally swept porous cylinder with its axis parallel to plate. At the rear of the cylinder was a Thwaites flap and by suction into the cylinder and with a suitable flap angle it was possible to develop sufficient circulation to induce on the plate a pressure distribution that was initially favourable and then adverse enough to cause the plate boundary layer to separate. The plate chord was 118 cms, the tunnel velocity about 20 m/s, the cylinder was 40 cms downstream of the plate leading edge and transition was tripped close to the leading edge. Results were obtained with two different Thwaites flap setting producing slightly different pressure distributions (C1, the more severe, illustrated in Fig. 1, and C2). Also investigated was the form of adverse pressure distribution produced by a spoiler flap 6.4 cm high, swept at 45 degrees, 58.5 cm downstream of the plate leading edge. This induced ahead of itself an almost monotonic adverse pressure distribution with flow separation, see Fig. 2.

Traverses were made across the boundary layer at various streamwise and spanwise stations both before and after separation using a Chu type of single tube yawmeter from which were derived velocity magnitude and direction distributions, and in the spoiler flaps experiments crossed hot wire probes were also used to determine turbulence distributions. For the cylinder induced separation the traverses were mainly confined to station A. With the spoiler induced separation the cases SA, SC and SG were examined in some detail, here A, C, and G refer to the spanwise station concerned, see Fig. 2. Oil flow pictures were taken (see Plate I).

For the second kind of separated flow, i.e. that in a wing body junction, a sketch of the model is shown in Fig. 3. The leading edge of the wing was a semi-circular cylinder of radius 10.8 cm and span about 92 cm, and the rest of the wing was formed by parallel sides of length 1.25 m. The wing was at zero incidence normal to a plate which extended about 70 cm upstream of the wing leading edge, and the tripped turbulent boundary layer on the plate simulated that of a fuselage. As is well known, separation of the latter ahead of the wing takes the form of a number of reversed flow vortices which wrap around the leading edge and then trail downstream. The main region investigated was therefore from just upstream of separation to where the vortices had largely turned downstream, and again measurements consisted of yawmeter traverses, surface pressure measurements and oil flow visualisation (see Plate II).

---

* X is distance downstream from the leading edge measured parallel to the wind tunnel axis. It is generally quoted in inches unless otherwise stated.

2. Main Results

2.1 Swept plate tests

2.1.1 Mean flow characteristics

The main coordinate system used has the x axis normal to the leading edge in the plane of the plate, the y axis is spanwise (positive when upstream) parallel to the leading edge, the z axis is normal to the plate. The corresponding mean velocity components are U, V and W outside the boundary layer and u, v, and w inside it. The corresponding turbulence velocity components are u', v' and w'. U_∞ is the undisturbed stream velocity.

In planes normal to the leading edge the flow patterns (see Fig. 4) were not very different qualitatively from what one would infer from unswept wing experience, i.e. there was either one or two reverse flow vortices and with the latter there was evidence of a small vortex of opposite sense between them. The latter was the pattern for C1 and all the spoiler induced separations. The oil flow patterns for the flow with the spoiler showed portions of black (scoured) spanwise line in the separated flow region which are believed to be characteristic of the region just below the small counter-flowing vortex (see Plate I).

Some mean velocity component profiles for C1 and SA are illustrated in Fig. 5. If we identify separation as occurring where u close to the surface begins to reverse in direction then we find that for C1 separation started at about X = 20, in accordance with the flow visualisation studies. By X = 30 the flow had almost reattached. A particular noteworthy feature was the development of a strong spanwise flow component in the downstream sense (negative v) in the separated flow close to the surface. Towards the rear of the separated flow the magnitude of v inside the boundary layer exceeded that outside and was of the same order as U . This was also evident in the case C2. For the case SA separation occurred at about X = 18.5 and the presence of the spoiler at X = 23 clearly influenced the shape of the u component profile nearby. Again we note the development of a strong spanwise velocity component in the inner region of the separated flow, here -v attained a value of about 1.3 U .

Fig. 6 shows the angle β between the flow direction just outside the boundary layer and the direction inside as a function of z for different values of X for the case SA. We see that for X = 20 and 21 β reached a maximum above the surface and then decreased as the surface is approached. At the other values of X investigated maximum β was attained at the surface. This is consistent with a small counter flowing vortex at X = 20 ∿ 21 since u/v may be expected to reach a minimum in the vortex core region. The oil flow pictures support this conclusion. Similarly for C1 the counter flowing vortex was inferred to be in the region of X = 23 - 24. It is interesting that the less severe C2 case showed no evidence of a counter flowing vortex.

Fig. 7 presents the boundary layer thicknesses as functions of X along station A for both the spoiler and cylinder induced (C1) pressure distributions. The corresponding two-dimensional results of Chu and Young[1] are shown for comparison. It will be seen that for C1 the rate of growth of the boundary layer thickness after separation was much smaller than for the corresponding two-dimensional flow, but one must bear in mind that for the swept plate flow the velocity component and downstream distance in a plane normal to the leading edge are important scaling quantities to consider in making such a comparison. The dashed curve labelled 3D (C1) Effective has therefore been derived based on these quantities, however, the overall picture of a much smaller rate of growth than for the two-dimensional case still remains. It would seem that the strong spanwise flow in the case of C1 helps to reduce the rate of growth of the boundary layer after separation. On the other hand, with the spoiler the rate of growth of the boundary layer after separation was somewhat greater in the swept plate case than in two-dimensions and the distance between the separation point and the spoiler was much less. However, it seems likely that here the scale of the flow pattern was dominated by the height of the spoiler and the upstream boundary layer thickness. These were 2 - 2½ times as large for the two-dimensional flow as for the three-dimensional flow and the distance the separation point was ahead of the spoiler scales similarly. It seems that in this case the transverse flow did not play such a key part as in the relatively unconstrained flow induced by the cylinder.

Fig. 8 shows plots of Head's entrainment form parameter $H_1 = (\delta-\delta^*)/\theta$ as a function of the form factor $H = \delta^*/\theta$ as measured in planes normal to the leading edge, i.e. δ^* (displacement thickness) and θ (momentum thickness) are derived in terms of u and w. For Head's integral prediction method for two-dimensional turbulent boundary layers an empirical relation between H and H_1 is used based on attached flow data. However, for separated flows this relation needs to be extended by appropriate experimental data. There is a great paucity of such data so the present results although derived from swept plate measurements should prove useful. We note that the results do not suggest a universal curve for separated flows, although a mean curve may prove adequate for many applications. Also, at separation $dH_1/dH = 0$.

This latter condition has been inferred from the integral momentum and entrainment equations by Dr. Horton in an unpublished communication.

2.1.2 Turbulence Components and Eddy Stresses

Fig. 9 shows distributions of the Reynolds stresses at X = 19, 20 and 21 for case SA. The distributions of the stress $\overline{u'w'}$ followed qualitatively those of $\partial u/\partial z$ (see Fig. 5) but there is no evidence of a simple quantitative relationship. Similarly the stress $\overline{v'w'}$ reflected broadly but not closely the corresponding distributions of the mean velocity gradient $\partial v/\partial z$. The stress $\overline{u'v'}$ was not insignificant in magnitude, and its behaviour was not readily correlated to the mean velocity distributions. At all stations in the separated flow the normal stresses $\overline{u'^2}$ and $\overline{v'^2}$ were relatively large through almost the entire width of the boundary layer, $\overline{w'^2}$ was considerably smaller but not insignificant.

Fig. 10 shows the ratio of τ/ρ to twice the turbulence kinetic energy per unit mass i.e. $\tau/\rho(\overline{u'^2} + \overline{v'^2} + \overline{w'^2})$, where τ is the resultant shear stress in planes parallel to the surface. In two-dimensional attached flow this ratio (usually denoted as a_1) has been found to be nearly constant across the boundary layer and about 0.15. Here it is seen to vary widely and to range from about 0.02 to about 0.34.

2.2 The Simulated Wing-Body Junction

The notation is illustrated in Fig. 3. The origin is in the plane of the plate at the centre of curvature of the semi-circular section of the wing leading edge. The x axis is along the centre line, the y axis is in the plane of the plate and the z axis is normal to the plate. The traverse planes are parallel to the wing centre plane and are identified by their distance y from it in centimetres.

Based on oil flow pictures and other data, Fig. 11a shows a sketch delineating the main features of the limiting streamlines and the positions of the separation and reattachment lines. Fig. 11b illustrates the inferred flow pattern in the centre plane and the singular points of separation and reattachment and we see that again the separated flow involved two main reverse flow vortices and two small vortices of opposite sign to the main ones, one between the main ones and the other deep in the junction. The oil flow pictures showed a black line suggesting a local scoured region between R_1 and S_2 and we infer the intermediate counter flowing vortex was just above it (see Plate II).

Fig. 12 shows the plate pressure distributions along the streamwise traverse lines. We see that in the region of the junction each had a maximum as the wing leading edge was approached followed by a minimum and then a further rise. The peak generally occurred at or a little ahead of the first separation line, and the minimum was between the main reverse flow vortices in the region of reattachment close to the small counter flowing vortex.

Fig. 13 illustrates u, v, w profiles for a few values of x on the traverse line -y = 3cm. Here separation was at about -x = 18.5 cm. We again note in the separated flow large lateral velocity components near the surface in the direction of the vortex axis and outwards from the centre plane.

Fig. 14 shows the distributions of the boundary layer thickness for the traverse lines y = 3, 5, 10 and 20 cm. The rate of thickening of the boundary layer as it approached the wing leading edge and separated was relatively small compared with the corresponding rate in two-dimensional flow ahead of an obstacle or in the flow on the swept plate referred to above. Indeed, in the junction flow the boundary layer thickness near the centre line grew at a rate after separation that was little more than before separation and away from the centre line the changes in thickness were even smaller and non-monotonic. The region of reversed flow was generally not more than a fifth of the total boundary layer thickness, this is considerably smaller than in corresponding laminar boundary layer experiments (see, for example Ref. 2). The strong lateral flow component presumably plays a major part in relieving the tendency of the boundary layer to thicken after separation.

3. Further Comments

The above measurements were relatively easy to make as compared with the earlier two-dimensional tests. This was because of the relative steadiness and ordered nature of the former. The reverse flow vortices with their strong axial velocities appear to impose a defined structure on the flow and so inhibit the large scale chaos of two-dimensional separation.

The reasons for this are not clear and there is a need for an appraisal of the stability of vortical flows. At this stage one can only speculate that since for wing trailing vortices an important stability parameter is the ratio of the axial velocity to the transverse peripheral velocity the development of a strong axial velocity may well be stabilising for the flows that we are here considering. More crudely it may be argued that the axial motion provides an escape for entrained fluid that would otherwise cause the separated flow to 'blow up'. In the case of the simulated wing body junction the strong curvature of the flow around the convex wing surface may also be expected to be a stabilising factor.

References

1. Chu, J.K., Young, A.D. Measurements in Separating Two-Dimensional Turbulent Boundary Layers. AGARD CP 168, Conference on Flow Separation, 13, 1975.

2. Norman, R. On Obstacle Generated Secondary Flows in Laminar Boundary Layers and Transition to Turbulence. Ph.D Thesis, Illinois Institute of Technology, 1972.

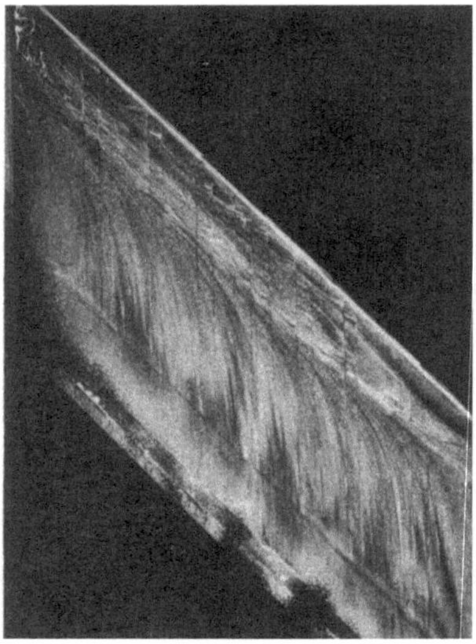

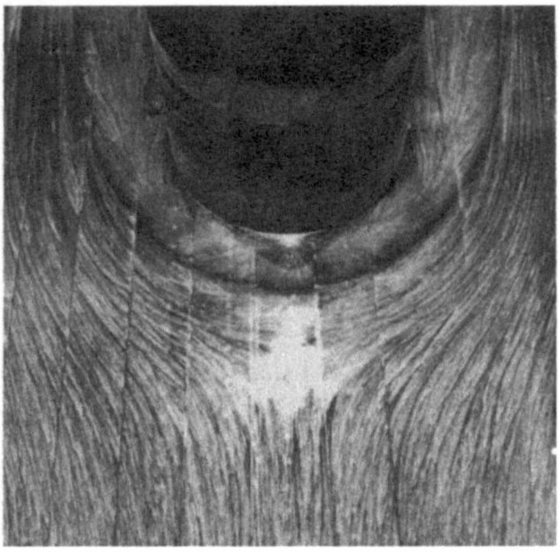

Plate I ↑ Plate II

wind
direction

Oil Flow Pictures: (I) Swept Plate with Spoiler
(II) Simulated Wing-Body Junction

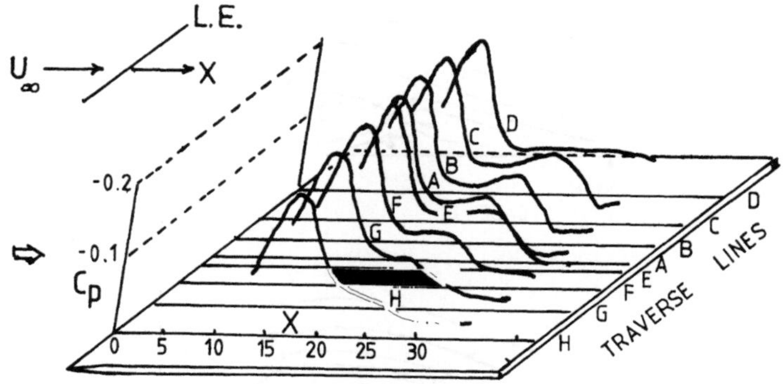

FIG.1 Pressure Distributions for Cylinder Induced Separation (C1)

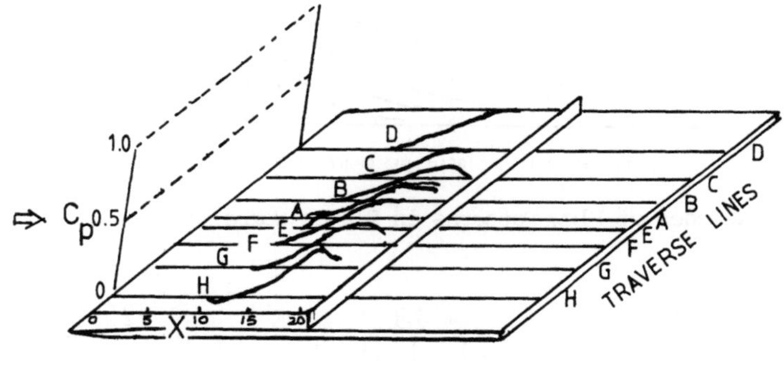

FIG.2 Pressure Distributions for Spoiler Flap Separation

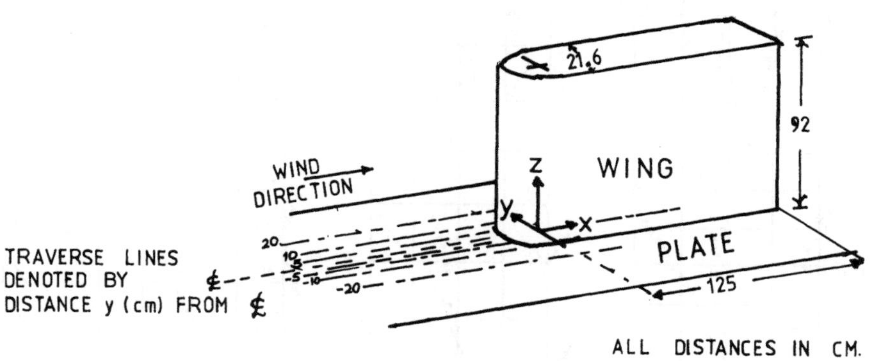

FIG.3 Sketch of Model for Wing-Body Junction Tests

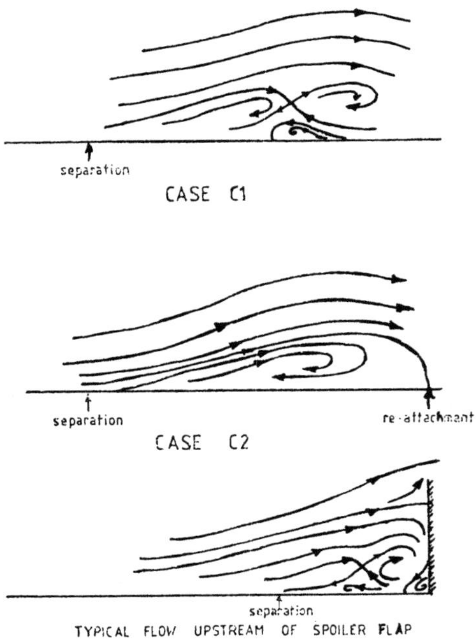

FIG.4 Rough Sketches of Flow Patterns in Planes Normal to Swept Plate L.E.

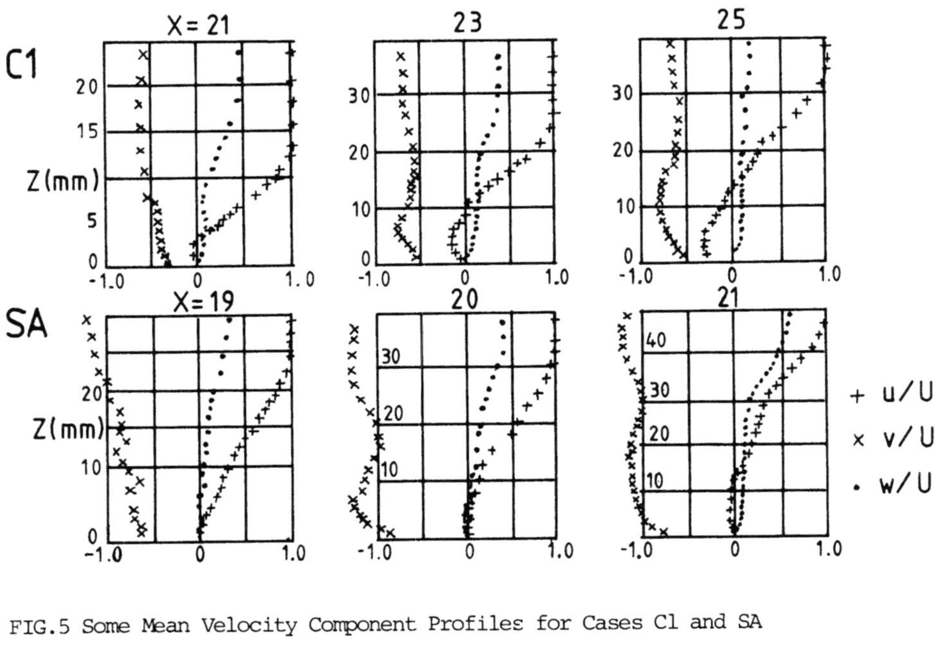

FIG.5 Some Mean Velocity Component Profiles for Cases Cl and SA

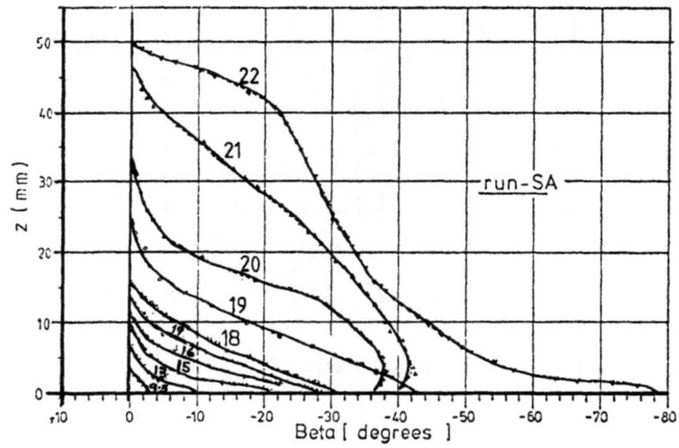

FIG.6 Beta Distributions for Case SA.

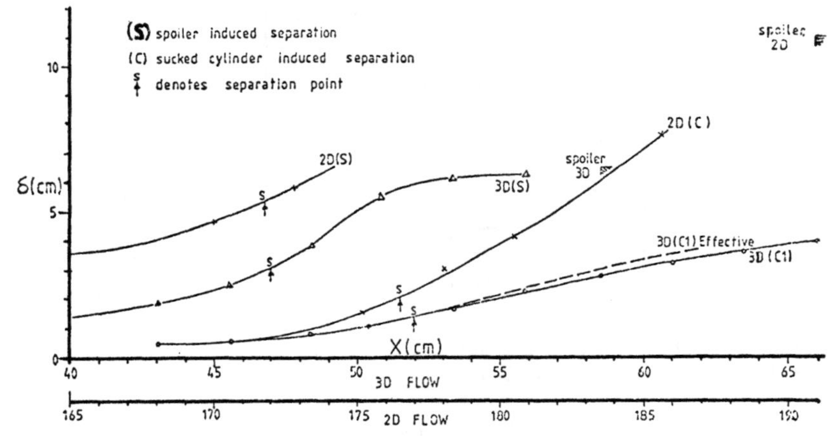

FIG.7 Comparisons of Growth of Boundary Layer Thicknesses for Separating B.L. in 2- and 3-Dimensions.

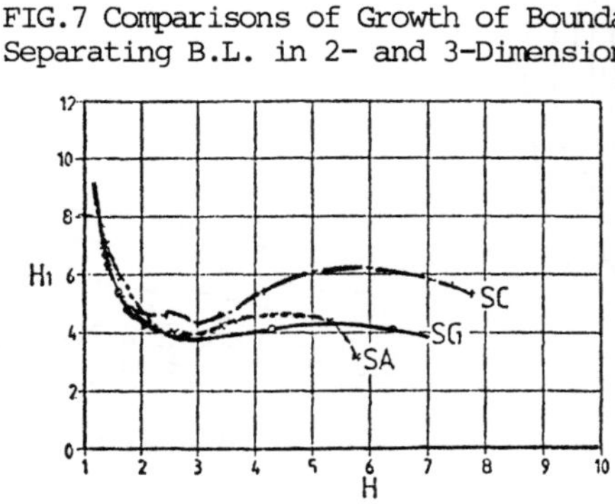

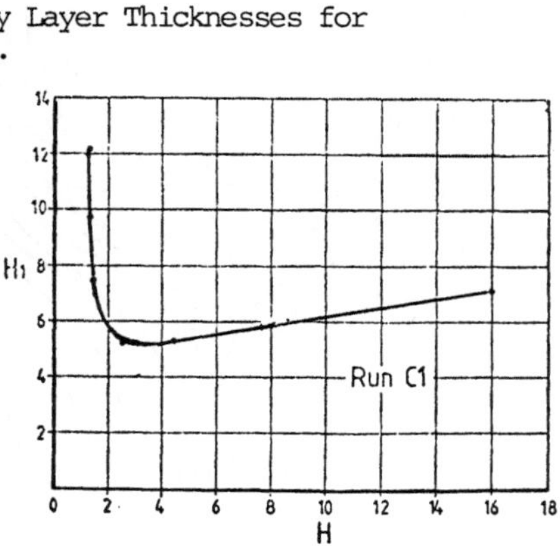

FIG.8 Head's Entrainment Factor H_1 v H in Planes Normal to L.E.

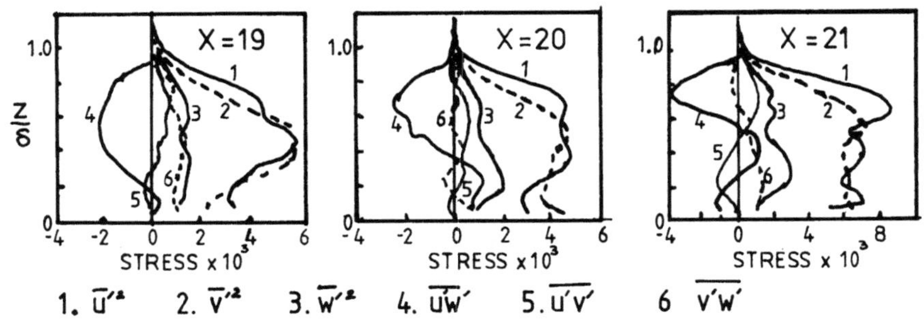

FIG.9 Reynolds Stress Components/U_∞^2. Case SA

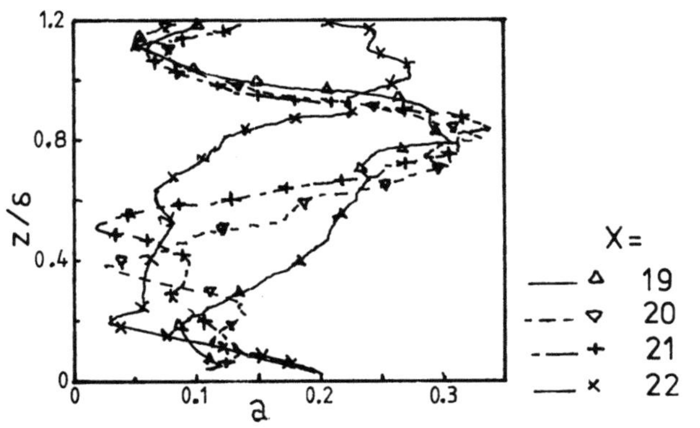

FIG.10 Some Distributions of a_1 (Ratio of τ to Turbulent Kinetic Energy), Case SA.

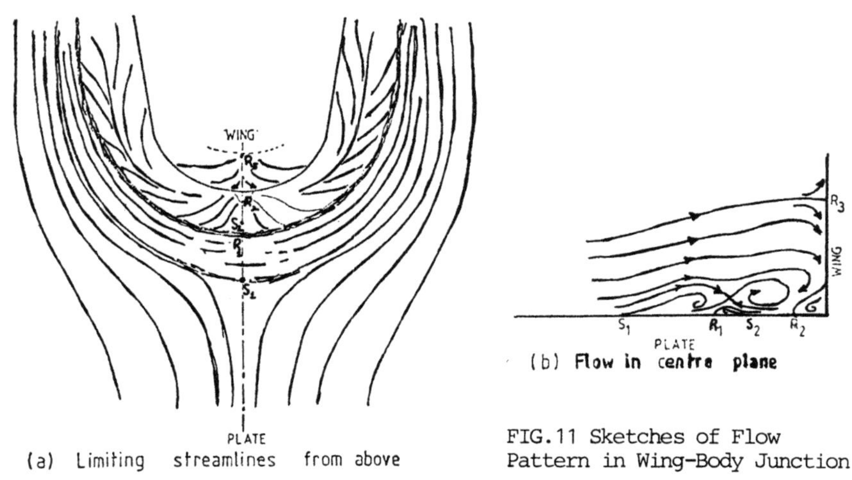

FIG.11 Sketches of Flow Pattern in Wing-Body Junction

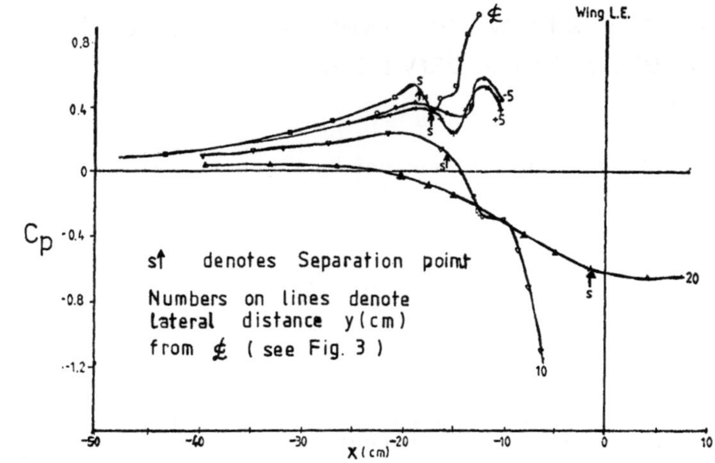

FIG.12 Pressure Coefficient Distributions Along Streamwise Traverse Lines in Wing-Body Junction

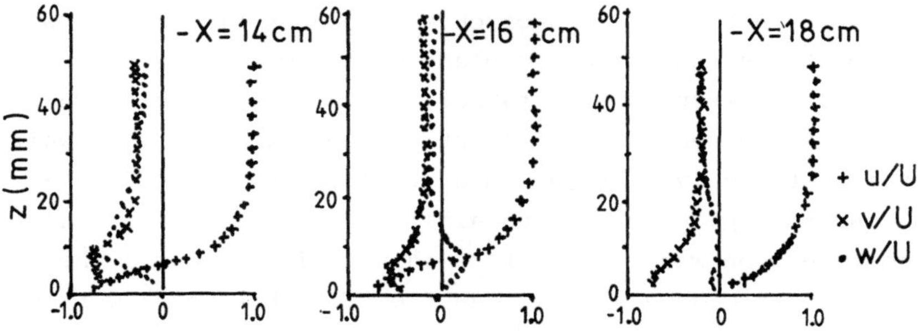

FIG.13 Some Mean Velocity Component Profiles Along a Traverse Line (-y = 3cm) in Wing-Body Junction.

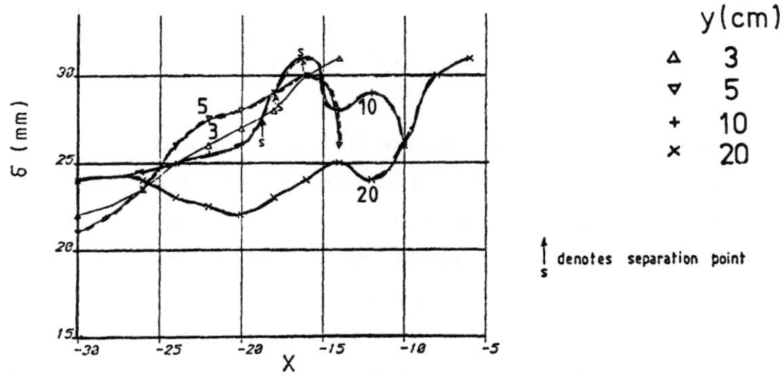

FIG.14 B.L. Thickness Distributions Along Streamwise Traverse Lines in Wing-Body Junction.

Experiments on a Flow With Swept Separation and Reattachment of a Boundary Layer

E.P. SUTTON

Engineering Department, Cambridge University, England

Summary

Flows with boundary-layer separation from swept salient edges and with subsequent turbulent reattachment have been studied experimentally. Some mainly qualitative observations of the types of flow which occurred are presented here.

Details of the Experiments

A plate with uniform cross-section as shown in Figure 1(a) was mounted on the vertical centre plane of the working section of a 1.7 x 1.2 m low-speed wind tunnel, as in Figure 1(b). The flat nose of the section had slightly extended corners at which separation of the (very thin) boundary layer was fixed. Reattachment took place on the parallel flat sides, beyond which the cross-section was tapered to a thin trailing edge. The plate passed through a slot in the floor of the tunnel working section and was clamped and sealed there; at its upper end it was attached to a bracket on the ceiling of the working section. The flat sides were always aligned with the free-stream direction, and the angle of sweep of the edges of the plate, Γ in Figure 1(b), was varied from 30° to 60°.

The wind speed was in the range between 25 and 50 m/s. The corresponding range of Reynolds number based on the width 2h of the nose of the section was about 40,000 to 80,000, and the range of Reynolds number based on the distance from separation to reattachment (measured normal to the upstream face of the strut) was about 2×10^{5} to 4×10^{5}. It was found that there was very little variation in the observed characteristics of the flow in that range; the separated shear layer may be assumed to have become turbulent well upstream of reattachment in all cases.

The experiments consisted mainly of surface oilflow observations on the sides of the plate, supported by measurements of surface static pressure along three rows of tappings aligned perpendicular to the separation edges.

Results

In a region adjacent to the floor of the tunnel there was a vigorous flow downstream of separation, with a length scale which increased roughly linearly with distance from the floor, bearing some resemblance to separated flows at the leading edges of thin delta wings at incidence (Figure 2(a)).

Further away from the floor the flow between separation and attachment was less vigorous, and the oilflow pattern appeared to show a narrow attachment region, (Figure 2(b)) rather than an attachment line. In some cases, at the higher angles of sweep, the distance from separation to attachment became approximately constant over a substantial part of the span, and the surface pattern suggested that the flow was of a swept two-dimensional kind with a length scale determined by the cross-sectional dimensions of the plate. Figure 2(c) shows the shape of the attachment line (or region) when the angle of sweep was 45°. In other cases, at the lower angles of sweep, the distance from separation to attachment continued to increase over the whole span of the plate, though at a much slower rate than near the floor. At sweep angles of 45° or more the influence of the tip and the tunnel ceiling appeared to extend over only a small part of the span.

Pressure-distribution measurements showed that even in cases where the surface oilflow pattern revealed no significant spanwise variation over part of the span, the chordwise pressure distribution still varied significantly. An example is shown in Figure 3(b), again for 45° sweep, corresponding to Figure 2. Chordwise pressure distributions at three stations whose positions are given in Figure 3(a) are compared. Each one is qualitatively similar to the two-dimensional separation-bubble type of pressure distribution observed when the plate is unswept, but the minimum pressure increases with distance along the span, and the subsequent steep pressure rise occurs progressively further from separation.

In some cases the separated flow region appeared to shed vorticity locally, at one or more positions along the span, into trailing vortices aligned approximately in the free-stream direction. When that happened, the feeding of vorticity into the recirculating region from the shear layer appeared to begin afresh from the separation edge, and for some distance further out along the span the flow near the surface between separation and reattachment became more vigorous. These occurrences seemed to be triggered and located by very small imperfections in the separation edge, sometimes barely detectable even by feeling along the edge.

Further experiments were carried out at 45° sweep with small disturbances applied deliberately at the separation edges, in the form of thin self-adhesive metal foil tapes attached tightly round them and across the forward-facing surface between them. Each disturbance consisted of an upward step of height approximately 0.05 mm, normal to the spanwise direction at the edge, followed by an equal downward step 12.7 mm further along; the upward step was the more effective part.

Figure 3(c) shows the pressure distributions at the outermost station with and without one such tape at the spanwise position of the middle station. The effect of the disturbance was to lower the pressure in the region close downstream of separation and bring forward the subsequent pressure rise.

Figure 3(d) shows the pressure distributions at all three stations when there were 8 tapes equally spaced along the separation edges at the positions given in Figure 3(a). At the two outer stations, which were subject to effects of disturbances at corresponding relative positions, the minimum pressure was considerably lower and the pressure rise occurred earlier than in the absence of separation-edge disturbances. The pressure distribution at the innermost station was changed less, but the beginning of the pressure rise became steeper and the end of it less steep.

The surface oilflow pattern in the vicinity of the two outer stations is reproduced in Figure 4. A possible interpretation of it is given in Figure 5, where the approximate locations of vortices initiated at the separation-edge disturbances have been inferred from the characteristic inflected shapes of the limiting streamlines. The main attachment region is the rather wavy band of more-or-less spanwise flow from which the limiting streamlines turn either towards the separation edge* or downstream. Its mean distance from the separation edge is only about three quarters of the corresponding distance in the absence of edge disturbances (Figure 2(c)). The vortices are shown crossing the attachment region, downstream of which they produced identifiable regions of cross-flow alternating with regions of approximately parallel flow and regions of converging flow (where "open" separation lines were seen in some cases).

Discussion and Concluding Remarks

The surface oilflow on the plate without separation-edge disturbances suggested a gradual spanwise change from a highly three-dimensional separated flow, near the upstream corner of the plate, towards a two-dimensional type of separated flow, further out along the span. The pressure measurements showed that even where the pattern of limiting streamlines had ceased to change appreciably in the spanwise direction the chordwise pressure distribution had not, though its shape was generally similar to that for a two-dimensional flow with a separation bubble.

Whether the spanwise variations can be accounted for by the finite aspect ratio of the plate remains to be investigated. The ratio of the length of the separation edges to the maximum

---

*Note that gravity had a significant effect on the pattern close to the separation edge, where the flow towards the edge separated again before reaching the edge and the surface shear stress was very small. Details there should be ignored.

distance from separation to reattachment, measured normal to the edges, was about 12 in the 45° swept case described. It is intended to do more experiments with that ratio increased to about 20.

The extreme sensitivity of the flow to small disturbances at the separation edge suggests that the quasi-two-dimensional flow is in any case not a very stable one. On the other hand, though the step height of the applied disturbances was very small indeed in relation to the overall dimensions of the separated flow, it was not small in relation to the shear-layer thickness at separation.

When the flow was interrupted by a row of disturbances applied to the separation edge, the mean distance from separation to attachment was reduced. The flow obviously became less uniform with respect to distance in the spanwise direction, on a scale of the order of the distance between the disturbances. In another sense, however, the flow became more uniform spanwise; the flow pattern repeated itself, approximately, with variations about a spanwise mean, over a large part of the span.

Experiments with roughness applied to the surface between the separation edges were reported briefly in Reference 1. The roughness supplied a large number of more-or-less random disturbances. The resulting flow was much the same as with the equally spaced disturbances, apart from the less uniform spacing of the vortices observed. In both cases it is easy to see qualitative analogies with two-dimensional flow about an unswept plate of the same cross-section; periodic or quasi-periodic variations from a spanwise mean flow in the one case correspond to periodic or quasi-periodic variations from a time-mean flow in the other case. The analogy provides some help towards the construction of physical models of the flow in both cases.

Both in the three-dimensional flow and in the two-dimensional flow regarded as an unsteady flow, incidentally, there are of course problems over definitions of separation and attachment.

It may be noted that the flows described here could be considered to be wholly attached in a frame of reference in which the downstream direction is spanwise; the flow near the surface would then nowhere be seen as a "reversed" flow.

Finally, it should be pointed out that the possible existence of trailing vortices in a boundary layer downstream of a region with swept separation and reattachment lines must be borne in mind when comparisons are made between calculations and measurements of such a boundary layer.

References

1. Sutton, E.P., Evans, G.P., McGuinness, M.D., Svehla, K.M. Influence of wall vibrations on a flow with boundary layer separation at a convex edge. IUTAM Symposium on Unsteady Turbulent Shear Flows, (Michel, R., Cousteix, J., Houdeville, R. (eds.)). Berlin, Heidelberg, New York: Springer-Verlag 1981; 285-293.

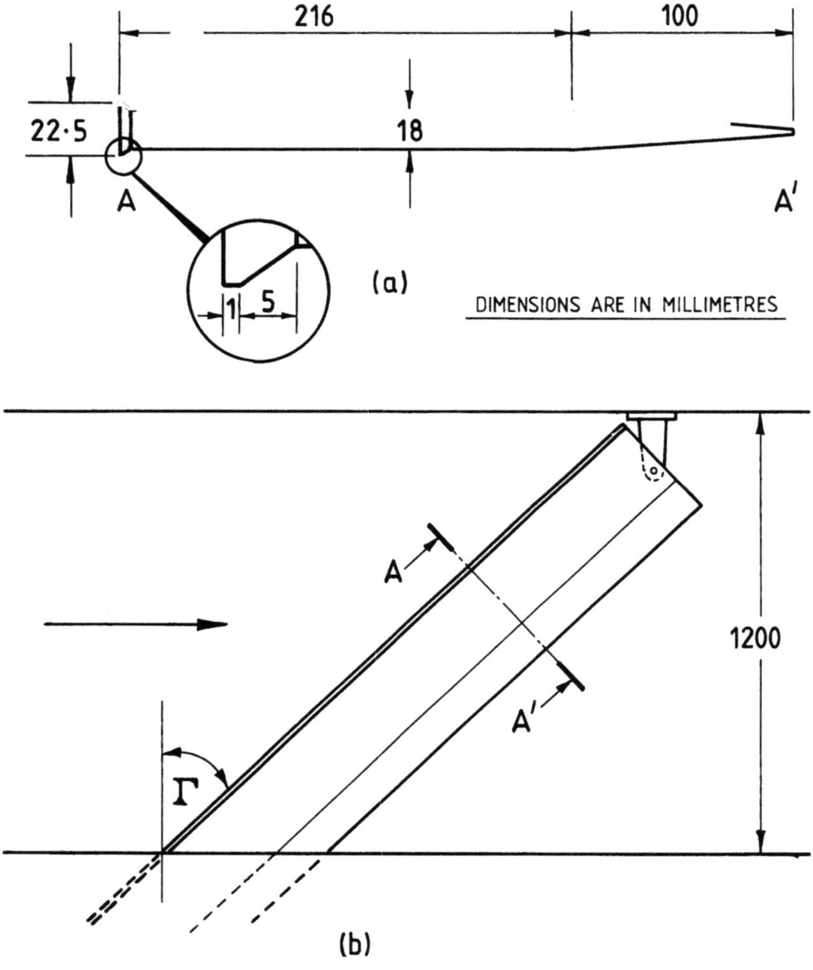

Figure 1 Details of the plate: (a) cross section AA' normal to the separation edges; (b) side elevation installed in the tunnel

(a) (b)

Sketches of limiting streamlines
(a) close to the tunnel floor
(b) further out along the span

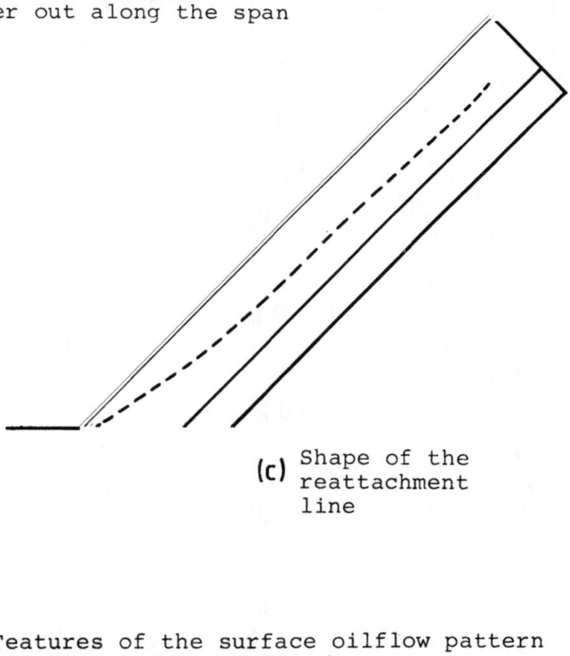

(c) Shape of the reattachment line

Figure 2 Features of the surface oilflow pattern at 45° sweep with wind speed 30 m/s

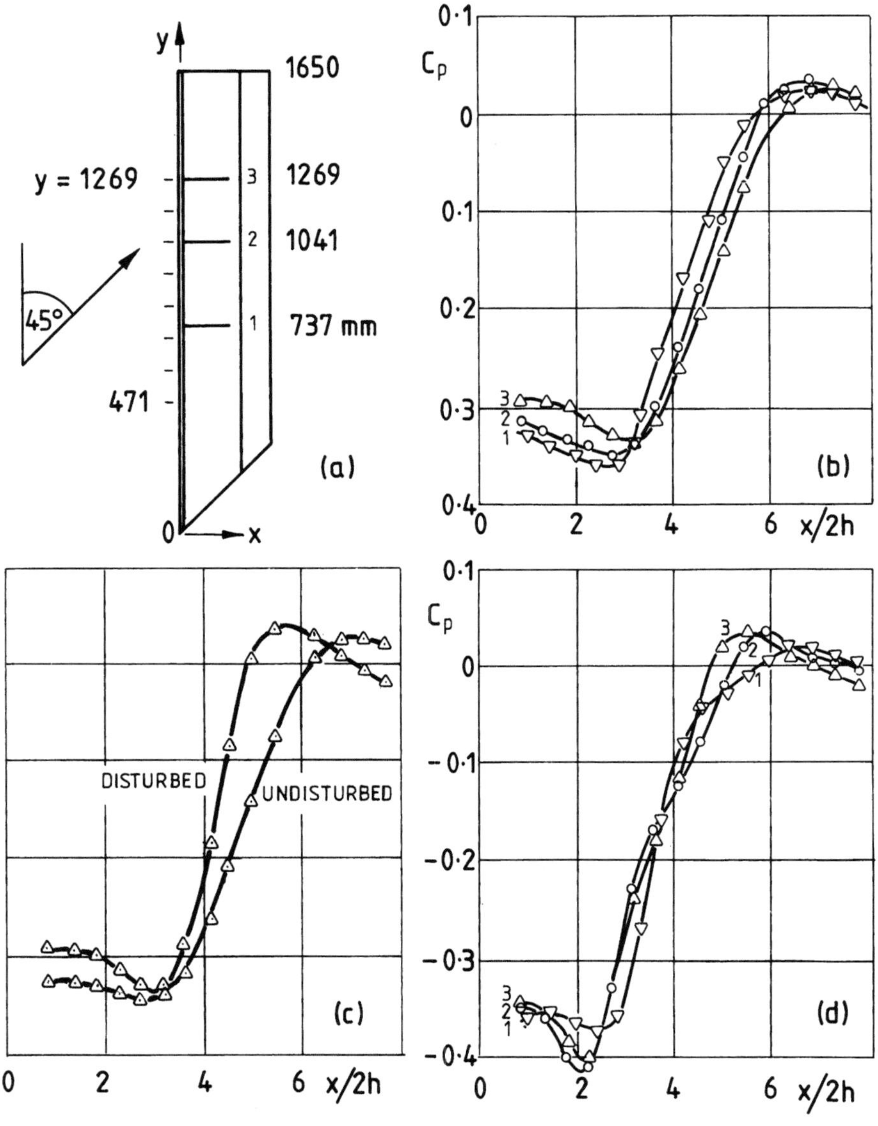

Figure 3 Pressure distributions chordwise at 45° sweep with wind speed 30 m/s: (a) Positions of stations 1, 2 and 3 at y = 737, 1041 and 1269 mm, and positions of edge disturbances equally spaced from y = 471 to 1269 mm; pressure distributions (b) at all 3 stations on 'clean' plate; (c) at station 3 with and without an edge disturbance at station 2; (d) at all 3 stations with 8 edge disturbances

Figure 4 Surface oilflow pattern between y = 927 and 1269 mm with edge disturbances as in Figures 3(a) and (d)

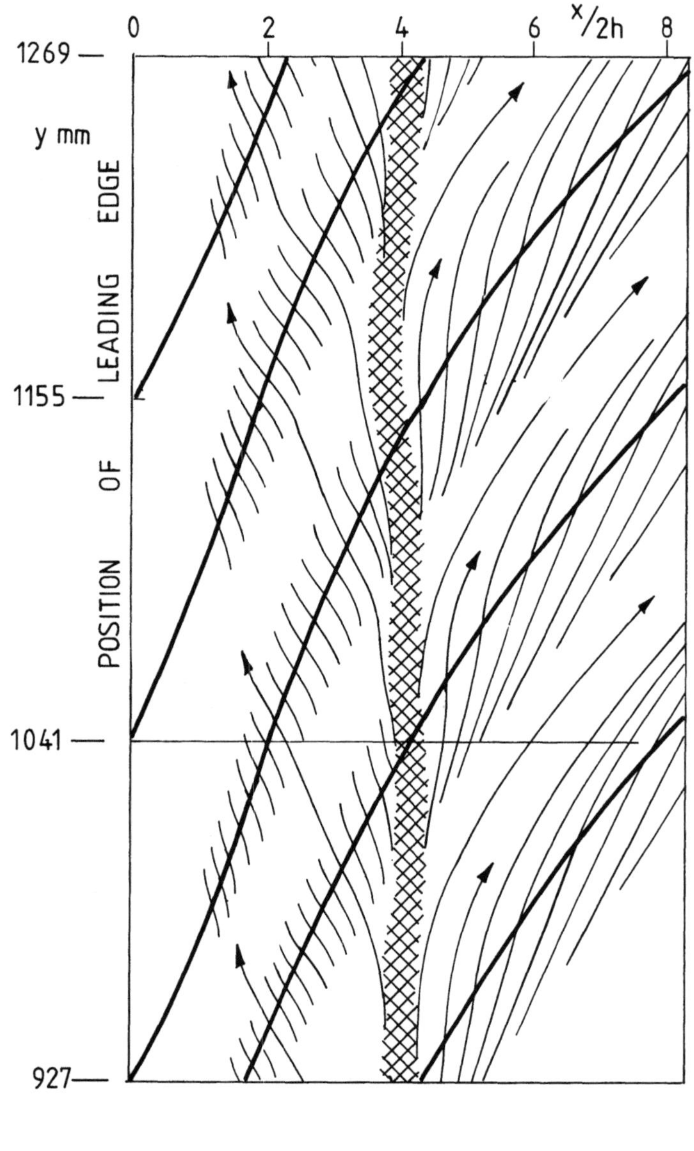

Figure 5 Suggested interpretation of the oilflow pattern reproduced in Figure 4

Traces of vortices from edge disturbances

Main region of flow reattachment

Three-Dimensional Boundary Layers in Turbomachines

Y. SENOO

Research Institute of Industrial Science
Kyushu University
Kasugashi, Fukuoka 816, Japan

Summary

Various kinds of three-dimensional boundary layer are observed in turbomachines. They are induced by relative motion of a rotor and a stator, by lateral pressure gradient based on curvature of streamlines and by revolution of a rotor. In many cases, secondary flow in boundary layer increases the friction-loss and induces flow separation, as a result the performance of turbomachine is reduced. In some cases separation of boundary layer is delayed by secondary flow and the performance of turbomachine is improved including widening the range of stable operation.

Introduction

Power is mostly generated from heat using fluid as the medium, and in a power plant tremendous amount of liquid and gas is transported, compressed and expanded to compose a heat cycle. At utility plants, chemical plants and many other kinds of plants a lot of fluid is handled. In these plants turbomachinery is the only kinds of machine which can handle a large amount of fluid with a machine of relatively small dimensions. Therefore, enormous amount of power is produced or consumed by turbomachines all year around, and improvement in the efficiency of turbomachines is one of the most effective way for energy saving.

In the case of electric power generating stations, the flow rate in the turbine must be adjusted to meet the power demand, while the shaft speed must be kept constant so that the electric frequency does not change. Therefore, the behavior of flow inside the turbine varies in a wide range, yet the operation must be stable and the efficiency must be high for the whole range.

A turbomachine consists of many components, and a lot of variety in design is conceivable to achieve a specific requirement. At the stage of planning and preliminary design, it is necessary for a design engineer to acquire necessary data for comparing various designs and to make fair judgement for the best design of turbomachines which would demonstrate good performance at the required wide range of operation. For that purpose, it is essential to predict the flow behavior within a turbomachine without spending too much time and labour.

In many cases actual pressure loss in a component of a turbomachine is considerably larger than the prediction based on a two dimensional flow analysis and the difference is often attributed to secondary flow loss without understanding secondary flow in the boundary layer. In some cases separation of boundary layer is delayed or removed by secondary flow, and a considerably better performance is achieved than the prediction based on a two-dimensional flow analysis. Therefore, better understanding of three-dimensional boundary layer flow is essential to improve the design technology of turbomachinery. In this paper it is intended to review various aspects of three-dimensional flow in components of turbomachines.

Macroscopic Flow in Turbomachines

In cases of axial compressors and turbines there are many stationary and moving blade rows, and it looks as if boundary layers along the blades and the casing walls fill up the flow passage and influence the flow pattern significantly. In reality, since the stationary blade row and the moving blade row are located alternately, there is a mixing process of the blade wake with the core flow for the following blade rows.

The relative velocity to the moving blade row is the vector sum of the absolute velocity vector out of the stationary blade row and the circumferential velociy vector of the moving blade row as shown in Fig.1. Consequently relative to the rotating blade row, the direction of wake flow of the stationary blade is different from that of the core flow, or the flow is periodic and it may be a cause of noise and vibration .

The angle of incidence of the wake flow to the succeeding blade row is so large that the blades would stall if it were steady flow. The frequency is so high that the boundary layer along the blades does not have time to develop and to separate, consequently the periodic nature is disregarded and only the circumferential mean flow out of the front blade row is assumed as the inlet condition to the succeeding blade row. In each blade row the blade chord is not long compared with the passage width and the boundary layer along the blade does not much influence the core flow. Therefore, inviscid flow analysis may be sufficient to evaluate the core flow except near the hub and casing.

Near the hub and casing of compressors, a wall boundary layer develops and the flow between blades in this region is a complicated three-dimensional boundary layer flow. At the tip of rotor-blade, the stationary wall moves circumferentially relative to the blade and furthermore a leakage flow through the tip clearance makes the flow complicated. For the standpoint of designners, only the time mean flow is the problem, that is, the blade loading and the distribution of pressure loss which are influenced by the end-walls and by the tip clearance come in their design, and experimental relations have been derived [1]. There are some theoretical and experimental study on the flow and further study is desired for better understanding of flow and for generalization of experimental results.

<u>Three-Dimensional Boundary Layer on the Endwall of Cascade</u>

The flow changes the circumferential component of velocity a lot while it passes through the blade rows of nozzle and the runner of turbines, that is, the blades are heavily loaded and there is a large pressure gradient between the blades or across a channel, which induces a strong secondary flow in the end-wall zone. The flow is made even more complicated by the distorted inlet flow due to the relative motion of stationary and rotating blade rows. That is, even though circumferential variation of velocity is disregarded by circumferentially averaging the incoming flow, the mean relative velocity near the end-wall varies the magnitude and the direction considerably from the main flow zone to the endwall as shown in Fig.1. That is, at the inlet of each blade

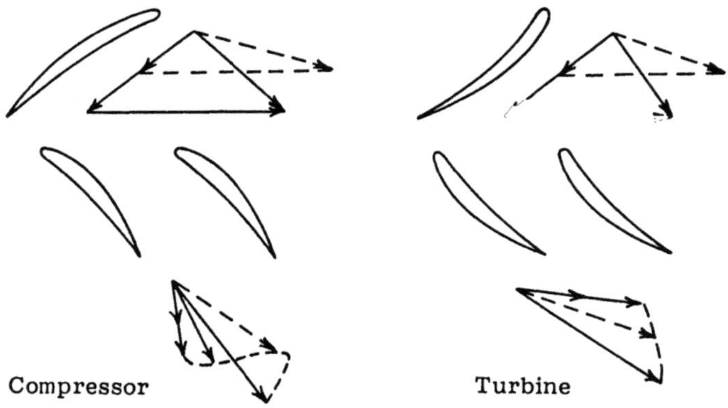

Fig.1 Relative velocity in the endwall boundary layer

row of turbomachine, the flow is hardly collateral.

The velocity vectors of skewed boundary layer at the inlet of turbine blade row and those of compressor are shown in Fig.1 together with the geometry of blades. It should be noted that the direction of blade loading is opposite for the two cases. Since secondary flow is induced in the end-wall boundary layer from the pressure surface of a blade to the suction surface of the adjacent blade, in cases of turbine blade row, the direction of induced secondary flow agrees with the secondary flow of the skewed inlet boundary layer. In cases of compressor blade row the secondary flow induced in the blade row is opposite to the secondary flow of the skewed inlet boundary layer.

The polar plot of velocity vectors in the end-wall boundary layer of compressor blade row is considerably more complicated than that of turbine blade row, and more parameters are necessary to express the velocity distribution properly for the case of compressor than for the case of turbine. For prediction of boundary layer development based on momentum integral method, the velocity distribution must be expressed well with a few parameters. If more parameters are required to express the velocity distribution, number of simultaneous equations to be solved increases and the merit of momentum integral method is reduced.

There are several types of turbine blade row: in cases of turbine nozzles a large pressure drop occurs due to a large turning of flow from the axial direction, in cases of impulse blade row a large turning of flow occurs with little pressure drop, in cases of reaction blade row a medium amount of turning of flow occurs with a medium amount of pressure drop, and there are many intermediate types of blade row. Among these turbine blade rows, nozzles and impulse blade rows are the cases where strong secondary flow is induced in the end-wall boundary layer due to the large turning angle. The cross flow in the end-wall boundary layer is obstructed by the blade suction surface and rolls up at the corner [2]. In cases of impulse blade row where the flow is little accelerated through the blade row, the upstream boundary layer conspicuously influences the flow in the end-wall boundary layer. That is, a zone of three-dimensional separation is recognized on the end-wall near the front edge of each blade. The separation zone may be represented by a horse-shoe vortex, and the direction of rotation of the vortex along the suction surface of the blade is opposite to that of the vortex induced by the accumulated end-wall boundary layer.

Bailey [3] tested a linear cascade of turbine nozzle and measured the velocity and turbulence in the end-wall boundary layer using a hot-wire anemometer and a laser velocimeter. The Reynolds number was 4.6×10^5 based on the axial chord and the inlet velocity. The maximum turbulence intensity within the longitudinal vortex was, as shown in Fig.2, four percent of the main

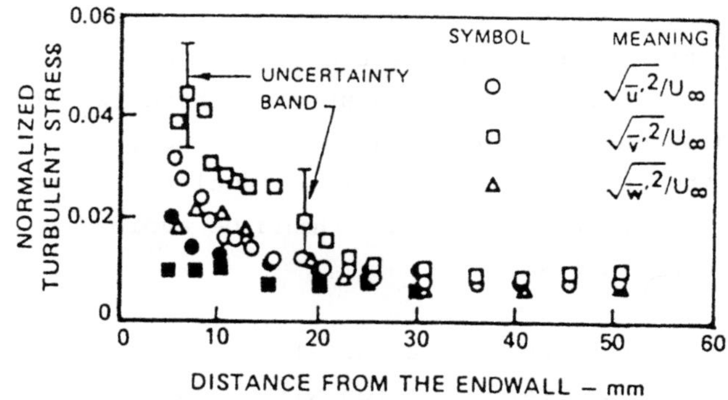

Open symbols: near suction surface. Closed symbols: middle

Fig.2 Turbulence at the throat of turbine nozzle [3]

flow velocity. Large region in the vortex was non-turbulent, with intensity approaching free stream value. Since a turbulent wall boundary layer was the source of vorticity that produced the passage vortex, such low turbulence level was not anticipated by the author. According to Senoo [4], although the end-wall boundary layer was turbulent at the inlet of a turbine nozzle cascade, the end-wall boundary layer became laminar inside the nozzle. He proved that it was laminar firstly by comparing the measured velocity distributions at several points across the nozzle with the prediction based on a laminar flow analysis, secondly by turbulence measurement and thirdly by recognition of transition point downstream of the nozzle where mass transfer rate increased discontinuously. From the view-point of blade cooling as well as pressure loss, it is important to clarify the factors and conditions which relaminarize the end-wall boundary layer.

Boundary Layer on Stationary Blade

Both in cases of axial compressors and turbines, there is a strong pressure gradient in the radial direction due to the swirl flow induced by a rotating impeller or stationary nozzles and the radial pressure gradient induces inward secondary flow in the boundary layer and the wake of stationary blades. In cases of multistage turbines with a large expansion ratio, the blade length increases downstream, and the hub and shroud are conical surface as shown in Fig.3 and the incoming flow is not perpendicular to the span of blades. Boundary layer flow along a swept wing is a typical case of three-dimensional boundary layer, and the secondary flow is stronger as the blade loading increases. If the hub diameter is reduced downstream, these two effects induce inward secondary flow along the stationary blades near the hub, and the boundary layer fluid accumulates at the root. Therefore, a converging conical hub should be avoided as much as possible as shown in Fig.3.

In cases of turbo-fan engines, an axial compressor of small diameter is located downstream of a fan with a large hub diameter, and the connecting annular passage has a converging conical hub. If the blades are mounted radially on the hub, the flow is not

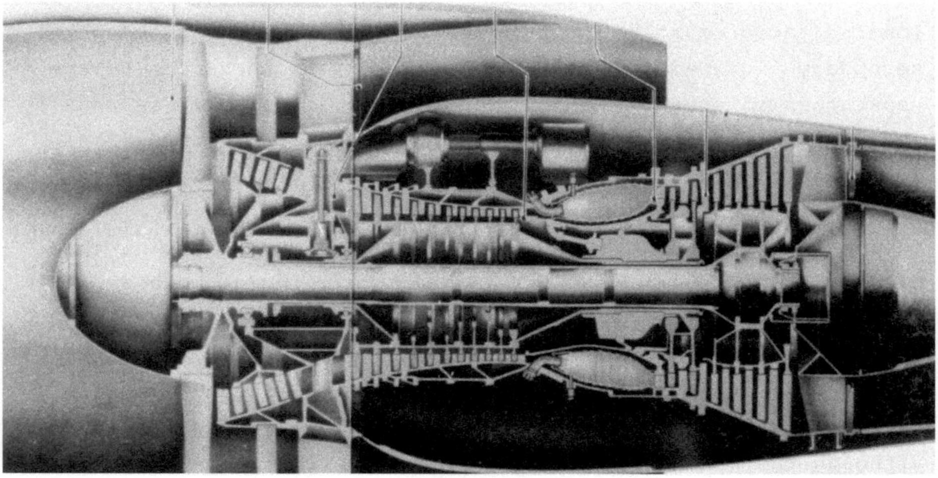

Fig.3 Cross section of turbo-fan engine JT9D

normal to the span, i.e. the blades are oriented with a swept angle relative to the flow. In order to avoid deterioration of compressor performance, in some engines the blades are mounted, as shown in Fig.3, normal to the conical hub surface despite difficulty with respect to strength of rotating blades.

Boundary Layer on Rotating Blades

Himmelskamp [5] demonstrated that the maximum lift coefficient of airfoil section of a propeller was larger than that measured in a wind tunnel, and the largest value 3.2 was observed near the hub. The improvement in the maximum lift coefficient is achieved by delay of boundary layer separation due to radial migration of low velocity fluid in the boundary layer. The radial driving force is approximately proportional to the radius while the force to generate boundary layer is proportional to the square of relative velocity which is approximately proportional to the square of radius. Therefore, the ratio of removing force to generating force of boundary layer is larger near the hub.

In cases of axial pump the pressure rise across an impeller stage must be about equal regardless of the radial position. Since the relative velocity is lower near the hub, the blade loading factor must be higher there. Therefore, even though secondary flow reduces the boundary layer thickness and delays separation of boundary layer near the hub, separation is usually observed near the hub section at off-design conditions. It is important to predict the behavior of boundary layer along the blades of axial pumps and compressors. Arakawa et. al. [6] were successful in predicting development of boundary layer on the blades of axial pump by means of momentum integral method. According to their calculation the boundary layer thickness is only slightly thinner than the thickness based on a two-dimensional prediction. The predicted separation line agrees fairly well with the observed separation line by means of oil film trace although the predicted shape of separation line somewhat depends upon how to express the velocity distribution in the analysis.

Accumulation of Boundary Layer

Secondary flow in boundary layer moves the fluid of low velocity laterally, as a result in the major part of blade surface and endwalls the boundary layer is thinner than the prediction based on two-dimensional flow. On the other hand the migrated fluid with low velocity must be disposed somewhere. If it is nicely dispersed into the main flow the turbomachine behaves well as shown in the case of propeller. However, in many cases, the fluid of low velocity accumulates at the corner or on the surface of minimum pressure. In the case of adverse pressure gradient, reverse flow may occur there unless the accumulated low velocity fluid mixes with the main flow. Therefore, from the standpoint of design engineers, it is important to examine the behavior of accumulated low velocity fluid. Flow visualization is a handy mean to grasp the overall picture. Since smoke diffuses quickly in turbulent flow, in some cases good smoke pictures are taken in laminar flow. Smoke picture of corner vortex in a turbine nozzle is well known [2].

At the rotor blade tip, the flow in the boundary layer along the stationary shroud casing is made complicated by three factors, they are the pressure gradient between blades, the leakage flow through the tip clearance and motion of the shroud relative to the blades. It was demonstrated that the flow along the shroud

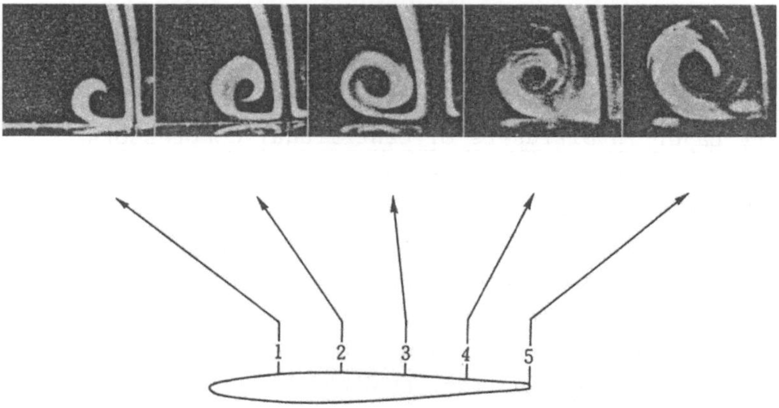

Fig.4 Boundary layer of casing is scraped by a moving blade and rolls up

accumulated at the corner of blade due to motion of the shroud relative to the blades. That is, the blades were unloaded so that there was no cross-flow based on the tip leakage and based on the pressure gradient between blades. Smoke was introduced to the casing boundary layer, which was scraped by the blade and rolled up. The smoke pictures at five sections in Fig.4 clearly demonstrate rolling up of a smoke sheet, which represents the boundary layer on the shroud casing. It should be noted that in the area of rolled smoke sheet a lot of fluid in the main flow is contained, i.e. it is sandwiched in the rolled smoke sheet. Therefore, the accumulated fluid of low velocity mixes well with the main flow at the corner and it may be possible to flow against an adverse pressure gradient.

In cases of turbines, secondary flow is induced in the boundary layer by the pressure gradient between blades, and it has the same direction as the relative motion of shroud casing. Therefore, the end-wall boundary layer rolls up at the corner of blade suction surface unless leakage through the tip clearance is significant. In cases of compressors, the direction of secondary flow induced by the pressure gradient is opposite to the relative motion of casing. It is said that the secondary flow induced by the pressure gradient between blades is intercepted by circumferential motion of casing and by the leakage flow through the tip clearance, consequently fluid with low velocity accumulates somewhere away from the corner.

Boundary Layer in Diffusers of Centrifugal Compressors

In cases of centrifugal compressors and radial turbines where a large change in pressure per stage is accomplished, a long passage is required compared to the passage width, consequently the boundary layer becomes thick, and in some cases the passage is filled up by the boundary layer flow. Due to the wakes of blades the flow out of an impeller is not axisymmetric and the pattern is rotating with the impeller, therefore, the flow to the diffuser is periodic. However, the circumferential distortion decays quickly and it is usual to assume that the flow out of an impeller is steady and axisymmetric.

Similarly to the case of axial compressors, the direction of inlet flow relative to the diffuser varies as the flow rate is changed. In cases of vaned diffuser, the change of angle of incidence often results in separation of flow and the range of flow rate for stable operation is very narrow. Therefore, in order to operate a centrifugal compressor in a wide flow range, it is common to use a vaneless diffuser which consists of two nearly

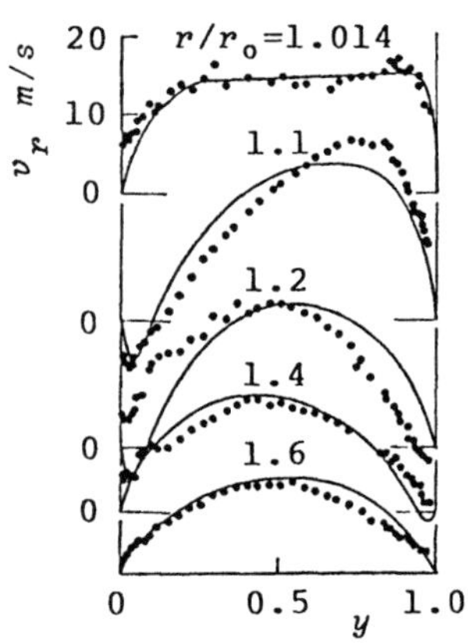

(a) Radial component of velocity

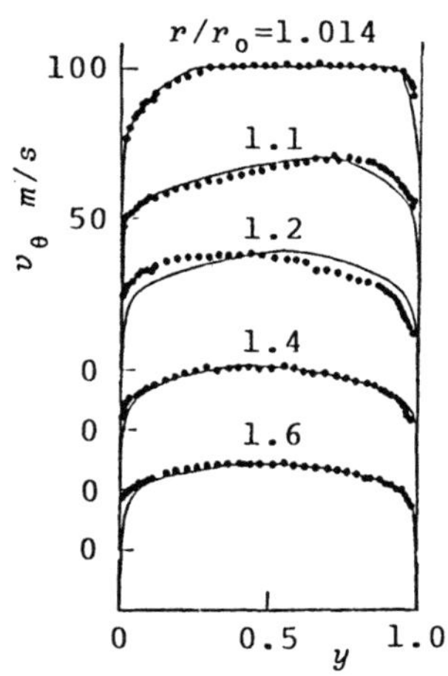

(b) Circumferential component of velocity

Fig.5 Local reverse flow in a vaneless diffuser of centrifugal blower [7] parallel disks.

As the flow rate is reduced and the direction of flow becomes close to circumferential, an axisymmetric reverse flow zone is observed on one wall of the vaneless diffuser. In cases of narrow diffusers the reverse flow zone on the wall disappears downstream while another reverse flow zone appears on the other wall [7] as shown in Fig.5. As the flow rate is reduced the reverse flow zone becomes thicker and larger, and eventually the flow becomes unstable, that is, the axisymmetric flow pattern is destroyed and a few patches of reverse flow zones propagate circumferentially. A boundary layer flow analysis based on momentum integral method predicted the location of axisymmetric reverse flow zone and the velocity distribution correctly, and the predicted critical inlet flow rate agreed well with experimental data. Therefore, the analysis is useful as a design tool.

In order to decelerate the flow out of an impeller quickly within a limited space, it is necessary to guide the flow with vanes. In such cases the velocity at the throat varies in proportion to the flow rate while the magnitude of velocity out of the impeller is not changed much by the flow rate. Therefore, at a reduced flow rate a large pressure rise is expected from the leading edge to the throat of blade row, and the flow along the side wall in this triangular zone is a complicated three-dimensional flow. The downstream portion of the throat is a relatively simple channel diffuser, and the pressure recovery coefficient in this zone mostly depends upon the boundary layer thickness at the throat section. Therefore, study on flow at the inlet triangular zone is quite important.

A vaned diffuser which consisted of a circular cascade of small solidity demonstrated a good pressure recovery for a wide range of flow rate at a low inlet Mach number. In spite of the extreamely small aspect ratio of 0.12 the maximum lift coefficient of the airfoil was comparable to or even better than the critical value of stall for the two-dimensional isolated airfoil.

The pattern of oil-film on the side wall in Fig.6 demonstrates

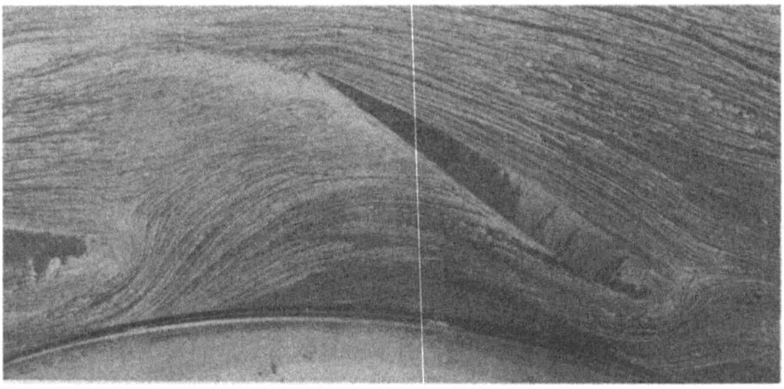

Fig.6 Oil film on a side-wall of vaned diffuser, near stall

that the side-wall boundary layer moves from the suction side of a blade to the pressure side of the adjacent blades, consequently the boundary layer on the suction surface of a blade is sucked by the secondary flow along the side walls and separation of boundary layer on the blade suction surface is delayed.

The oil film pattern at the inlet section indicates that all incoming wall-streamlines near the suction side of a blade flow back near the pressure side of the adjacent blade and none of the wall streamline passes through the blade row to the exit. Therefore, it looks as if the entire boundary layer flows back. However, the velocity vector in the side-wall boundary layer continuously varies from the wall-streamline to the main flow and major part of fluid in the boundary layer flows toward the pressure side of the adjacent blade. In the present case since the solidity of the blade row is small the pressure rise along the blade is not large and the accumulated wall boundary layer can flow downstream without separation, thus the blade does not stall. Such lateral movement of side-wall boundary layer is a feature of flow in a cascade with a large stagger angle, and the stall limit of airfoil is not penalized by the side-wall boundary layer.

As the angle of incidence to the blade increases, the layer of reverse flow on the side wall-triangular zone becomes thicker, and when it exceeds a critical value the performance of the blade row drops. The exit/inlet velocity ratio across the blade row was 0.37 at the critical condition of diffuser stall. This value is smaller than the critical value of stall for many diffusers. That is, the secondary flow in the side-wall boundary layer improved the behavior of the wall boundary layer and delayed the stall limit.

When two rows of small solidity cascade are arranged to form a tandem cascade, the blade boundary layer of the front cascade is sent into the main-flow and a new boundary layer is developed on the blade of the rear cascade, consequently, flow has little chance to separate from the blade surface. However, the side walls are continuous for the two cascades, therefore, it looks as if it is not possible to flow against a very high pressure rise across the tandem cascade. In reality at the critical condition of diffuser stall, each blade row achieves almost as much pressure rise as the critical pressure rise of respective single blade row, consequently the overall exit/inlet velocity ratio is very small and a large pressure rise is achieved by a properly designed tandem cascade.

As it is mentioned earlier, the oil-film pattern on the side wall demonstrates that reverse flow in the side wall boundary layer is limited to the inlet triangular zone, and at the exit of the blade row the boundary layer flows downstream, that is, the inlet flow to the rear blade row is not bad and the performance of the rear blade row is not penalized by the front blade row. It is presumed that the low energy fluid in the side wall boundary layer is nicely swept laterally and flows along the pressure side of a blade and it is eventually shoveled into the main flow through the gap between the blade and one of the blades in the rear blade row, consequently the side-wall boundary layer at the inlet to the next blade row is thin. That is, the three-dimensional nature of flow in the side-wall boundary layer prevents the tandem cascade from stalling against a high pressure rise. At present the optimum arrangement of the two blade rows is examined only experimentally, and further theoretical examination on three-dimensional boundary layer on the side wall is desired.

Impeller

The flow passage in an impeller of centrifugal compressor is three-dimensional, and flow changes the direction in the meridional plane as well as in the surface of revolution, and the boundary layer flow within the passage is three-dimensional due to curvature. Flow is made even more complicated by rotation of the impeller. Since the flow passage inside the impeller is long compared to the passage width, the boundary layer makes the effective passage narrower and as a result the velocity distribution of core flow is significantly modified.

Moore [9] calculated three-dimensional turbulent boundary layer along the hub, shroud and blade surfaces using a momentum integral method, assuming that at a corner of two surfaces the secondary flow in the boundary layer along a wall continues to the secondary flow in the boundary layer along the other surface.

Eckardt [8] measured the velocity distribution inside a centrifugal impeller with a laser velocimeter and demonstrated that at the corner of shroud and blade suction surface there is an area where the velocity is about one half of the velocity of core flow and that the area occupies about 1/5 of the cross-sectional area at the exit of impeller as shown in Fig.7-a.

Several groups analysed the flow within Eckardt's impeller by different ways assuming inviscid shear flow and predicted the

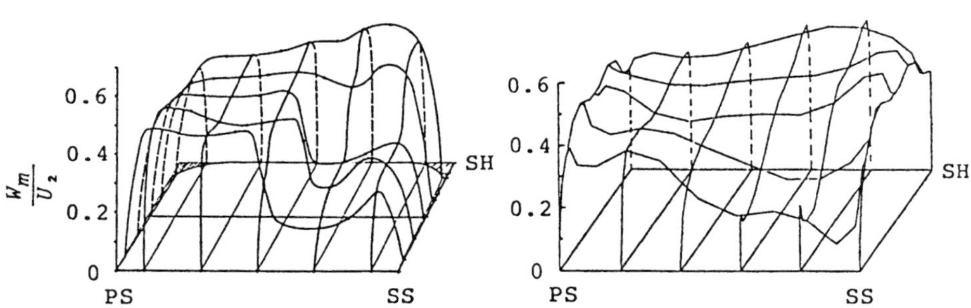

(a) Experiment [8]
Stationary shroud

(b) Prediction [11]
1% tip gap, stationary shroud

Fig.7 Velocity distribution at the exit of impeller

low velocity zone correctly. Since inviscid flow analysis is not simple for the complicated geometry, it is not practical to iterate a boundary layer analysis and an inviscid flow analysis. These days several groups try to solve two-dimensional viscous flow within a complicated passage of a centrifugal impeller.

Moore et. al. [10] developed a partially-parabolic calculation procedure to calculate three-dimensional viscous flow in a centrifugal impeller. The three-dimensional pressure field within the impeller is obtained by first performing a three-dimensional inviscid flow calculation and then adding a viscosity model and a viscous-wall boundary condition to allow calculation of the three-dimensional viscous flow. The mixing-length model is used as the turbulent viscosity. Calculated wake development in Fig. 7-b and pressure distributions at the exit are compared [11] with measurements. The results are satisfactory predicting impeller efficiency within one percent.

Conclusion

At the inlet of a blade row the velocity distribution in the boundary layer is skewed due to relative motion of blade rows, and the boundary layer inside a blade row is made even more complicated by curvature of the passage and rotation of the impeller. Due to the cross-flow component of velocity within the boundary layer, the boundary layer becomes thin while low velocity fluid accumulates at the corner of low pressure.

The zone of accumulated fluid of low velocity is susceptible to adverse pressure gradient, and it may become a cause of separation of flow from the wall. If the accumulated low velocity fluid is managed to flow downstream, separation of flow is delayed considerably because the wall boundary layer is made thin by the cross-flow, and it is possible to operate a turbomachine stably in a wide range of flow rate.

The three-dimensional effect of boundary layer is qualitatively understood but quantitative prediction is difficult for complicated geometry of flow passage. Behavior of accumulated low velocity fluid is beyond flow analysis based on the boundary layer assumption because of the extraordinary thickness, but the behavior is crucial for the performance of turbomachines.

References

1. Deruyck, J.; Hirsch, D.: Investigations of an Axial Compressor End-Wall Boundary Layer Prediction Method. Journal of Engineering for Power, Trans. ASME 103 (1981) 20-33.

2. Herzig, H.Z; et. al.: A Visualization Study of Secondary Flows in Cascades. NACA Rep. 1163, 1954.

3. Bailey, D.A.: Study of Mean and Turbulent-Velocity Fields in a Large-Scale Turbine-Vane Passage. Journal of Engineering for power, Trans. ASME 102 (1980) 88-95.

4. Senoo, Y.: The Boundary Layer on the End-Wall of a Turbine Nozzle Cascade. Trans. ASME 80 (1958) 1711-1720.

5. Himmelskamp, H.: Profiluntersuchungen an einem umlaufenden Propeller. Reports of the Max-Plank-Institut fur Strömungsforschung, Göttingen, No.2 (1950).

6. Arakawa, C.; Tagori, T,; Shirakura, M.: Analysis of Three-Dimensional Turbulent Boundary Layer on the Rotating Blade of Axial Flow Pump. Proceedings 10th Symposium Tokyo, The International Association for Hydraulic Research, Section Hydraulic Machinery, Equipment and Cavitation (1980) Vol.1 341-352.

7. Senoo, T.; Kinoshita, Y.; Ishida, M.: Asymmetric Flow in Vaneless Diffusers of Centrifugal Blowers. Journal of Fluids Engineering, Trans. ASME 99 (1977) 104-114.

8. Eckardt, D.: Detailed Flow Investigations Within a High-Speed Centrifugal Compressor Impeller. Journal of Fluids Engineering, Trans. ASME 98 (1976) 390-402.

9. Moore, J.: A Wake and an Eddy in a Rotating, Radial-Flow passage, Part 1; Experimental Observations, Part 2; Flow Model. Journal of Engineering for Power, Trans. ASME 95 (1973) 205-219.

10. Moore, J.; Moore, J.G.: Calculations of Three-Dimensional, Viscous Flow and Wake Development in a Centrifugal Impeller. Journal of Engineering for Power, Trans. ASME 103 (1981) 367-372.

11. Moore, J.; Moore, J.G.: Three-Dimensional Viscous Flow Calculation for Assessing the Thermodynamic Performance of Centrifugal Compressor. AGARD Conference Proceedings No.282 (1981) 9.1-9.19.

Experimental Study of the Boundary Layer on a Turbomachinery Rotor Blade

B. LAKSHMINARAYANA
T. R. GOVINDAN
C. HAH

Department of Aerospace Engineering
The Pennsylvania State University
University Park, Pennsylvania, U.S.A.

Abstract

The development of the turbulent boundary layer on a turbomachinery rotor blade is very complex. This paper is concerned with the measurement, analysis, and prediction of the blade boundary layer on a rotor blade. A comprehensive measurement of the blade boundary layer on a fan rotor blade was carried out in the axial flow fan facility at the Applied Research Laboratory. The measurements were taken with a miniature "x" configuration hot-wire probe. The streamwise and the radial velocity profiles as well as the turbulence intensities at one radius are presented and interpreted.

Introduction

Lack of knowledge of three dimensional turbulent boundary layers, especially those with rotation and curvature, has hampered technological progress in areas such as turbomachinery (e.g., compressors, pumps, propellers, turbine, etc.). The rotation induced coriolis forces generate spanwise flows inside the blade boundary layers, the magnitude of which depends on the angular velocity of the blade, the flow coefficient, and the rotor blade geometry. The nature of the boundary layer is schematically illustrated in Fig. 1. The spanwise migration of the flow results in a decrease of the boundary layer growth near the hub and an increase near the tip, an increase in the skin friction coefficient and other structural changes in the turbulence characteristics. A knowledge of the characteristics of the three dimensional boundarv layers on rotor blades is essential for the analysis and design of these machinery as well as prediction of flow properties, losses and noise generated. Very few investigators have attempted to measure or analyze the boundary layer developing on a turbomachinery rotor blade. A very brief review of these investigations is given below.

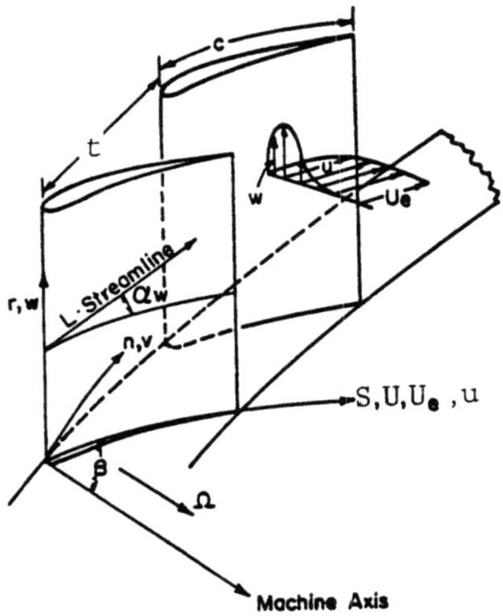

Fig. 1. Nature of boundary layer and notations used

Turbulent Boundary Layers on Axial Flow Rotor Blades

To date, very little work has been reported on the characteristics of rotating turbulent boundary layers on the blades of axial flow fluid machinery. Banks and Gadd [1] investigated the turbulent boundary layer growth near the leading edge of a single helical blade of large stagger, rotating in a stationary fluid. They approximate the blade by a flat sector of a disk with leading and trailing edges. The two momentum integral equations were solved by assuming a series solution, small cross flow and 1/7th profile for the mainstream velocity. Later, Lakshminarayana et al. [2] conducted an extensive experimental and analytical investigation of the three-dimensional turbulent boundary layer growth on a single rotating helical blade of large chord length. Using the momentum integral approach, they predicted the boundary layer growth and the limiting streamline angle of the flow over the entire blade. They also obtained the nature of mean velocity profiles in the outer part of the turbulent boundary layer and a skin friction correlation to show the direct effects of rotation. The skin friction stress was found to increase with rotation according to the formula

$$c_f/c_{f_{\Omega=0}} = 1 + c_1\sqrt{(\varepsilon_w\ x\ /r)}$$

where ε_w is the limiting streamline angle and x is the distance from the leading edge, r is the radius, and c_f is the skin friction coefficient, subscript $\Omega=0$ represents its value for a stationary case, c_1 is a constant.

Analytical and experimental investigations of the characteristics of three-dimensional turbulent boundary layers in a rotating helical passage of an axial flow inducer rotor were reported in refs. 3, 4, and 5. Expressions were developed for the velocity profiles in the inner layer, where the viscous effects dominate, in the outer layer, where the viscous effects are small, and in the interference layer, where the endwalls influence the flow. The prediction of boundary layer growth was based on the momentum integral technique. The experimental investigations were carried out in a flat plate inducer 3 ft. in diameter. The mean velocity profiles, turbulence intensities and shear stresses, wall shear stress, and the limiting streamline angles were measured at various radial and chordwise locations by using rotating pitot, and three sensor hot wire probes. The measurements are in general agreement with the predictions. The radial flows are found to be large. The collateral region was found to exist only very near the blade surface. The radial component of turbulence intensity was found to be higher than the streamwise component because of the effect of rotation. A wall shear stress correlation, which includes the effects of both the Reynolds number and rotation parameter, was derived from the measured data.

Lakshminarayana and Govindan [6] extended the momentum integral technique of ref. 3 to include the entrainment equation as a closure model and showed good agreement with the measured boundary layer data on cascade and rotor blades.

Detailed measurement of the boundary layer properties on an axial flow compressor or fan type blade, where the coupled pressure gradient and rotation effects make the boundary layer highly complex, is lacking and this prompted the present investigation.

Facility, Measurement Technique, and Program

Measurements reported in this paper were carried out using the Axial Flow Fan Research Facility (AFRF) at the Applied Research Laboratory. The facility and the rotor are described in detail in ref. 7. A brief description follows.

A nine-bladed rotor with cambered blades was installed in the facility and had the following specifications: tip diameter = 0.54 m, hub/tip diameter ratio = 0.44, blade chord (const) = 15.24 cm, maximum thickness (const) = 1.5 cm. Blade spacing, stagger angle and incidence angle at mid-radius were, respectively, 13.72 cm, 50°, 2.94°. Other relevant data are tabulated in Table 1 of ref. 8. The blade was designed for the free vortex distribution with substantial radial twist in the blade geometry. The performance characteristic of the rotor is given in ref. 8. The rotor alone (no IGV or stator) was operated at the design condition, 1150 rpm. The required flow coefficient was achieved by means of an auxiliary fan at the exit of the facility. The axial velocity at inlet was 13.6 m/sec. The Reynolds number based on the inlet relative velocity and chord length at the midspan was 2.8×10^5.

All the measurements were taken with a hot wire probe rotating with the rotor. This is accomplished by driving an internal probe traverse assembly with a stepping motor and a gear reduction unit that permits increments of 0.083°. Details of this rotating probe traverse assembly are given in ref. 7.

A miniature cross flow "X" hot wire probe, TSI 1247, with sensor diameter of 3 μm and sensor length equal to 1 mm was used in this program. Both the sensors are within a diameter of approximately 1.5 mm. The sensors were located in the (sr) plane with its axis at 45° to the s axis (Fig. 1) and was traversed normal to the blade. The mean and the fluctuating voltages from each of these wires were noted. Since the flow traverse was done close to the blade surface, the component of velocity in the n direction (V) is assumed to be small. The present configuration and the relevant hot wire equations will provide the values of the streamwise velocity U, radial velocity W and the resultant turbulence intensity in the (sr) plane, T_{sr}. These are plotted and interpreted in this paper.

Experimental Results and Interpretation

As mentioned earlier, the data was acquired at five radial and several chordwise locations. The data at R = 0.86, where R is the local radius normalized by the blade tip radius, is presented and interpreted in this paper. The data at R = 0.54, 0.72, and 0.94 is given in ref. 9. The data is presented at a streamwise distance S = 0.38 and 0.79, where S is the streamwise distance (Fig. 1) from the blade leading edge, normalized by the blade chord. All velocities are normalized by the freestream relative velocities (U_e), which are 28.1 and 23.8 m/sec, respectively, at S = 0.38 and S = 0.79. Other notations are shown in Fig. 1.

Streamwise Velocity Profile (U)

The non-dimensional streamwise velocity (U) is shown plotted in Fig. 2. N in this figure is the normal distance n (Fig. 1), normalized by the blade spacing (t). No data is available very close to the wall. The boundary layer thicknesses are about 2 to 3 percent of the blade spacing. The velocity profiles are typical of those that exist in an adverse pressure gradient. The experimental data is compared with the power law profile,

$$U = \left(\frac{n}{\delta}\right)^{(H-1)/2} \quad (1)$$

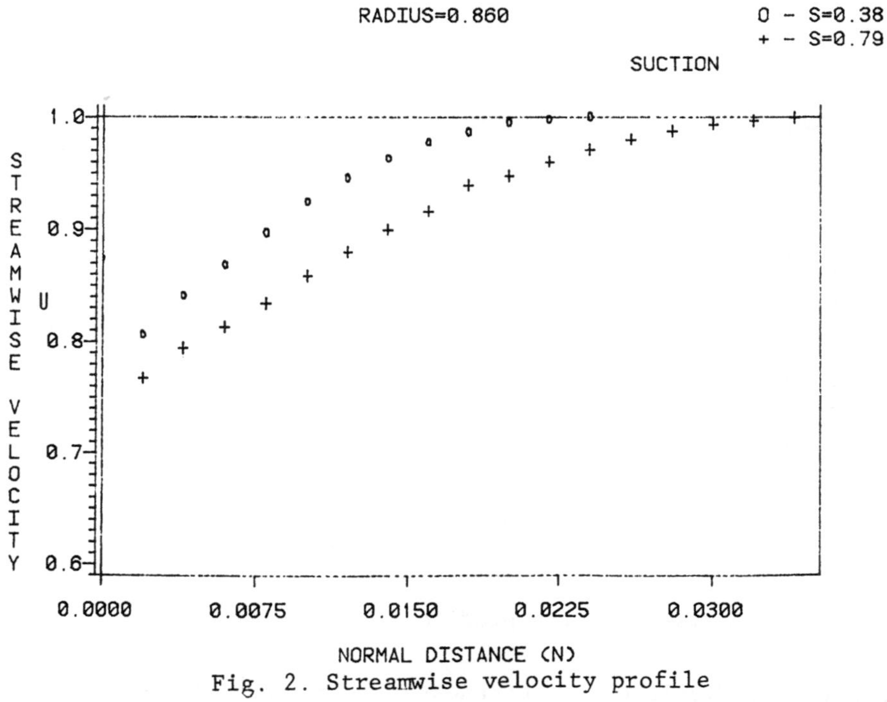

Fig. 2. Streamwise velocity profile

in Fig. 3. The shape factor (H) and the value of the boundary layer thickness (δ) were derived from the experimental values of

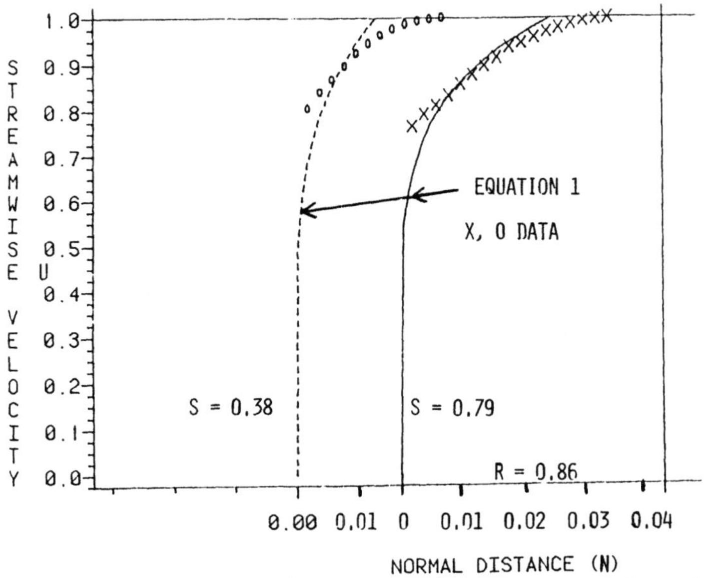

Fig. 3. Streamwise velocity profile in power law form

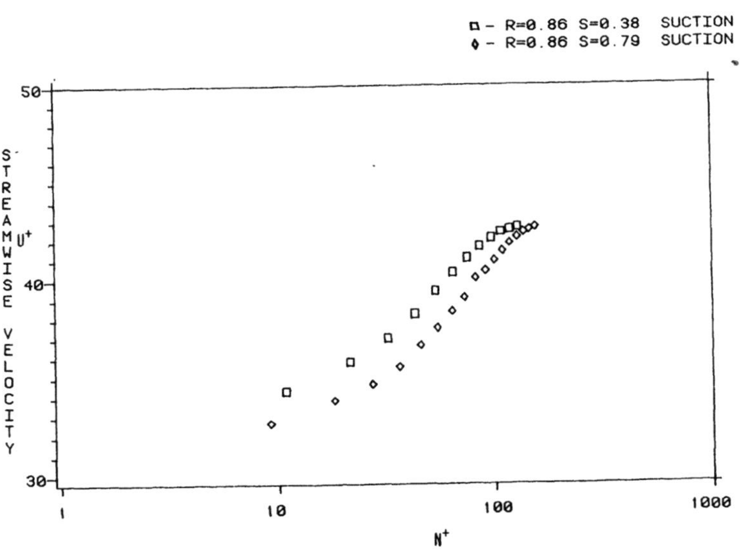

Fig. 4. Streamwise velocity profile in "law of the wall and wake" form

momentum and displacement thicknesses. The data at S = 0.79 shows excellent agreement with the power law profile, except in the wall region. The agreement at S = 0.38 is not as good. The boundary layer is very thin (2 to 3 mm) in this region. The profile is also plotted in a logarithmic form in Fig. 4. Here U^+ is velocity normalized by the friction velocity (U^*) derived from the experimental data and $N^+ = U^*n/\nu$, ν is the kinematic viscosity. It is evident that the "law of the wall" region is very small, but the "wake" region is substantial with a large amplitude. The effect of adverse pressure gradient is clearly evident.

In summary, the streamwise velocity profiles seem to follow the power law profile as well as the empirical law of the wall and wake.

Radial Velocity Profile (W)

The radial velocity normalized by the freestream relative velocity (U_e) is shown in Fig. 5. The radial velocities (W) are

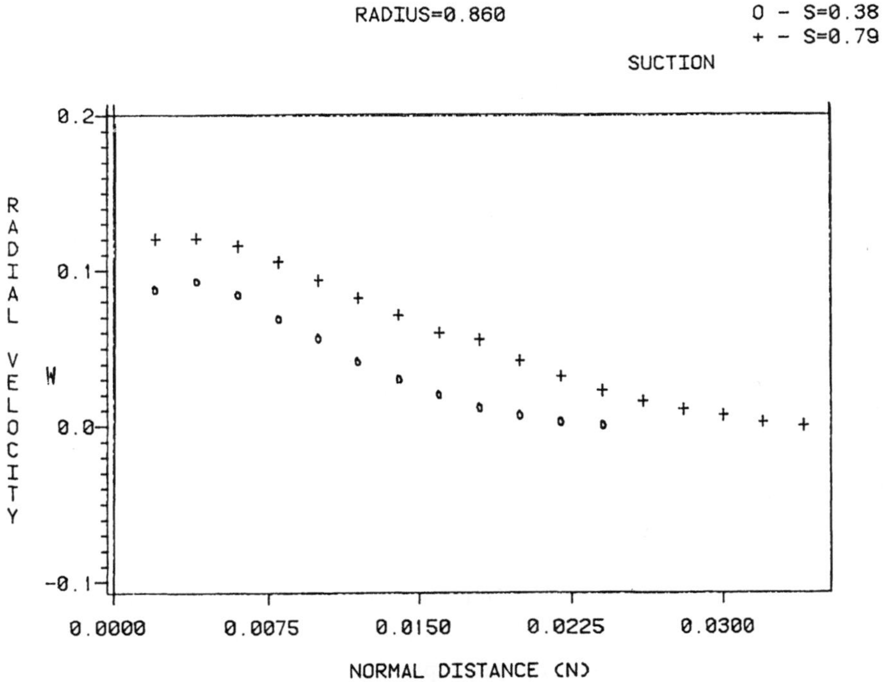

Fig. 5. Radial velocity profile

maximum close to the blade surface, decreasing to zero in the outer edge. The radial velocities are about 10 to 20 percent higher at S = 0.79 than at S = 0.38. The plot of the radial velocity in the hodograph plane, shown in Fig. 6, clearly reveal the linear region in the outer layer with a limiting streamline angle (angle of the streamline as the blade surface is approached, measured from the free stream direction) of 9° at S = 0.79 and 7° at S = 0.38.

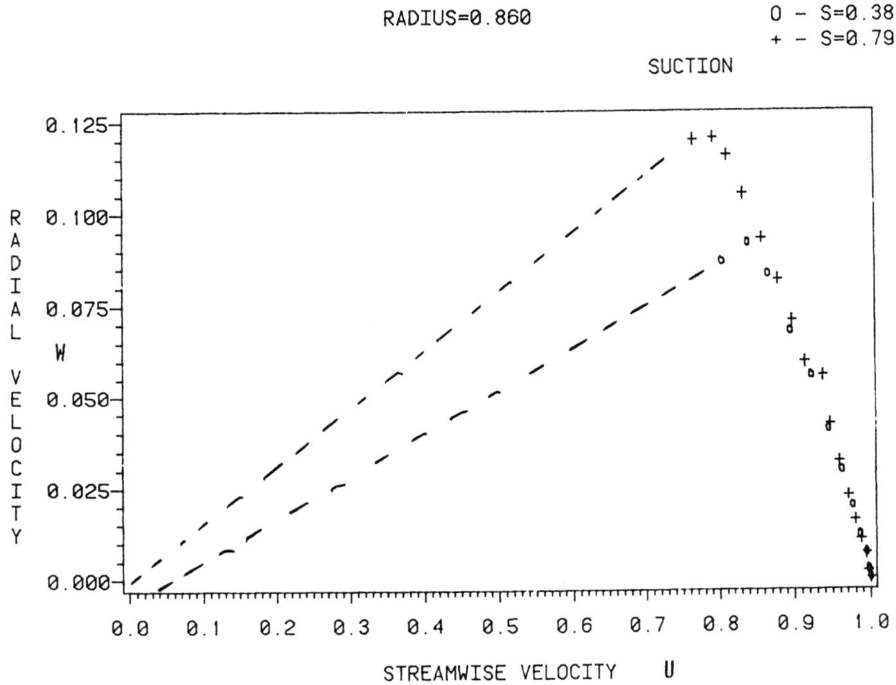

Fig. 6. Hodograph plot of the streamwise and radial velocity

Turbulence Intensity (T_{rs}, T_1)

The turbulence intensity $T_{rs} = \sqrt{\overline{u^2} + \overline{w^2}}/U_e$ where u and w are fluctuating components of velocity in s and r directions, respectively, is shown plotted in Fig. 7. The maximum turbulence intensity near the wall is 16 percent, decreasing to about 8 percent in the free stream. The freestream intensities generally are high in rotor passages. The turbulence intensity (T_1) in a direction normal to one of the wires is shown in Fig. 8. This shows a freestream intensity of 7 percent increasing to about 13 percent close to the blade surface.

Momentum and Displacement Thicknesses and Comparison with Predictions

The momentum and displacement thicknesses and the shape factor

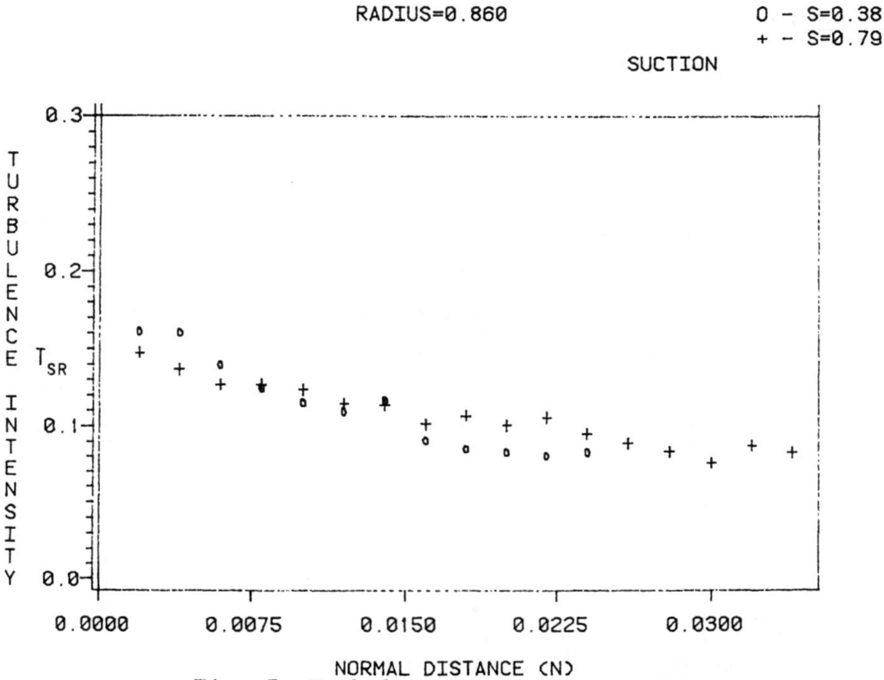

Fig. 7. Turbulence intensity profile, T_{sr}

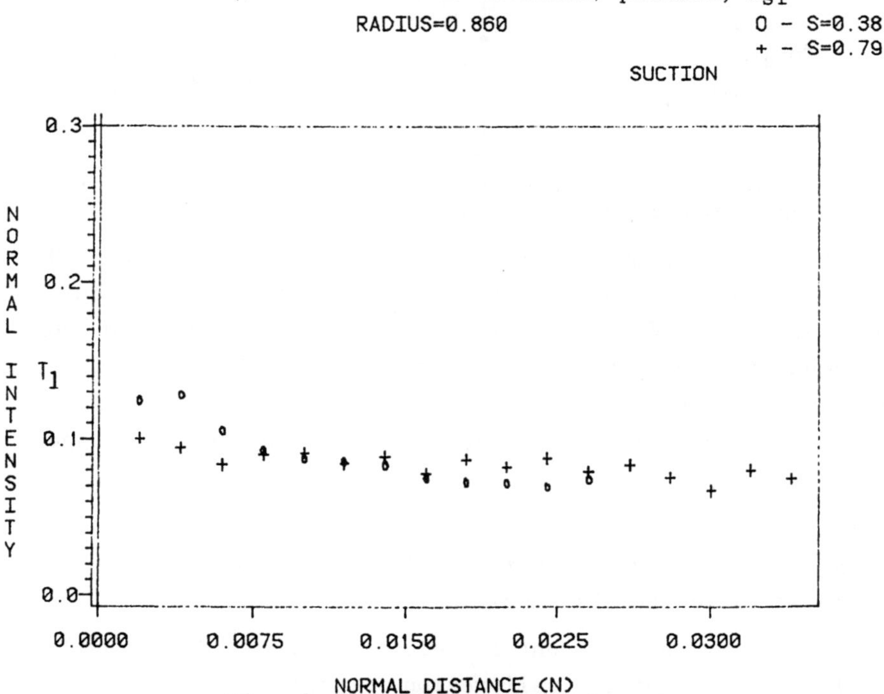

Fig. 8. Turbulence intensity, T_1

are defined, respectively, by

$$\theta_{11} = \frac{1}{C} \int_o^h U\left(1 - U\right) dn \qquad h > \delta \qquad (2)$$

$$\delta_1 = \frac{1}{C} \int_o^h \left(1 - U\right) dn \qquad (3)$$

$$H = \delta_1 / \theta_{11} \qquad (4)$$

The momentum and displacement thicknesses and the shape factor calculated from equations 2-4 are plotted in Fig. 9. Both δ_1 and θ_{11} increase almost linearly in the chordwise direction and the shape factor varies from 1.33 to 1.4, indicating a moderate to large adverse pressure gradient.

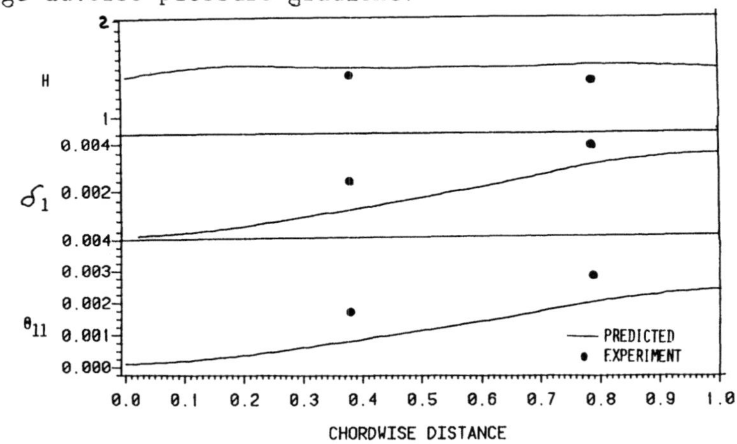

Fig. 9. Displacement and momentum thicknesses, shape factor, and comparison with predictions (ref. 6)

The data is also compared with the predicted values in Fig. 9. The predictions are based on the momentum integral technique developed in ref. 6. The inviscid flow field was derived from the computer code by Katsanis and McNally [10,11]. Considering the complexity of the flow, the agreement between the theory and experiment is good for δ_1 and θ_{11} and excellent for the shape factor.

Conclusions

1. The boundary layer on a rotor blade is three dimensional, with appreciable radial velocities.

2. The streamwise component of velocity follows the power law profile. The law of the wall region is small and the wake region is substantial.

3. The outer part of the hodograph plot is linear and the limiting streamline angles are of the order of 7 to 10°.

4. The free stream turbulence intensities as well as the intensities near the blade surface are high.

5. Momentum and displacement thickness vary almost linearly with the chord length. The values of these and the shape factor are predicted reasonably well.

Acknowledgment

This work was supported by the U. S. National Aeronautics and Space Administration through grant NSG 3266 with P. M. Sockol as the Technical Monitor. The experimental work was carried out at the Applied Research Laboratory in their Axial Flow Research Fan Facility.

References

1. Banks, W.H.; Gadd, G.E.: A preliminary report on boundary layers on screw propellers and simpler rotating bodies. National Physical Laboratory (England) Report SHR 27/62 (1962).

2. Lakshminarayana, B.; Jabbari, A.; Yamaoka, H.: Three dimensional turbulent boundary layer on a single rotating helical blade. J. Fluid Mech. 51 (1972) 545.

3. Anand, A.K.; Lakshminarayana, B.: Three-dimensional turbulent boundary layer in a rotating helical channel. J. Fluids Engr. 97 (1975) 197.

4. Anand, A.K.; Lakshminarayana, B.: An experimental and theoretical investigation of three dimensional turbulent boundary layer inside an axial flow inducer. NASA CR 2888 (1977).

5. Anand, A.K.; Lakshminarayana, B.: Three dimensional turbulent boundary layer in a rotating helical channel. J. Engr. Power 97 (1978) 197-210.

6. Lakshminarayana, B.; Govindan, T.R.: Analysis of turbulent boundary layer on cascade and rotor blades of turbomachinery. AIAA Journal 19:10 (1981) 1333-1341.

7. Pierzga, M.J.: Experimental verification of the streamline curvature analysis method applied to the flow through an axial flow fan. M.S. Thesis, Dept. of Aerospace Engr., The Pennsylvania State Univ. (1981) also as ARL TM 80-181 (1980).

8. Lakshminarayana, B.; Govindan, T.R.; Reynolds, B.: Effects of rotation and blade incidence on properties of turbomachinery rotor wake. AIAA Journal 20:2 (1982) 245.

9. Lakshminarayana, B.; Hah, C.; Govindan, T.R.: Three-dimensional turbulent boundary layer development on a fan rotor blade. AIAA Paper 82-1007 (1982).

10. Katsanis, T.; McNally, W.D.: Revised Fortran program for calculating velocities and streamlines on the hub-shroud mid channel stream surface of an axial-, radial-, and mixed-flow turbomachine or annular duct. NACA TND 8430 (1977).

11. Katsanis, T.: Use of arbitrary quasi-orthogonals for calculating flow distribution on a blade to blade surface in a turbomachine. NASA TND 2809 (1965).

Boundary Layers on Bodies of Revolution Spinning in Axial Flows

Ikuo NAKAMURA and Shintaro YAMASHITA

Department of Mechanical Engineering, Nagoya University
Nagoya, Japan

Summary

A brief survey is made of the various aspects of the boundary layer which develops on a spinning axisymmetric body in an axial flow. At first some studies of laminar and transitional flow are reviewed. General features of this boundary layer are discussed by use of the basic equations in which two notable terms represent the effects of centrifugal force and Coriolis-like force. Similar or universal turbulent velocity distribution laws are discussed and a new law of the wall is presented. The log-law shows considerable effects of rotation and transverse curvature. Finally, future problems are discussed from the authors' viewpoint.

Introduction

Since rotating motion is one of the fundamental factors in fluid flow as in the dynamics of rigid bodies, various rotating flow patterns have drawn the attention of many researchers. For example, Taylor [1] published his monumental study in 1923, and Kármán [2] proved the similarity of the flow on a rotating disc. There have been a number of reports dealing with problems of rotating flows such as a spinning body in a quiescent fluid, a swirling flow in a pipe, the rotation of a pipe [36] or a duct, a swirling jet, not to mention geophysical flows. Among these various rotating flows, boundary layers developing on bodies of revolution spinning in various axial main flows are rather a basic type, as seen in the example in Fig.1.

Even in the laminar state, this flow is complex compared with a thin boundary layer on a stationary axisymmetric body, because the latter can be reduced to a two-dimensional problem through Mangler transformation, whereas there is no such effective transformation for the spinning body case. If the flow is tur

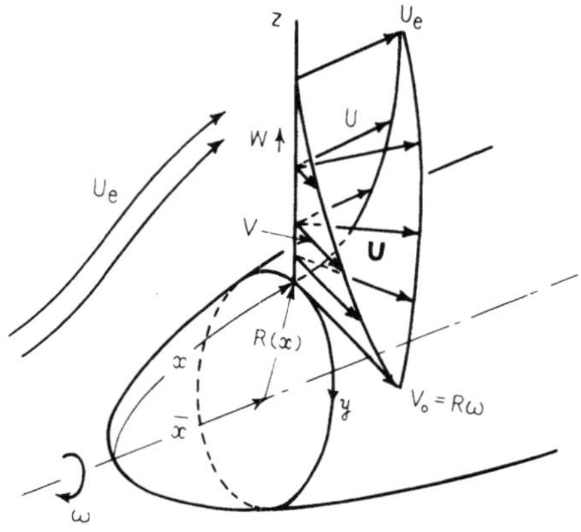

Fig.1. Representative flow field and nomenclature

bulent, the centrifugal force produced by rotation and the transverse curvature effect complicate the structure of this boundary layer. Inclination of the bodv surface generates particular body forces, which are described in what follows, and they control the development of the boundary layer. In his stimulating study, Bradshaw said of these flows: "There are a few experiments of swirling flows that are not easy to classify" [3]. In spite of its complexity, this boundary layer has very good characteristics, axisymmetry, and it makes this flow tractable in comparison with other three-dimensional boundary layers.

From an engineering point of view, this problem concerns the flow around bosses of various fluid machinery and spinning projectiles. For example, in the case of a propeller water turbine the radius of the rotating boss decreases downstream; furthermore, in the off-design operation, the main flow has the swirl component. Because of the many rotating flow problems, this brief review is limited to the indicated flow and involves no discussion of either the Magnus effect or the main flow swirl.

Laminar Flows and Their Stability

Although the turbulent boundary layer is the subject of discussion, so many eddy viscosity models are found in this field nowadays that it seems relevant to make some remarks on the laminar flow case. It also may serve to clarify its fundamental properties.

Schlichting [4] calculated laminar boundary layers of this type by use of the Kármán-Pohlhausen method and showed the movement of the separation point by rotation. Hannah's exact solution of the Navier-Stokes equation [5], Ishizawa's consideration on similar solutions [6], Nakamura's universal series solution [7], and the solution of Furuya et al.[8] for the thick laminar boundary layer are basic ones. The most important facts revealed by these studies are the movement of a separation point due to the meridian component of the centrifugal force [4,7] and self-induced pressure gradient in the case of a rotating thin cylinder [8].

Very recently, a stability analysis of this boundary layer was performed by Kobayashi [9]. He envisaged Taylor-Görtler type vortex system in the boundary layer on a cone spinning in an axial flow and calculated it by use of linear theory. Also, a visualization study carried out by Kohama and Kobayashi [10] showed very clearly the predicted spiral vortices. Two-types of these vortices are classified by them.

Experimental Studies

Drag and Torque In 1926 Wieselsberger [11] published a report concerning drag exerted on three types of spinning bodies in a wind tunnel. One of the results presented in Fig.2 shows clearly increased drag acting on a bullet-type body with spin. From his experimental results, Wieselsberger conjectured a vortex system somewhat like Taylor vortices. As mentioned in the previous section, his hypothesis was verified by Kobayashi and Kohama [9,10] after about a half century.

Luthander and Rydberg [12] measured the drag experienced by a sphere spinning in axial flows. Some of their results are re

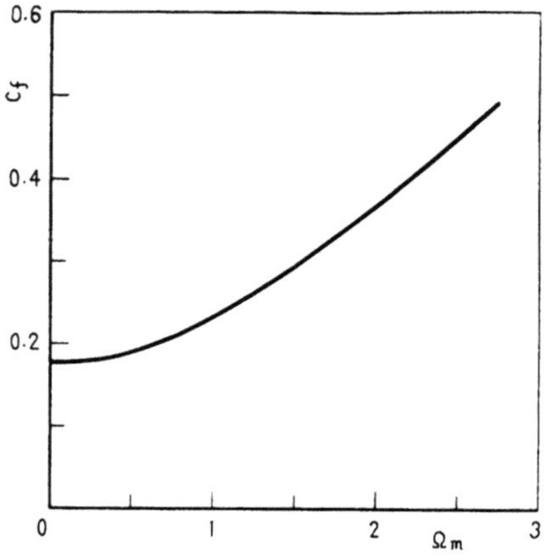

Fig.2. Drag coefficient measured by Wieselsberger [11]. $\Omega_m = R_m\omega/U_m$: speed ratio

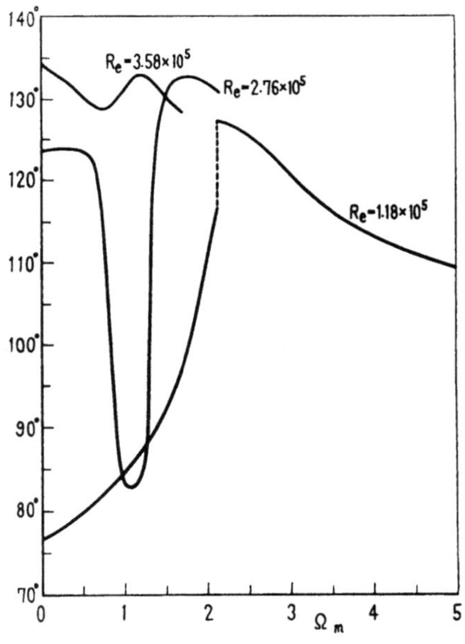

Fig.3. Movement of separation point (line) with speed ratio. Ordinate indicates the angle of separation point from the pole (forward stagnation point).

arranged in Fig.3 in order to show the complex movement of the separation point. The most probable explanation for this change in the range $0.5 < \Omega_m \lesssim 1$ may be the relaminarization of the boundary layer. Parr [13] measured the torque of two types of rotating body and also discovered increased torque with spin.

Measurements of base pressure of a spinning bullet type body, performed by Badrinarayanan and Kangovi [14], showed a definite trend with rotating speed.

<u>Mean Velocity Field</u> No measurement of velocity was carried out in earlier studies. Skewed velocity profiles on a nose part and on a tail body were measured by Parr [13]. He obtained various integrated thicknesses of the boundary layer without taking account of the transverse curvature effect. Development of the integrated thicknesses showed clearly the effect of the rotation speed. Truckenbrodt's proposal [15] as to a relationship between the meridian velocity component U and the azimuthal component V, may be stated

$$V/V_o = 1 - U/U_e, \tag{1}$$

where V_o and U_e are the circumferential speed of the body surface and main flow speed respectively. This was checked by Parr [13], and the present authors, who found Eq.(1) to be valid except at high rotating speed.

Many experimental studies have been performed by the present authors on these boundary layers developing on spinning bodies of various shapes and under pressure gradients: 1) a streamlined body [16], 2) a hemispherical nose body [16], 3) a conical nose body [16], 4) a thin cylinder and one with a ring or a backward facing step [17,19], 5) a rotating cylinder in adverse and favourable pressure gradients [18,19], 6) a conical tail body [20], and 7) a rotating thin cylinder in a uniform stream [21].

Some discussion will be made using basic equations deduced from the thin boundary layer approximation;

$$U\partial U/\partial x + V\partial U/\partial z = -\rho^{-1}dP/dx + \lambda V^2/R + R^{-1}\partial\{R(\nu\partial U/\partial z - \overline{uw})\}/\partial z, \tag{2a}$$

$$U\partial V/\partial x + V\partial V/\partial z = -\lambda UV/R + R^{-2}\partial\{R^2(\nu\partial V/\partial z - \overline{vw})\}/\partial z, \tag{2b}$$

$$\partial(RU)/\partial x + \partial(RW)/\partial z = 0, \tag{2c}$$

where $\lambda = dR/dx$ represents the inclination of the body surface and other notation is evident (see Fig.1). When the boundary layer thickens in relation to the body radius, more terms must be included in the basic equations; but in the present context, Eqs.(2a,b,c) are sufficiently accurate.

As clearly seen from the above equations, an important non-dimensional parameter can be constructed as the ratio of a representative peripheral velocity of the body surface $V_m = R_m\omega$ to a representative main flow velocity U_m as follows:

$$\Omega_m = R_m\omega/U_m = V_m/U_m\text{: speed ratio.} \qquad (3)$$

A comparison of the basic equations of the present problem to the ones for the stationary body case serves to distinguish two important terms,

$$C_n = \lambda V^2/R, \quad C_o = \lambda UV/R, \qquad (4)$$

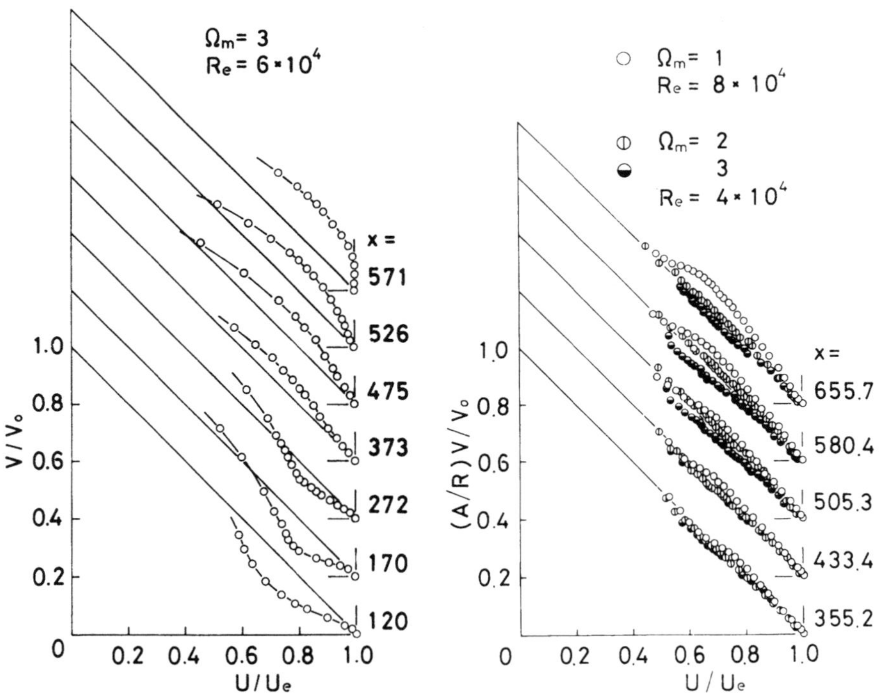

a) Radius increasing toward downstream, $\lambda > 0$

b) Radius decreasing toward downstream, $\lambda < 0$

Fig.4. Deflection of the polar plots from the quasi-collateral condition in case of conical bodies which can be called the centrifugal force term and the Coriolis-like force term [20]. The speed ratio Ω_m may be taken for the ratio of C_n to C_o.

From Eqs.(2a,b), the role of C_n and C_o in the velocity distribution can be classified qualitatively as follows.

$\lambda > 0$, then C_n accelerates U but C_o decelerates V;

$\lambda < 0$, then C_n decelerates U but C_o accelerates V.

This phenomenon produces a deviation of the polar plots from the quasi-collateral condition Eq.(1), as shown in Figs.4a, b. The pressure gradient term, when it exists, will have an appreciable effect on the quasi-collateral condition. When the flow field shows a quasi-collateral condition, an important similarity exists between the meridian direction and the azimuthal direction. Equation (1) is generalized [8] for the thick boundary layer as

$$(a/r)V/V_o = 1 - U/U_e. \tag{5}$$

A different polar plot was proposed by Bissonnette and Mellor [22]. Arzoumanian, Fulachier and Dumas [23] expressed polar plots using $\{(a+z)/a\}V/V_o$ and U/U_e for the sublayer region.

<u>Turbulent Field</u> The turbulence measurements in this boundary layer were performed in detail at first by Bissonnette and Mellor [22], who investigated the axisymmetric turbulent boundary layer which enters from a stationary cylinder into the rotating one that follows and encounters a sudden circumferential strain. Mean flow measurements of the same flow configuration were performed by Ikui et al. [24,25]. Bissonnette and Mellor's experiment showed that the redistribution process of turbulent energy responded very rapidly to the applied straining motion. This flow was worthy of attention, and the apparatus somewhat easier to construct than the present authors'. Several experiments on the same flow configuration used by Bissonnette and Mellor were reported by Lohmann [26], a Marseille group under Dumas and his co-workers [22,27]. The present authors have investigated in detail the turbuler ·ɘld in the boundary layer on a rotating conical body of r reasing toward the downstream in which $C_n < 0$ and C_o

Bradshaw [3,28] has pursued an interesting discussion on the rotation effect, or the general effect of streamline curvature or extra rate of strain, on turbulence. Many Richardson numbers introduced by him are employed in various studies.

Universal Velocity Distribution

Since the logarithmic layer in two-dimensional turbulent boundary layers is one of their most conspicuous features, there are many proposals for the logarithmic law valid for this turbulent boundary layer. The present authors have investigated what type of law of the wall and logarithmic law should exist for the peripheral velocity component in the thin and thick turbulent boundary layers [16,18,19]. Also, Bissonnette and Mellor [22] considered the logarithmic law, while Lohmann [26] and Arzoumanian et al.[26] applied the usual logarithmic law. The results of various authors are sometimes contradictory.

In a recent study [21] the present authors discussed the wall layer similarity. For the sublayer, the velocity distribution is given as

$$(U_{R0}/U_R)U_1/U_{1\tau} = z^+ = (U_{1\tau}a/\nu)\{(r^2 - a^2)/2r^2\}, \qquad (6)$$

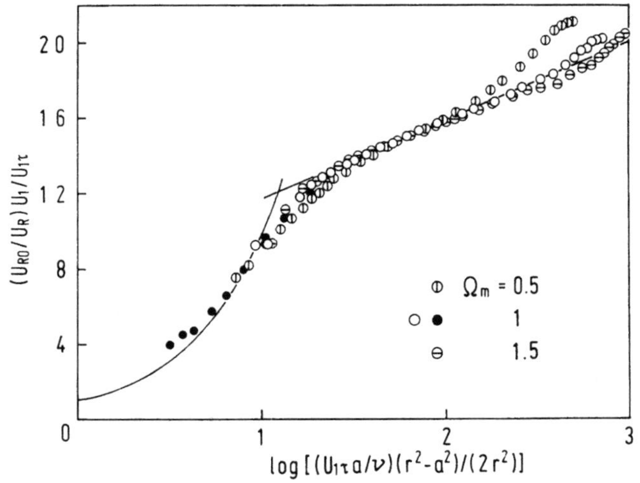

Fig.5. Logarithmic profile representing Eq.(7) with A_1 = 4.2 and B_1 = 7.5 [21]. Near the wall the curve shows Eq.(6).

where U_R is relative inviscid flow velocity, U_{R0} is its value on the wall, U_1 is the mean velocity component in the direction of the vector U_R, and $U_{1\tau}$ is the friction velocity in the U_{R0}-direction. The logarithmic law is written

$$(U_{R0}/U_R)U_1/U_{1\tau} = A_1 \log z^+ + B_1. \quad (7)$$

These equations describe well the experimental velocity distribution shown in Fig.5.

Prediction of the Turbulent Boundary Layer

Historically, the calculation of this boundary layer began with the momentum integral method of Truckenbrodt [15]. Parr [13] completely revised it and obtained reasonable agreement. The present authors extended Parr's method to the more general case [29].

The application of various field methods recently developed for this turbulent boundary layer has been aggressively undertaken by an Imperial College group under Launder and his co-workers [30-33]. The agreement between the predictions and the data is fairly good, as shown in Fig.6. Higuchi and Rubesin's contributions to this problem are also notable [35]. Although further development of Reynolds stress equation model is desired,

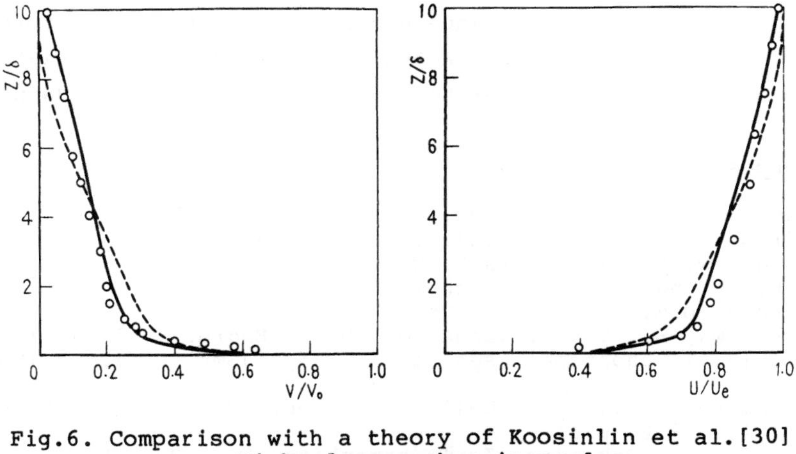

Fig.6. Comparison with a theory of Koosinlin et al.[30].
———, Richardson number is used.
-----, Richardson number is not used.

it cannot be overlooked that the proof of Speziale [34] shows that the Reynolds stress transport equations are frame-dependent.

Future Research

The following areas deserve future study.

(1) Laminar flow Can a multi-scaled solution be obtained for the laminar boundary layer that encounters a sudden circumferential strain (i.e., the same flow configuration studied by Bissonnette and Mellor for the turbulent flow)?

(2) Stability How are the spiral vortices and transition involved in the changes in curves shown in Fig. 3?

(3) Turbulence i) Direct measurement of local wall shear stresses τ_{ox} and τ_{oy} is extremely valuable. ii) Consideration of the classical self-preservation concept of this turbulent boundary layer is not a hackneyed analysis. iii) Turbulent energy balance is essential to understand the physical process in this boundary layer. Dissipation should be measured. iv) How does the rotation of a body modify the coherent structure observed in the flat-plate boundary layer?

Acknowledgement

The authors extend their sincere gratitude to Emeritus Professor Dr. Yoshimasa Furuya of Nagoya University for initiating and encouraging these researches and for his overall guidance.

References

1. Taylor, G.I., Phyl. Trans., A 223, 1923, p.289
2. Kármán, T. von, ZAMM, Vol.1, 1921, p.233
3. Bradshaw, P., AGARDograph, No.169, 1973
4. Schlichting, H., Ing.-Arch., Bd.21, 1953, p.227
5. Hannah, D.M., ARC Rept., No.2772, 1947
6. Ishizawa, S., Trans. Japan Soc. Mech. Engrs., Vol.28, 1962, p.1512, in Japanese; see also Bull. JSME, Vol.6, 1963, p.487
7. Nakamura, I., Res. Bull. Dept. Gen. Ed., Nagoya Univ., Vol.16, 1972, p.31
8. Furuya, Y., Nakamura, I. & Yamashita, S., Mem. Fac. Eng., Nagoya Univ., Vol.30, No.1, 1978, p.1

9. Kobayashi, R., Trans. Japan Soc. Mech. Engrs., Vol.46, 1980, p.1900, in Japanese
10. Kohama, Y. & Kobayashi, R., Prep. Japan Soc. Mech. Engrs., No.810-15, 1981, p.159, in Japanese
11. Wieselsberger, C., Phy. Zeit., Bd.28, 1927, p.84
12. Luthander, S. & Rydberg, A., Phy. Zeit., Bd.36, 1935, p.552
13. Parr, O., Ing.-Arch., Bd.32, 1963, p.393
14. Badrinarayanan, M.A. & Kangovi, S., Ind. Inst. Sci. Rept., 71FM5, 1971
15. Trukenbrodt, E., Ing.-Arch., Bd.22, 1954, p.21
16. Furuya, Y. & Nakamura, I., J. Appl. Mech., Vol.37, 1970, p.17
17. Nakamura, I., Yamashita, S. & Furuya, Y., 2nd Int. JSME Symposium, Tokyo, 1972, Vol. 2, p.41
18. Furuya, Y., Nakamura, I., Yamashita, S. & Ishii, T., Bull. JSME, Vol.20, 1977, p.191
19. Furuya, Y., Nakamura, I. & Yamashita, S., 5th Conf. Fluid Machinery, Budapest, 1975, p.297
20. Nakamura, I., Yamashita, S. & Furuya, Y., Turbulent Shear Flows 2, Ed., J.S. Bradbury et al., Springer, 1980, p.99
21. Nakamura, I., Yamashita, S., Watanabe, T. & Sawaki, Y., 3rd Symposium on Turbulent Shear Flows, U. C. Davis, California, 1981, p.2.7
22. Bissonnette, L.R. & Mellor, G.L., J. Fluid Mech., Vol.63, 1974, p.369
23. Arzoumanian, E., Fulachier, L. & Dumas, R., 2nd Symposium on Turbulent Shear Flows, Imperial College, London, 1979, p.4.28
24. Ikui, T., et al., Prep. Japan Soc. Mech. Engrs., No.87, 1963, p.43, in Japanese
25. Ikui, T., et al., Prep. Japan Soc. Mech. Engrs., No.115, 1964, p.33, in Japanese
26. Lohmann, R.P., J. Fluids Eng., Vol.98, 1976, p.354
27. Arzoumanian, E., Leborgne, M. & Fulachier, L., 3rd Symposium on Turbulent Shear Flows, U. C. Davis, California, 1981, p.2.13
28. Bradshaw, P., J. Fluid Mech., Vol.36, 1969, p.1
29. Nakamura, I., Yamashita, S. & Yamamoto, K., Mem. Fac. Eng., Nagoya Univ., Vol.32, 1980, p.282
30. Koosinlin, M.L., Launder, B.E. & Sharma, B.I., J. Heat Transf., Vol.96, 1974, p.204
31. Koosinlin, M.L. & Lockwood, F.C., AIAA J, Vol.12, 1974, p.547
32. Sharma, B.I., AIAA J., Vol.15, 1977, p.271
33. Launder, B.E., Priddin, C.H., & Sharma, B.I., J. Fluids Eng., Vol.99, 1977, p.231
34. Speziale, C.G., Phys. Fluids, Vol.22, 1979, p.1033
35. Higuchi, H. & Rubesin, M.W., AIAA J, Vol.17, 1979, p.934
36. Murakami, M., Kikuyama, K. & Nishibori, K., 3rd Symposium on Turbulent Shear Flows, U. C. Davis, California, 1981, p.2.1

Effect on a Developed Turbulent Boundary Layer of a Sudden Local Wall Motion

L. FULACHIER, E. ARZOUMANIAN, R. DUMAS

Institut de Mécanique
Statistique de la Turbulence
Laboratoire Associé au CNRS n°130
12, Avenue Général Leclerc
13003 Marseille, France.

Summary

An experimental investigation is carried out in the turbulent boundary layer on an axially rotating cylinder following a stationary one. Measurements are made up to the viscous sublayer which plays a crucial role in the generation of turbulence. The threedimensionality of this complex flow is studied through the mean and turbulent velocity field. In particular it appears the trend towards a new twodimensional equilibrium in a relative system linked to the moving wall. Various eddy viscosities have been considered and discussed. The gradient mean velocity models are irrelevant mainly for the transverse relative motion.

Introduction

The purpose of the present paper is to present an experimental investigation of a complex turbulent shear flow. The distorsion is due to a sudden change of the boundary conditions. The experiment is carried out in the boundary layer of an axially rotating cylinder following a stationary cylinder. So,the fully developed upstream two-dimensional turbulent boundary layer is suddenly submitted to a transverse strain. This generates complex three-dimensional effects which are studied through their influence on the mean and turbulent velocity field.

Such experiments have been performed by Bissonnette and Mellor [1] and Lohmann [2], but only within the fully turbulent part. However it appears that the very-close-to-the-wall region plays a crucial role. In particular, many difficulties arise in the modelling : the prediction of this flow by an anisotropic viscosity k - ε model, performed by Cousteix and Aupoix [3], does not agree the experimental results in this region.

In this set of experiments, the wall velocity w_0 has been chosen neighbouring the free-stream velocity in order to create a peripheral friction of the same order as the longitudinal friction. In these conditions we may appreciate the curvature effect by the ratio $(w/(a+y)/(\partial w/\partial y)$ [4] where w is the peripheral velocity, a the radius of the cylinder, y the distance to the wall. At the wall this value is about 0.005 which may be not negligible [5]. Besides using relative coordinates (see Fig.3, $w_r=w-w_o$)the sign of the Coriolis parameter [6] is négative, that is slightly destabilizing $(-2(w_o/a)((\partial w_r/\partial y)-2(w_o/a))/(\partial w_r/\partial y)^2<0)$.This might explain the inflexional feature of the hodograph ($\overline{u}$, $\overline{w}$)found [7] at distances y^+ inferior to about 11 ($y^+=(y\ u_o^*/\nu$, u_o^* friction velocity with no rotation).

As we had shown previously [8], the experimental results obtained in absolute coordinate system lead to misinterpretation and they must be examined in a relative coordinate system linked to the moving cylinder (see Fig.3). Then, we found that the flow tends to be twodimensional in respect with this relative system more especially as the distance to the leading edge of the spinning part increases. Close to this leading edge, the tendency to the twodimensionality applied only to the sublayer, at least for the mean velocity. But actually, the flow is never twodimensional, farther the wall even at the trainling edge of the rotating part. Moreover, the relative coordinate system is suitable only within the rotation layer, $\overline{w} \neq 0$.

One of the objectives of the present paper is to underline the lack of twodimensionality, or the level of threedimensionality, through the turbulent tangentiel stresses. Finally taking into account the mean velocities, we define various eddy viscosities in such a complex situation.

Experimental Conditions and Procedures

The experiment is carried out in the I.M.S.T. closed-return wind tunnel (800 x 800 mm),(Fig.1). The model is rigidly suspended so that very little wall vibrations are perceptible about 0,01 mm for the rotation speed generally chosen (w_o = 10.8 m/s).Noze form of the body is adjusted with a Pitot-tube profile. Transition to turbulence is triggered by means of scores, a stripe and roughnesses. The rotating portion is ended by a stationary profiled trailing edge. Whith no rotation, a good axisymmetry of

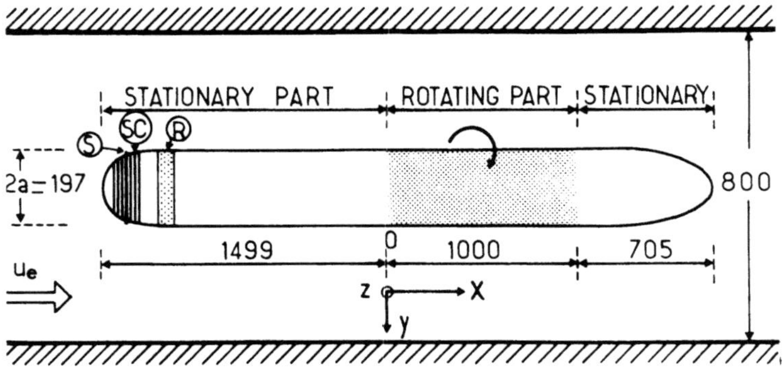

Fig.1. Schematic diagram of model (dimensions in mm)
(S) Stripe, (SC) Scores, (R) Roughnesses

the boundary layer has been obtained all along the rotating part. The streamwise pressure gradient is slightly negative, the Clauser parameter being small ($|\pi|$ = 0.26 for w_o = 0).

The aerodynamical conditions are summarized hereafter :
- Just before the rotating portion, the free stream velocity u_{e_o} = 12.34 m/s. The boundary layer thickness $\delta_o(\bar{u}=u_{e_o}) \simeq$ 28 mm. The Reynolds number R_{δ_o} = 19000 and the friction velocity u_o^* = 0.47 m/s.

Two stations have been investigated :
- Firstly, a region sufficently far from the origin of the spinning part (X/δ_o = 28, u_e = 12.38 m/s, δ = 34 mm), where the reorganization of the threedimensional boundary layer into a twodimensional one (in a relative coordinate system) is in an advanced phase.
- Secondly nearer the origin, where the threedimensional effects are stronger,(X/δ_o = 5.5, u_e = 12.36 m/s, δ = 29 mm).

As concerns the instrumentation, various static and total flattened pressure tubes (0.34 mm) have been used up to the wall vicinity ($y^+ \simeq 6$). Hot wire measurements have enabled to approach very close to the wall ($y^+ \simeq 2$). Corrections of wall effects have been examined. Very close to a stationary wall, measurements are altered by the vicinity of the wall. These effects have been taken into account. When the wall is moving with a speed sufficently strong as here, these corrections can no more

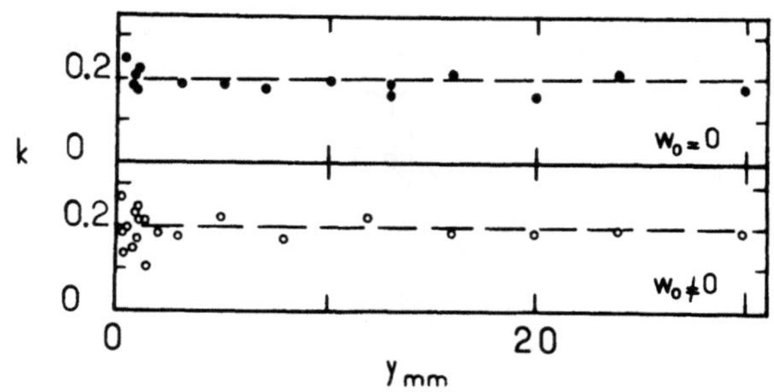

Fig. 2. Degree of sensitivity to the tangential velocity component along the wire.

be applied. It appears then, that data with no correction are in good agreement with the boundary conditions. Direct measurements of longitudinal $(\overline{u'^2})$ and transversal $(\overline{w'^2})$ variance and correlation $(\overline{u'w'})$ have been performed by using a single hot wire rotating in a plane parallel to the wall. A data acquisition system enables a digital treatment taking into account "locally" the influence of the velocity component tangential to the wire. This tangential sensitivity of the hot wire influences mainly the measurements of correlations $\overline{u'w'}$. So the value of the tangential coefficient k has been measured accuratly. In this measurement this coefficient can be defined from the effective cooling velocity

$$u_{eff} = [u_n^2 + k^2 u_t^2]^{1/2}$$

u_n and u_t are the velocity components respectively normal and tangential to the wire. Figure 2 shows an example of the results which are in good aggreement with the values found by Champagne [9] in the case of similar aspect ratio of the wire and for an angle of 40 degrees between the wire and the mean velocity.

Concerning the stress including the normal to the wall velocity component, $\overline{u'v'}$ and $\overline{w'v'}$, the size of X-wire probes or yawed probes does not enable to measure them close to the wall. So we have deduced their values from the integration of the corresponding momentum equations using the classical boundary layer simplifications, the remaining terms being deduced from velocity profiles. This method can be used if the azimuthal derivatives terms remain negligible in respect with the other terms. These

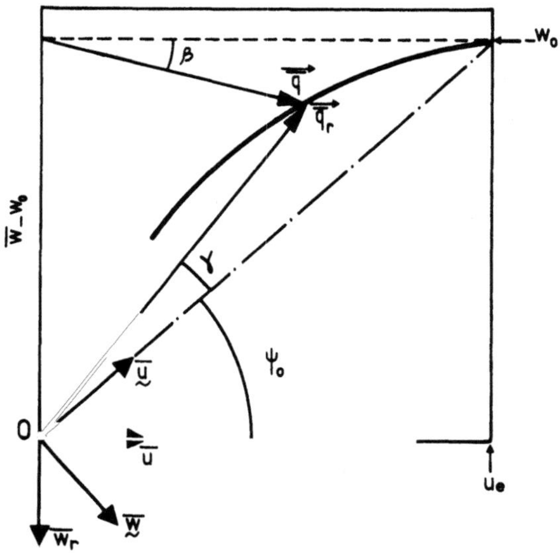

Fig.3. Definitions of angles and velocities

assumptions seem to be justified by the experimental results presented hereafter.

Experimental Results and Interpretations

As it is usual for threedimensional flows, we have considered a polar representation in the hodograph plane. Figure 3 shows the definitions of various angles and velocities. γ, defined from the direction of the relative mean outer velocity $\vec{q}_e(u_e,-w_o)$, is significant of the threedimensionality of the mean velocity field, as it is the case on a stationary wall. Concerning the turbulent field the results presented in the absolute system $S(\overline{u},\overline{w})$ yield misleading interpretations. So, we have considered the results in the relative coordinate system $\underset{\sim}{S}(\overline{\underset{\sim}{u}},\overline{\underset{\sim}{w}})$:

$$\underset{\sim}{u} = \frac{u_e}{q_e} u - \frac{w_o}{q_e} (w-w_o) \quad (1)$$

$$\underset{\sim}{w} = \frac{w_o}{q_e} u + \frac{u_e}{q_e} (w-w_o) \quad (2)$$

$$\underset{\sim}{v} = v$$

The turbulent values obtained in $\underset{\sim}{S}$ are weighted by q_e. Close to

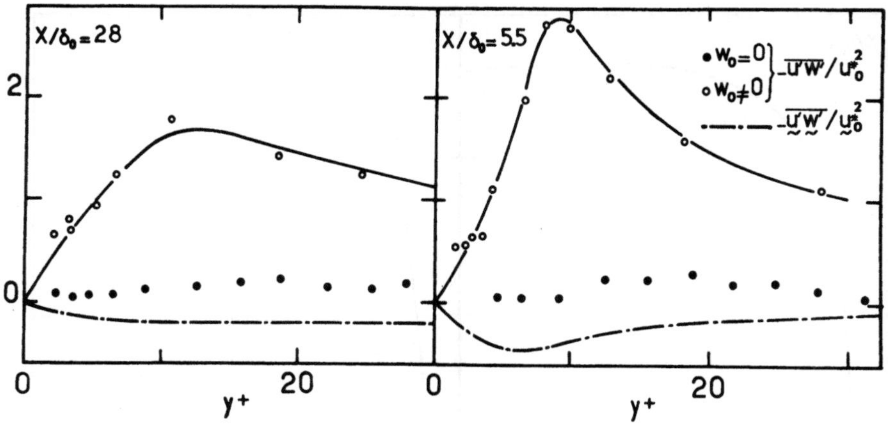

Fig.4. Correlations of longitudinal and transversal velocity components close to the wall.

the wall, it is more convenient to weight them by the friction velocity $\underset{\sim}{u}_o^* = (q_e/u_e)^{11/12} u_o^*$ -from a power law variation of the friction coefficient with the Reynolds number- which would exist on a stationary cylinder with an outer velocity q_e. In this representation, the results can be compared with the two-dimensional case. In a previous paper [8] it has been shown,at the farthest station (X/δ_o = 28),that the turbulence intensities $[\underset{\sim}{u}'^2]^{1/2}/q_e$ and $[\underset{\sim}{w}'^2]^{1/2}/q_e$ are practically coincident with the no rotation case values. On the contrary nearer the leading edge of the spinning part (X/δ=5.5) the trend to the twodimensionality of the turbulence intensities is in defect everywhere, except close to the wall in the inner part of the rotation layer.

Figure 4 presents the tangential stress $\overline{u'w'}/u_o^{*2}$ close to the wall with and without rotation at the two stations. Of course the values with rotation differ markedly from zero, but if we consider the $\underset{\sim}{S}$ system the deduced values of $\overline{\underset{\sim}{u}'\underset{\sim}{w}'}/\underset{\sim}{u}_o^{*2}$ are very much low. At the farthest station this quantity is very small, almost of the order of the measurement errors. At the position X/δ_o = 5.5. the numerical values are a little more high, but the threedimensionality remains weak.

In similar manner we see in figure 5 that in relative system $\underset{\sim}{S}$ the tangential stresses $\overline{\underset{\sim}{w}'\underset{\sim}{v}'}$ are practically null at X/δ_o = 28,

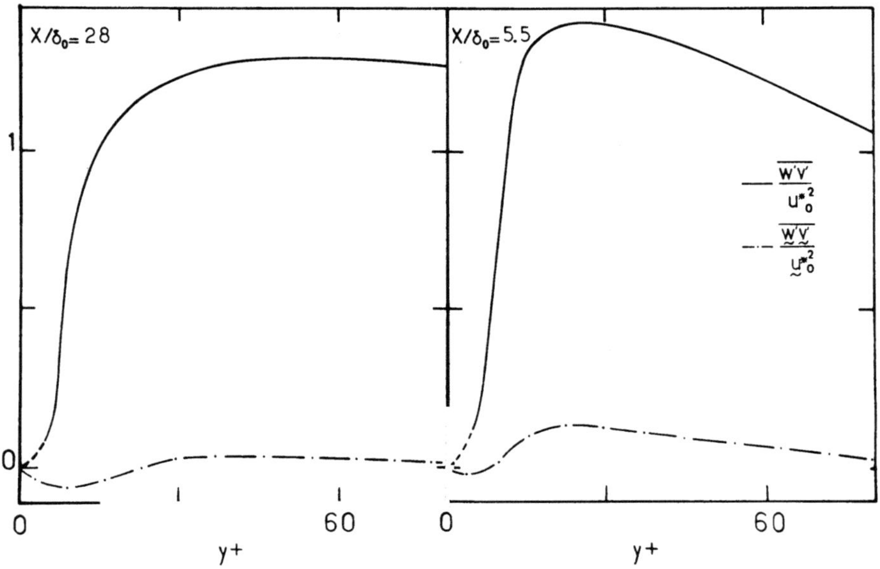

Fig.5. Correlations of transversal and normal velocity components close to the wall

Actually, the turbulent field is significantly threedimensional and this is revealed through the tangential stress $\overline{u'v'}$ (Fig.6). Indeed, in particular at the closer station to the leading edge of the spinning part (X/δ_o=5.5), the values of $\overline{\tilde{u}'\tilde{v}'}/q_e^2$ and $\overline{u'v'}/u_e^2$ are noticeably different. This difference is yet reduced at the downstream station X/δ= 28.

At this step of our investigation, we have considered the eddy viscosity concept and examined various definitions of ν_t. The first of them which came to mind has been defined in the absote system S:

$$-\overline{u'v'} = \nu_{t_x} \frac{\partial}{\partial y} \overline{u}, \quad -\overline{w'v'} = \nu_{t_z} \frac{\partial}{\partial y} \overline{w} \qquad (3)$$

Phisically, in the region affected by the rotation, these definitions are irrelevant as shown on the experimental results [10]. We have then turned to account the fact that the relative velocity intensity $|\vec{q}_r|$ follows very well the twodimensional laws -at least far from the leading edge of the moving part-. So we have introduced the corresponding eddy viscosity [10] :

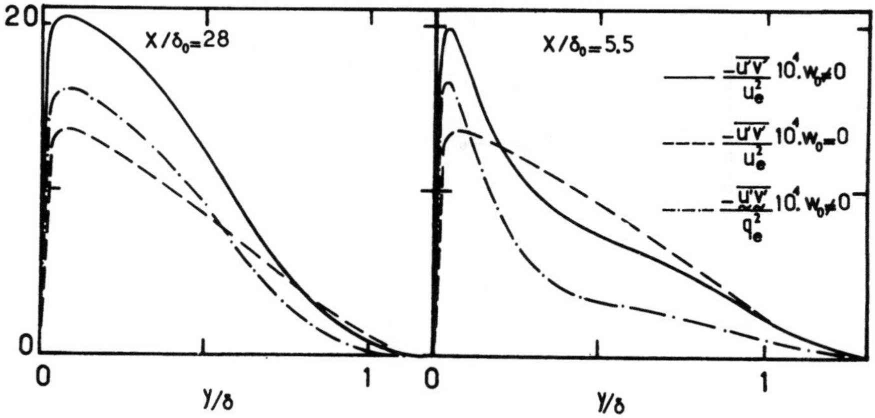

Fig.6. Reynolds stress $\overline{u'v'}$

$$-\overline{v'q_r'} = \nu_{tot}.\frac{\partial}{\partial y}\overline{q}_r \text{ with } q_r = (u^2+w_r^2)^{1/2} \qquad (4)$$

Figure 7 displays the results obtained at the two previous stations. Taking into account the Reynolds number effect, the values of $\nu_{tot.}(u_e/q_e)$ are close to the no rotation values at the farthest station $X/\delta_o = 28$ and even at $X/\delta_o = 5.5$, but close to the wall in the inner part of the rotation layer. Nevertheless it is

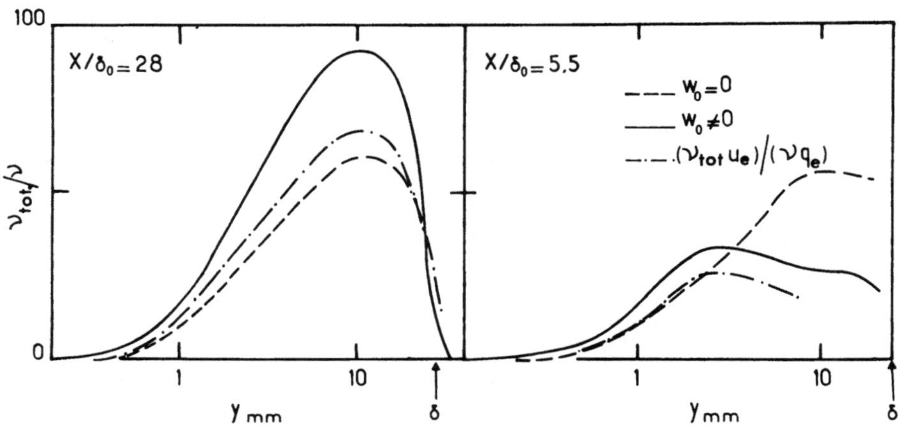

Fig.7. Eddy viscosity

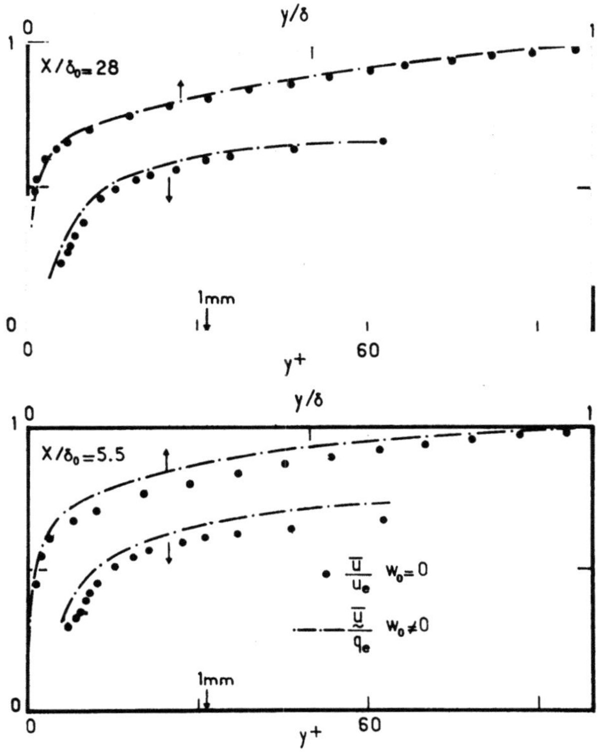

Fig.8. Mean longitudinal velocity profil

noteworthy that this eddy viscosity $\nu_{tot.}$ is related with the velocity intensity only. It does not take into account the velocity direction i.e. the deflecting angle γ. It would be necessary to define an other eddy viscosity from γ (y) gradient. Actually, instead to consider this polar representation it comes to the same thing to define eddy viscosities in respect with cartesian coordinates of $\vec{q}_r(\underset{\sim}{u}, \underset{\sim}{w})$ in the $\underset{\sim}{S}$ system with the only assumption to be positive :

$$-\overline{\underset{\sim}{u}'\underset{\sim}{v}'} = \nu_{t\underset{\sim}{x}} \frac{\partial}{\partial y} \overline{\underset{\sim}{u}} \quad , \quad -\overline{\underset{\sim}{w}'\underset{\sim}{v}'} = \nu_{t\underset{\sim}{z}} \frac{\partial}{\partial y} \overline{\underset{\sim}{w}} \qquad (5)$$

Figure 8 points out the behaviour of the longitudinal mean velocity $\overline{\underset{\sim}{u}}/q_e$ compared to $\overline{u}/u_e$ in the no rotation case. At the station X/δ_o = 28 the $\overline{\underset{\sim}{u}}$ profile is practically coïncident with the one of a twodimensional flow.Thus, taken into account the $\overline{\underset{\sim}{u}'\underset{\sim}{v}'}$ profile (Fig.6) the eddy viscosity $\nu_{t\underset{\sim}{x}}$ is not very different

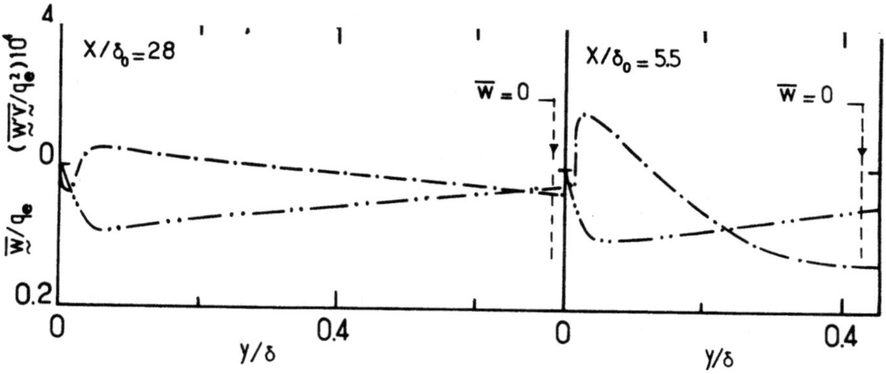

Fig.9. Transverse velocity and Reynolds stress in relative $\underset{\sim}{S}$ system

from ν_t, the eddy viscosity in the case of a boundary layer on a stationary wall with an outer velocity equal to q_e. At the station X/δ_o=5.5, the feature of $\underset{\sim}{\bar{u}}$ profile is quite different from the one of twodimensional flow. Concerning $\overline{\underset{\sim}{u}'\underset{\sim}{v}'}$ the diffrence is more pronounced (Fig.6). This implies a noticeable difference between $\nu_{t\underset{\sim}{x}}$ and $\nu_t(w_o=0)$. This must not be forgotten in any attempt of modelling using a concept of eddy viscosity.

As concerns the transverse mean velocity, its profile $\underset{\sim}{\bar{w}}(y)$ is asymmetrical (Fig.9), i.e. the slopes on each side of its maximum have numerical values strongly different. At the location of this maximum the tangent to the hodograph, that is to say the velocity gradient direction, is parallel to the outer velocity direction $\vec{q}_e$.On the other hand, there is a displacement between the zeros of the turbulent shear stress $\overline{w'v'}$ and the normal gradient $\partial\underset{\sim}{\bar{w}}/\partial y$ of the transversal relative mean velocity. The presence of a counter-gradient zone is so revealed, as it occurs in such flows, the mean profile of which being asymmetrical [11]. The size of this zone is larger and larger as the longitudinal distance X increases.However it must be remarked the level of $\overline{w'v'}$is lower and lower. Let us note that in the very-close-to-the-wall region the behaviour of this Reynolds stress is more complex.

In these conditions it appears the second relation (5) is irrelevant. It would be necessary to introduce for $\nu_{t\underset{\sim}{z}}$ an other definition :

$$-\overline{w'v'} = \nu_{tz} \frac{\partial}{\partial y} \overline{w} - (\overline{w'v'})^* \qquad (6)$$

The additional term would correspond to the one proposed by Townsend to describe the actions of the big eddies [12]. Besides, if the relation (5) are expressed in the absolute system it can be shown that each Reynolds stress $\overline{u'v'}$ and $\overline{w'v'}$ is a linear function of the velocity gradients $\partial\overline{u}/\partial y$ and $\partial\overline{w}/\partial y$. In fact, these relations take only into account the distorsion of the turbulent field by the mean flow, but not at all the counter gradient turbulent diffusion feature which hasbeen here specified.

References

1. Bissonnette, L.R.; Mellor, G.L. : J.F.M., 63 (1974) 369-413.
2. Lohmann, R.P. : The response of developed turbulent boundary layer to a local transverse surface motion. Ph.D. Connecticut Univ. 1973.
3. Arzoumanian, E.; Fulachier, L.; Cousteix,J.; Aupoix, B. : AGARD. Conf. n°271 The Hague 1979.
4. Bradshaw, P. : AGARD. Conf. n°169, 1973.
5. Cousteix,J.; Aupoix, B. : La Rech. Aero., 4 (1981) 275-285.
6. Bradshaw, P. : J.F.M., 36 (1969) 177-191.
7. Arzoumanian, E.; Fulachier, L.; Dumas, R. : 2nd Symp. Turbulent Shear Flows, 1979, London, 4.28-4.33.
8. Arzoumanian, E.; Leborgne, M.; Fulachier, L.: 3nd Symp. Turbulent Shear Flows, 1981, Davis.
9. Champagne, F.H. : Turbulence measurements with inclined hot-wires. Boeing Sc. Res. Lab. Rep. n°103, 1965.
10. Fulachier, L.; Arzoumanian, E.; Dumas, R.; Cousteix, J ; Aupoix, B. : XVth Int. Cong. of Theor. and Appl. Mech. Toronto, 1980.
11. Béguier, C.; Fulachier, L.; Keffer, J. : J.F.M., 89 (1978) 561-587.
12. Townsend, A.A. : The structure of turbulent shear flow. Cambridge Univ. Press. Second Edition (1976).

Some Measurements in the Intermittent Region of a Turbulent Boundary Layer along a Corner

I. NAKAMURA*, M. MIYATA**, T. KUSHIDA* and Y. KAGIYA*

*Department of Mechanical Engineering, Nagoya University, Nagoya, Japan

**Department of Mechanical Engineering, Yamanashi University, Kofu, Japan

Summary

An experimental investigation is reported of the turbulent boundary layer along a 90 degree streamwise corner. Conditional sampling measurements of the primary and the secondary velocity fields in the outer intermittent region of the boundary layer were made in a cross section 1.7 m downstream from the leading edge. Measurements were also made of the conventional mean flow and turbulent quantities.
The experimental results of turbulent zone averaged secondary velocity on the bisector clearly indicate a sign reversal in the outer part of the boundary layer and the whole secondary velocity field includes a closed vortex pattern.

Introduction

Turbulent flows along walls intersecting at right angles have long been of special interest because they present a particular flow situation, where the secondary current of Prandtl's second kind appears and distorts the primary flow field in a characteristic way. Although its obvious practical importance is to provide information about the distribution of the wall shear stresses or the heat transfer rates near the corner, the flow is of additional interest because of the usefulness to verify the validity of the proposed turbulence closure models. Since the time of Nikuradse many experimental studies have been carried out in the fully developed part of the duct flow with a non-circular cross section. However, there has been much less study of developing boundary layer flows. Important contributions were made by Gessner & Jones [1], Bragg [2], Eichelbrenner & Preston [3] and Gessner & Emery [4]. As for the origin of the secondary flow, several explanations were given by different authors. Gessner [5] reviewed and criticized these explanations, and proposed another plausible interpretation.

There is no doubt that the direct agency for the generation of the secondary current exists only when the flow is turbulent. It is, therefore, natural to apply the well established method of conditional sampling measurements to the outer intermittent region of the corner boundary layer which is absent in the fully-developed duct flow.

The aim of this research is to provide the results of the conditional sampling measurements as well as a comprehensive set of conventional mean flow and turbulence data taken in the zero pressure gradient boundary layer along a corner.

Experimental Apparatus and Technique

Experiments were conducted in a test section measuring 1.6 m wide by 0.32 m high by 2.6 m long, which was installed just downstream from the exit nozzle of an Eiffel type wind tunnel so that the boundary layer develops from the leading edge. An aluminum flat plate was mounted vertically on a floor of the test section, thus making a 90 degree streamwise corner. An opposing false roof was so adjusted to give a constant static pressure in the free stream. Tripping wires of 1 mm in diameter were placed 50 mm downstream of the leading edges. The sharp leading edges were adjusted carefully to minimize a flow asymmetry. A schematic flow configuration as well as some of the symbols used are shown in Fig. 1, where the value with a prime is refered to the coordinate 45 degree rotated about the x-axis, hence the y'-axis coincides with the corner bi-

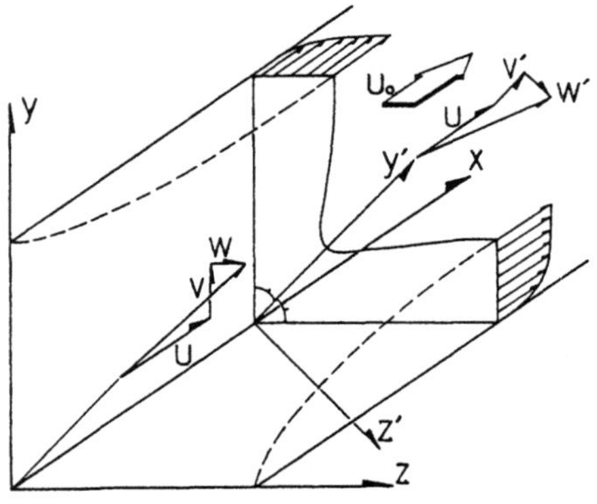

Fig. 1. Definition sketch of corner boundary layer

sector. Measured fluctuation level in the free stream was about 0.4 percent at $U_0 = 23$ m/s. All measurments were made under a constant unit Reynolds number of $U_0/\nu = 1.4 \times 10^6$ m^{-1}.

Constant temperature hot wire anemometers with conventional I- and X-probes were used for the mean flow and turbulence measurements. The sensing element of each probe was tungsten wire of 5 μm in diameter with a heated length of 1 mm. Analogue-digital circuits were devised for the measurements of the intermittency factor γ and the conditional measurements of streamwise and transverse secondary velocities.

The square of the first time derivative and the full wave rectified second time derivative of the fluctuating streamwise velocity u were fed separately into phase shifters, followed by level comparators and an or-gate. The out-put of the or-gate after smoothing gives an intermittent signal $I(t)$ which is 1 when either $(du/dt)^2$ or $|d^2u/dt^2|$ exceeds certain preset thresholds and otherwise 0. The random square wave $I(t)$ was modulated at 100 kHz, and γ could be read directly by a universal counter. The signal $I(t)$ drives an analogue switch and enables to generate signals proportional to $I\tilde{u}$ and $(1 - I)\tilde{u}$, where $\tilde{u}$ is an instantanuous streamwise velocity. All fluctuating out-put voltages were averaged over a time interval of 40 seconds after V-F conversion.

The results of conditional measurements were counter checked by the relation $U = \gamma U_t + (1-\gamma)U_p$, where U_t and U_p are turbulent and non-turbulent streamwise velocities, respectively. Independent measurements of U, γ, U_t and U_p satisfied this relation within an accuracy of 8 percent, where the experimental scattering of γ was less than 5 percent.

Experimental Results and Discussion

Contours of the streamwise mean velocity measured in the cross-section 1.7 m downstream from the leading edge are shown in Fig. 2. Other results given in this paper were obtained in the same location, where the Reynolds number based on the momentum thickness in the two-dimensional flow region was 4.80×10^3.

A characteristic distortion of isovels resulting from the secondary flow is familiar and resembles those reported so far. The

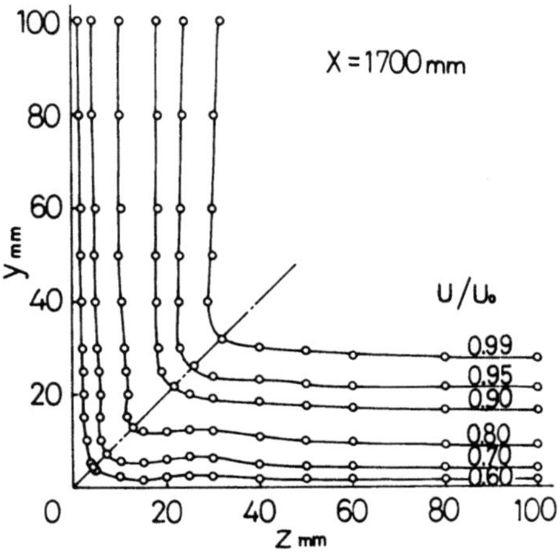

Fig. 2. Isovels of axial velocity

flow field shows a slight asymmetry regarding the bisector, which was found to be more pronounced for the distribution of wall shear stresses as shown in Fig. 3. The wall shear stress was measured by a Preston tube of 1 mm in outer diameter. The logarithmic straight part of the velocity profile could be observed close enough to the corner point. As can be seen in these two figures, the flow field is essentially two-dimensional in the region beyond the points $2\delta_\infty$ apart from the corner, where δ_∞ is the boundary layer thickness far from the corner.

The distribution of the secondary velocity resultants is shown in Fig. 4. It is clear that the secondary flow which is from the free stream toward the corner on the bisector has a spiral, vortex-like structure and is rather concentrated in the region near the corner. In the outermost region, however, careful measurements showed that the secondary flow moves opposite to that described

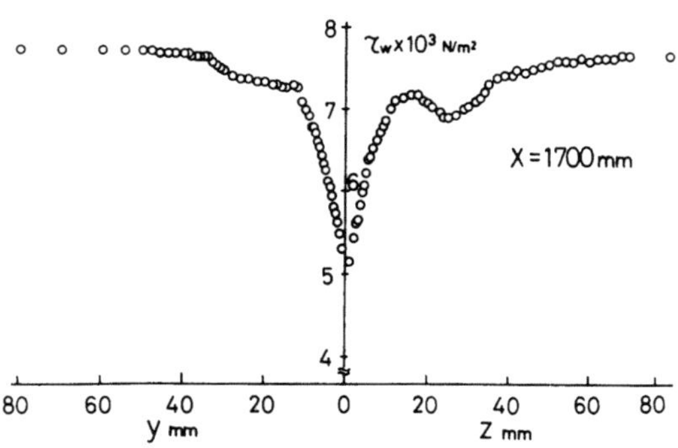

Fig. 3. Wall shear stress distribution

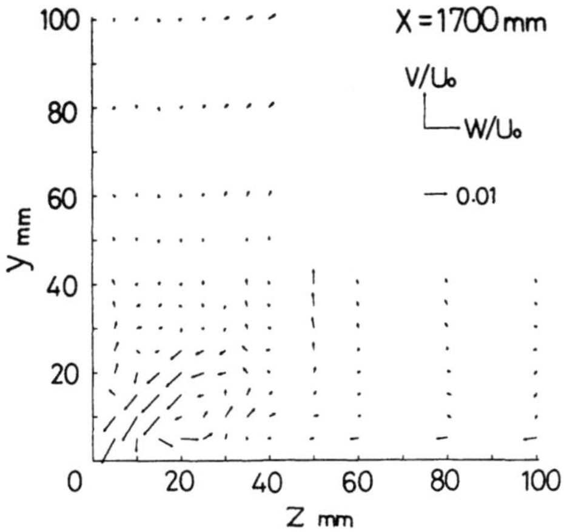

Fig. 4. Secondary flow vectors

just above, i.e., from the corner toward the free stream, although the magnitude is as small as the experimental uncertainty. More details of the secondary flow field will be discussed again later concerning the conditional sampling measurements. The secondary velocity reaches its maximum value of 1.5 percent of the free stream velocity near the middle between the corner and the boundary layer edge on the bisector.

Mean streamwise velocities and the Reynolds normal and shear stresses on the bisector are shown plotted with a data set for the flat plate boundary layer in Fig. 5. Of note is the almost constant distribution of each Reynolds stress in the deeper half of the boundary layer. Unlike the flat plate boundary layer, $\overline{v'^2}$ and $\overline{w'^2}$ are nearly the same, but the difference $\overline{v'^2}-\overline{w'^2}$ which

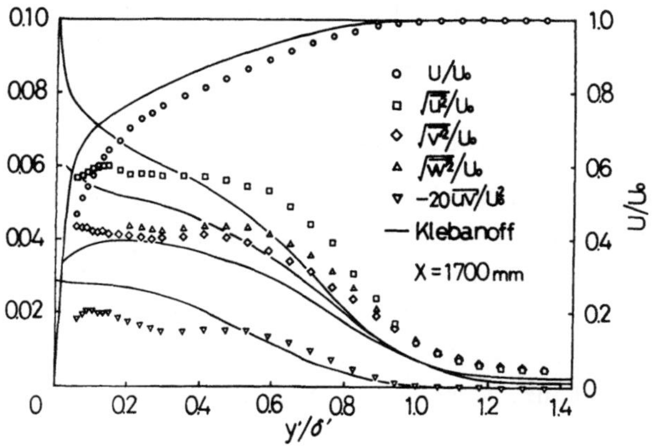

Fig. 5. Reynolds stress distribution on corner bisector

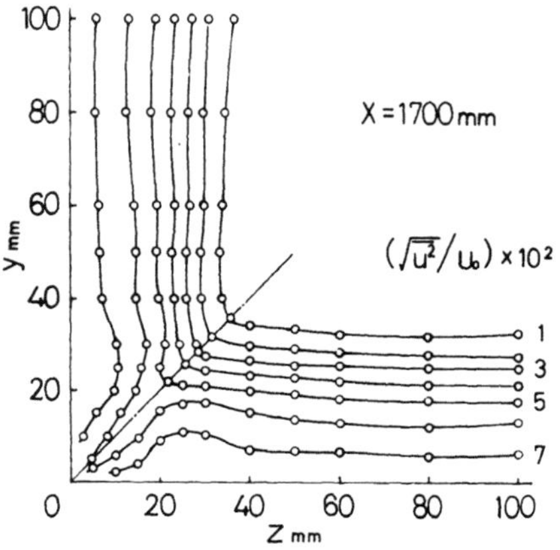

Fig. 6. Contours of streamwise fluctuation velocity

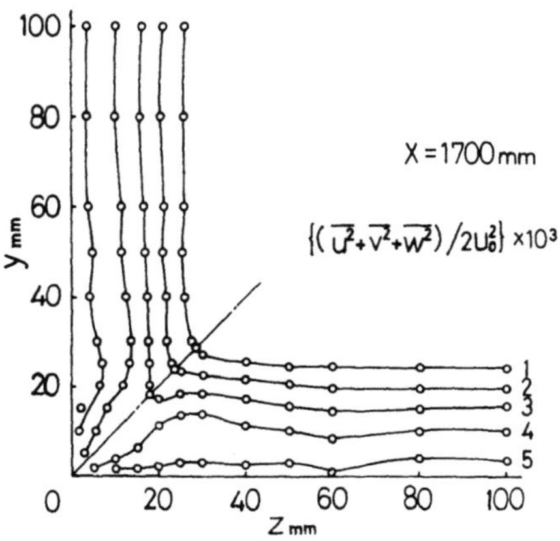

Fig. 7. Contours of turbulent kinetic energy

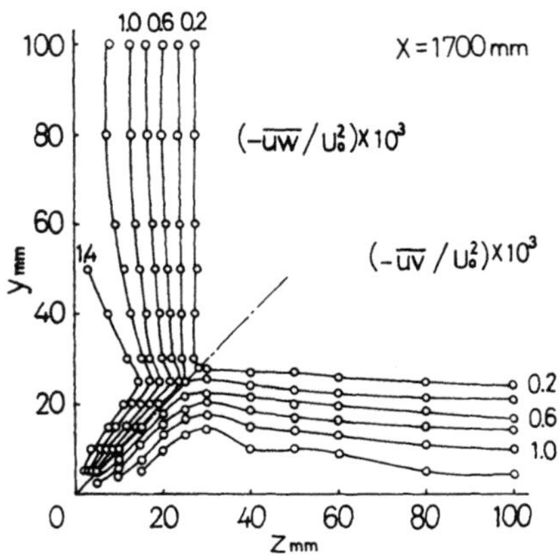

Fig. 8. Contours of Reynolds shear stresses

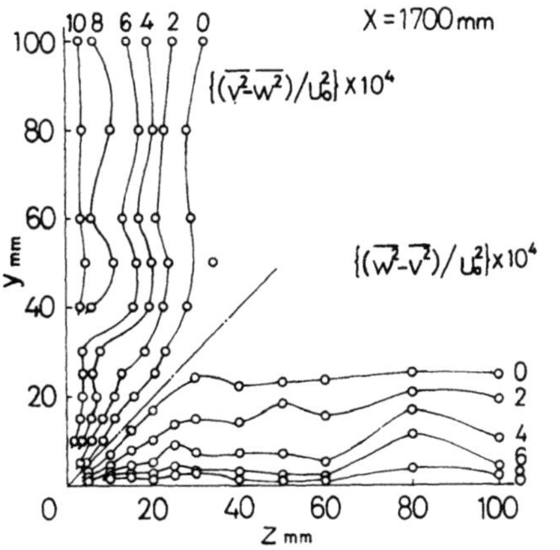

Fig. 9. Contours of difference of transverse Reynolds normal stresses

corresponds to $2\overline{vw}$ in the y-z coordinate has small negative value.

Contours of the streamwise fluctuation velocity are shown in Fig. 6. A deeper penetration toward the corner along the bisector corresponds to its almost constant distribution near the corner as shown in Fig. 5. The strong effect of the convective transport due to the secondary flow vortex results in a more significant distortion of the contour. The other two components of the fluctuating velocity show a similar behaviour, and so do the con-

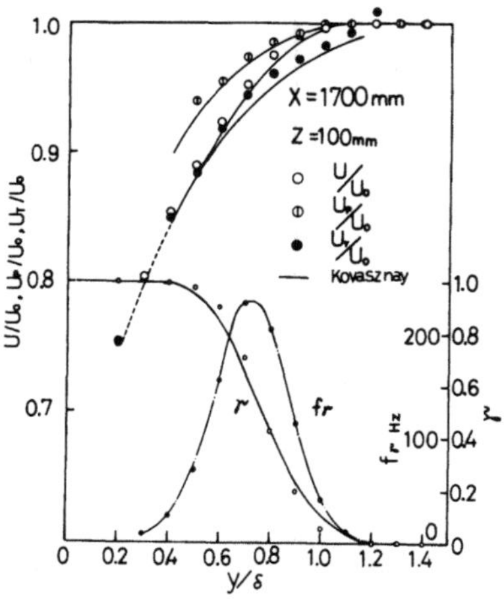

Fig. 10. Intermittency factor and regional mean velocity distributions in the two-dimensional flow region

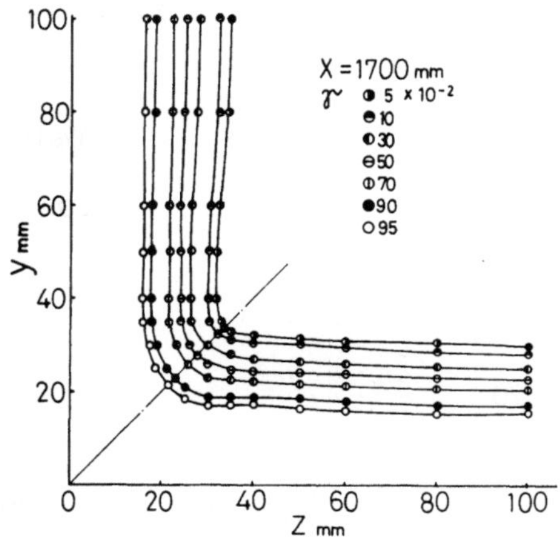

Fig. 11. Contours of intermittency factor

tours of the turbulent kinetic energy as shown in Fig. 7.

Figs. 8 and 9 show the lines of constant Reynolds shear stresses and the difference of the transverse Reynolds normal stresses, respectively. The mixed derivative of the latter with respect to y and z is the most dominant production term in the streamwise vorticity equation for the turbulent shear flow. The mixed derivatives were not evaluated because of uncertainty in obtaining them from the measured data. However, a deeper penetration toward the corner and a remarkable concentration along the bisector indicate that the significant production will take place in the region near along the bisector.

In obtaining the intermittent signal I(t), to make an adequate judgement from a fluctuating signal about whether the flow is turbulent or not is a most difficult task and may possibly depend on the experimenter's presuppositions. Measurements of γ and the conditional sampling were made in the two-dimensional flow region to confirm the reliability of the data. The results are shown plotted in Fig. 10 with Kovasznay's data for a flat plate, showing a reasonable agreement.

Fig. 11 shows the lines of constant intermittency factors in the

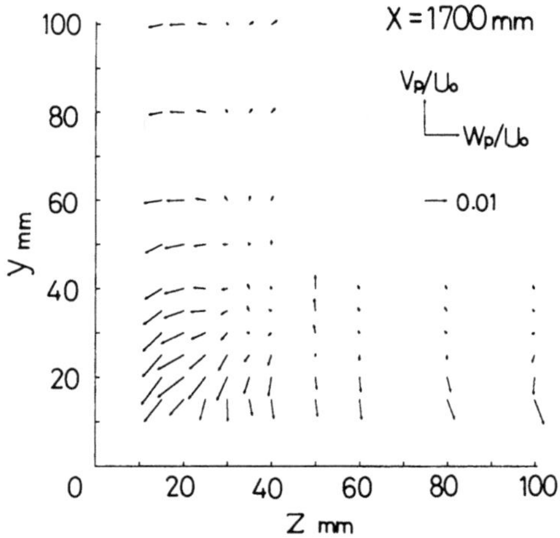

Fig. 12. Non-turbulent regional mean secondary flow vectors

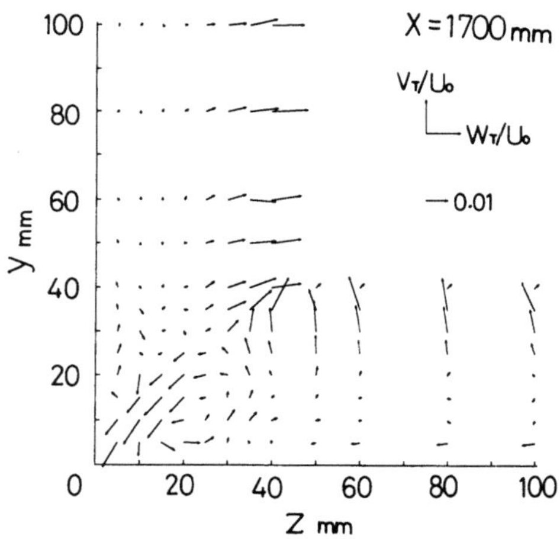

Fig. 13. Turbulent regional mean secondary flow vectors

corner region. One might think that the lines of constant γ will become distorted in a more complicated manner compared to those for the other quantities. For a large scale motion in the outer intermittent region and the secondary flow motion both have the same spatial scales of the boundary layer thickness and the same order of magnitude, therefore they interact strongly on each other and the turbulent non-turbulent interface is to be distorted considerably. The results of Fig. 11, however, show that this is not the case; the contours of γ show a simpler form than those for the turbulent quantities and are rather similar to the form of isovels. Indeed, the line of $\gamma = 0.5$ nearly coincides with the

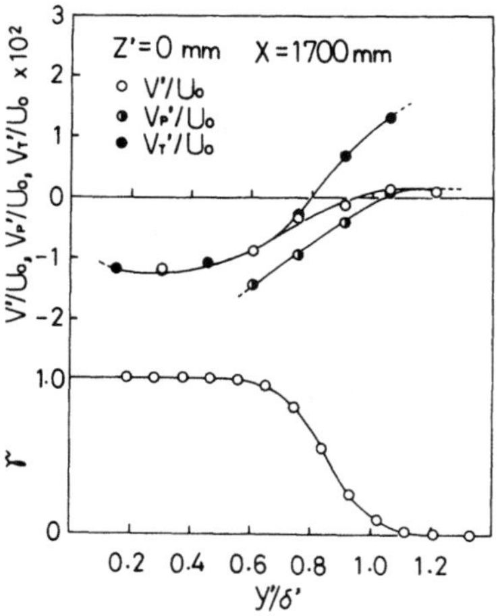

Fig. 14. Regional mean secondary velocity distribution on corner bisector

isovel of $U/U_0 = 0.965$.

Considering from the origin of the secondary flow discussed, the driving force to generate it exists only in the turbulent part of the flow. Fig. 9 indicates that such an agency is concentrated near the corner along the bisector. On the other hand, in the non-turbulent part, there is no such direct agency for the generation of the secondary flow. The transverse flow there is the results of advection and cntinuity. Figs. 12 and 13 show the conditional averaged secondary velocity resultants in the non-turbulent and the turbulent regions, respectively. It is clear that in the non-turbulent part, each transverse flow moves toward the wall, indicating that the flow is like that of an entrainment throughout the flow field. On the other hand, in the turbulent part the secondary flow clearly shows a closed vortex-like structure and is completely confined within the deeper half of the boundary layer. It is to be noted that in the outer region of the corner the velocity resultants are pointing in an opposite direction to those in the inner region, namely toward the free stream as in the two-dimensional flow region. This pattern of the secondary flow in the turbulent part is very similar to that proposed by Mojola [6] for the conventional secondary velocity field.

The distributions of the conditional averaged secondary velocities on the bisector, as well as the conventional mean values and the

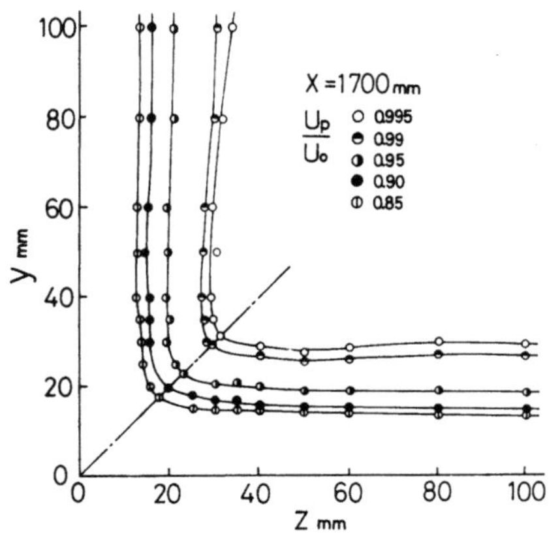

Fig. 15. Isovels of non-turbulent regional mean axial velocity

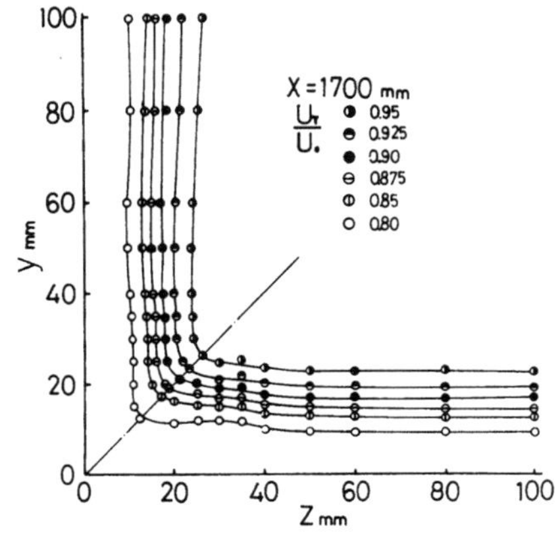

Fig. 16. Isovels of turbulent regional mean axial velocity

intermittency factors are shown in Fig. 14. The sign reversal of the turbulent zone averaged secondary velocity is beyond doubt. Of special interest is the fact that the reversal of sign takes place at nearly the statistical mean position of the boundary layer edge.

The contours of the regional mean streamwise velocity are shown in Figs. 15 and 16. No significant difference of the contour distortion from the conventional isovels was apparent, even in the outermost part of the corner boundary layer where the turbulent and the non-turbulent secondary flow move in the opposite direction.

Concluding Remarks

Conditional sampling measurements in the outer intermittent region of the turbulent boundary layer along a streamwise corner have shown that the non-turbulent zone averaged secondary flow moves toward the wall throughout the flow field and that the turbulent zone averaged secondary velocity changes its sign on the corner bisector at nearly the point $\gamma = 0.5$. The turbulent zone averaged secondary flow field thus shows a closed vortex pattern confined within the deeper half of the boundary layer and looks as if there was a source of turbulence production at the point where the secondary velocity changes its sign.

Acknowledgements

The authors gratefully acknowledge the valuable advice of Prof. Emeritus Y. Furuya for this study.

References

1. Gessner, F. B.; Jones, J. B.: A preliminary study of turbulence characteristics of flow along a corner. J. of Basic Engineering. 83 (1961) 657.

2. Bragg, G. M.: The turbulent boundary layer in a corner. J. of Fluid Mech. 36 - 3 (1969) 485.

3. Eichelbrenner, E. A.; Preston, J. H.: On the role of secondary flow in turbulent boundary layers in corners (and salients). J. de Mécanique 10 - 1 (1971) 91.

4. Gessner, F. B.; Emery, A. F.: The numerical prediction of developing turbulent flow in rectangular ducts. Proc. of 2nd Symposium on turbulent shear flows, London (1979) 17 - 1.

5. Gessner, F. B.: The origin of secondary flow in turbulent flow. J. of Fluid Mech. 58 - 1 (1973) 1.

6. Mojola, O. O.: On secondary flow in streamwise corner. J. de Mécanique appliquée. 4 - 2 (1980).

Three-Dimensional Turbulent Boundary Layer along a Concave Wall

Y. AIHARA and H. KOYAMA

Department of Aeronautics,
Faculty of Engineering,
University of Tokyo, Tokyo, Japan

Summary

Three-dimensional properties associated with longitudinal vortices in the turbulent boundary layer along a concave wall are investigated experimentally.
Ordered motion of horseshoe type vortices is revealed to be an important fluctuation not only in the transitional region but also in the turbulent region. The fundamental behaviors of the longitudinal vortices and the horseshoe type vortices seem to be common to those observed in the turbulent boundary layer over a flat plate.

1. Introduction

The three-dimensional behavior of the boundary layer along a concave wall is caused by a system of longitudinal vortices called Görtler vortices in the upstream laminar region [1]-[5]. Nonlinear development and secondary instability of Görtler vortices were investigated extensively and the process of the rapid transition of the boundary layer due to the vortices pointed out so far [2] has been revealed through these experimental investigations [3]-[5]. Therefore, a three-dimensional turbulent boundary layer thus formed is considered to be distinguished from those classified by Prandtl [6]. As the three-dimensional nature in the upstream region also applies to the downstream region, it would be worthwhile to deal with the turbulent boundary layer in connection with the behaviors of the transitional flow in the present work.

2. Experimental Apparatus and Procedure

Experiments were carried out in a rectangular sectional curved channel, 300mm×300mm, made of transparent acrylic resin having a concave wall with a radius of curvature r=1m (Fig.1). Probes

were inserted into the channel through the slits of the convex wall spanned 150mm in the lateral direction. A two-dimensional precise traversing mechanism was set at the center of the concave arc so that the probe could be moved in normal (y-) or spanwise (z-) direction to the concave wall surface. The streamwise (x-) position was set by adjusting the angle of the mechanism.
Three characteristic x-y sections, I, II and III, were defined as the most accelerated section, the intermediate one and the most decelerated one, respectively, in the spanwise periodicity of the streamwise velocity u due to longitudinal vortices.
The main flow velocity, U_0, defined at the outer edge of the boundary layer in section II was set at 1.7m/s throughout the present experiment. Turbulence level in the main flow was 0.13%.
The static pressure distribution was satisfactorily uniform along the center line of the concave wall in the present experimental range (x=605~1300mm).
Visual and hot-wire observations were simultaneously performed for better understanding of the flow field. For the further details of the apparatus and the procedure, refer to [4] and [5].

3. Result and Discussion

Reynolds number R_θ based on momentum thickness θ and Görtler parameter $G \equiv R_\theta (\theta/r)^{\frac{1}{2}}$ are 650 and 44.5 at x=1000mm. Therefore the turbulent boundary layer here is considered to belong to a low Reynolds number case strongly influenced by the system of longitudinal vortices.

3.1 Mean Flow Field

The development of the turbulent boundary layer is discussed by the measurements at x=605mm (breakdown region of horseshoe type vortices generated because of secondary instability of Görtler vortices [4] [5]), at x=1000mm (just downstream of the transition region) and at x=1300 mm (developed turbulent region). Streamwise mean velocity distributions, u/U_0, are shown in Fig. 2. At x=605mm, the boundary layer is seen to be remarkably three-dimensional due to a nonlinear development of Görtler vortices. However, as the boundary layer flow becomes more turbulent in the downstream, the spanwise nonuniformity in the outer layer appears to be reduced. Values of shape factor $H=\delta^*/\theta$ (δ^*; boundary-layer displacement thickness) in section II are 1.75, 1.53 and 1.50 at x=605, 1000 and 1300mm, respectively. Normal mean velocity distributions, v/U_0 (Fig.3), reveal that the longitudinal vortices tend to regain the previous symmetry with respect to section II, as δ^* becomes uniform in the spanwise direction [5]. The vortices are rather restricted to the inner region of the turbulent boundary layer in the downstream.

3.2 Turbulence

Figures 4(a) and (b) are respectively a plan view and a side view of the turbulent field revealing the important process of horseshoe type vortices from their generation to breakdown along section III.

Streamwise turbulence intensities, $\sqrt{\overline{u'^2}}/U_0$ (Fig.5), increase in the turbulent boundary layer, and after the transition the distributions are characterized by two peaks, i. e., one in the outer and the other in the inner layer. The same is true for Reynolds stress $\overline{u'v'}/U_0^2$ (Fig.6). The dominant fluctuations in the outer layer are considered to be due to the collapse of transported horseshoe type vortices which have been generated due to the secondary instability of Görtler vortices. It is interesting to note that the intensity is still increasing through the energy supply, $-\overline{u'v'}\partial u/\partial y$, from the mean flow field. However, new fluctuations generated in the inner layer become more important rapidly in the downstream both in turbulence intensity and in Reynolds stress. As the growth and the distributions of the latter appear to be again three-dimensional in relation to the longitudinal vortices, further measurements are carried out in conjunction with the bursting phenomenon over a flat plate [7] [8].

Figure 7 shows the time records of u', v' and u'v' at x=605mm in section III, that is, on the plane of symmetry of horseshoe type vortices. u' denotes a fairly periodic generation of the disturbance and the waveform shows a relatively sharp increase followed by a slow decrease. It is noted that instantaneous values of u'v' are essentially negative and over ten times the local mean have been observed. A reference hot-wire probe was set at x=1000mm and y=26mm in section III and another probe was

traversed in the normal direction in order to investigate the space-time correlation of fluctuations higher than 8 Hz (the effect of meandering motion with low frequency is excluded). Large scale motion is seen to affect the outer layer but the close correlation between fluctuations in the outer layer and those in the inner layer is not recognized in this region (Fig. 8). Therefore, the fluctuations in the inner layer are rather considered peculiar to the three-dimensional nature of reorganized longitudinal vortices in the turbulent boundary layer.
Figure 9 shows similar time records at x=1300mm to those in Fig.7, revealing that the fluctuations bring about same effects in the flow field as the horseshoe type vortices do in the upstream region. Here, stationary high shear layer is not recognized and the organized motion is considered to arise from instantaneous instability in the random environment.
Flow visualization in the developed turbulent region (Fig.10) also reveals the generation of horseshoe type vortices followed by an irregular breakdown in the random environment.

4. Concluding Remarks

Experimentally investigated are the three-dimensional properties of the transitional and the turbulent boundary layers along a concave wall with a radius of curvature 1 m. Longitudinal vortices and the characteristic fluctuations in the form of horseshoe type vortices play an important role in the development of the turbulent boundary layer. The fundamental behaviors seem to be common to the main constituents leading to the bursting phenomenon in the turbulent boundary layer along a flat plate [7] [8]. This may give a new insight in the study of steady longitudinal vortices in the boundary layer along a concave wall.

References

1. So, R.M.C.; Mellor, G.L.: Experiment on turbulent boundary layers on a concave wall. Aeron. Quart. 26 (1975) 25-40.

2. Tani, I.; Aihara, Y.: Görtler vortices and boundary layer transition. Zeit. Angew. Math. Phys. 20 (1969) 609-618.

3. Bippes, H.; Görtler, H.: Dreidimensionale Störungen in der Grenzschicht an einer konkaven Wand. Acta Mech. 14 (1972)

251-267.

4. Aihara, Y.; Koyama. H.: Secondary instability of Görtler vortices-Formation of periodic three-dimensional coherent structure. Trans. Japan Soc. Aeron. Astron. 24 (1981) 78-94.

5. Aihara, Y.; Koyama, H.: Nonlinear development and secondary instability of Görtler vortices. Proc. of second IUTAM symposium on stability in the mechanics of continua. Spriger-Verlag (to be published).

6. Fernholz, H.H: Bradshaw, P. (eds.) Turbulence. Berlin, Heidelberg, New York: Springer-Verlag (1976).

7. Blackwelder, R.F.; Eckelmann, H.: Streamwise vortices associated with the bursting phenomenon. J. Fluid Mech. 94 (1979) 577-594.

8. Cantwell, B.J.: Organized motion in turbulent flow. Ann. Rev. Fluid Mech. 13 (1981) 457-515.

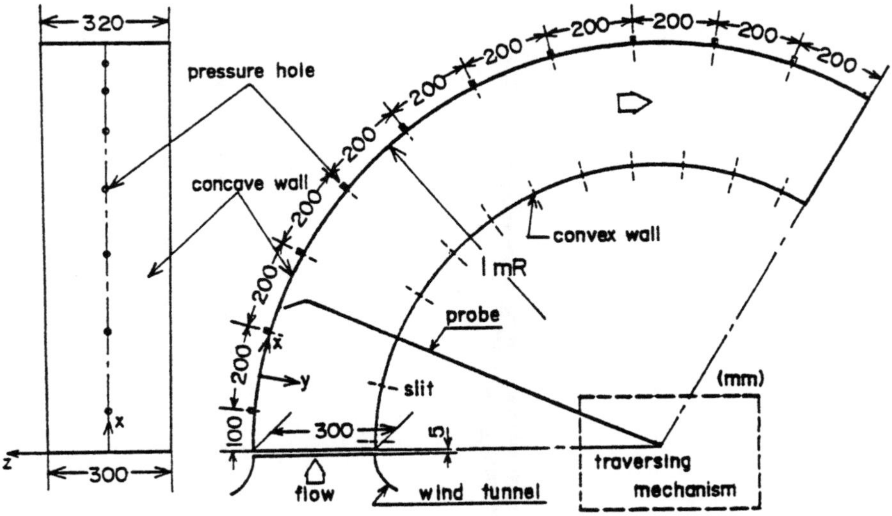

Fig.1. Curved channel

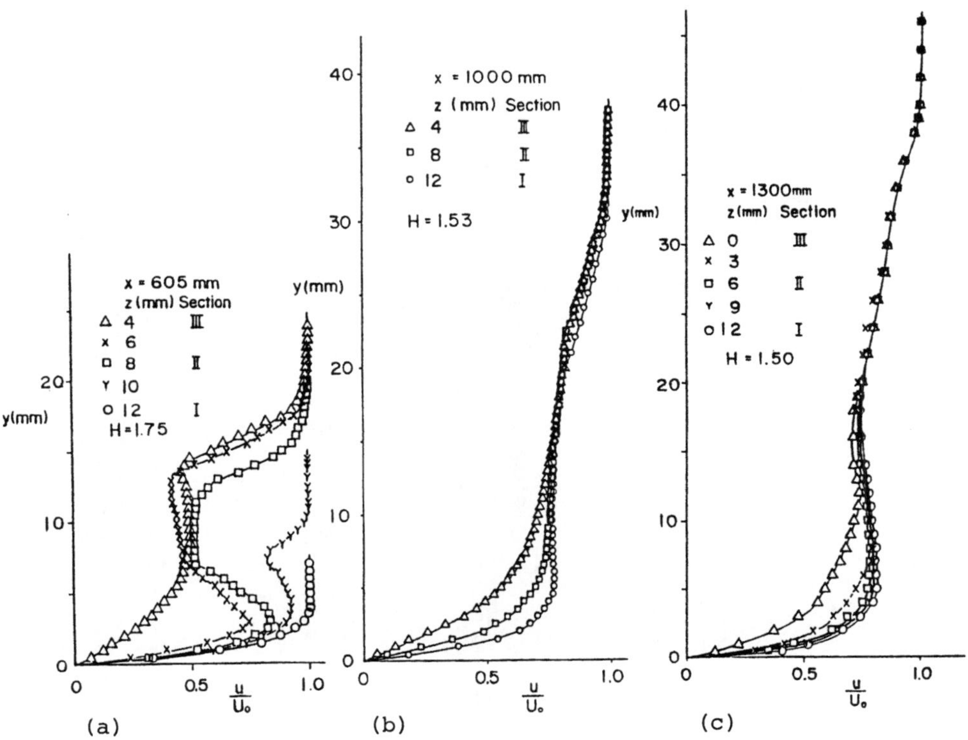

Fig.2. Streamwise mean velocity
(a) x=605mm (b) x=1000mm (c) x=1300mm

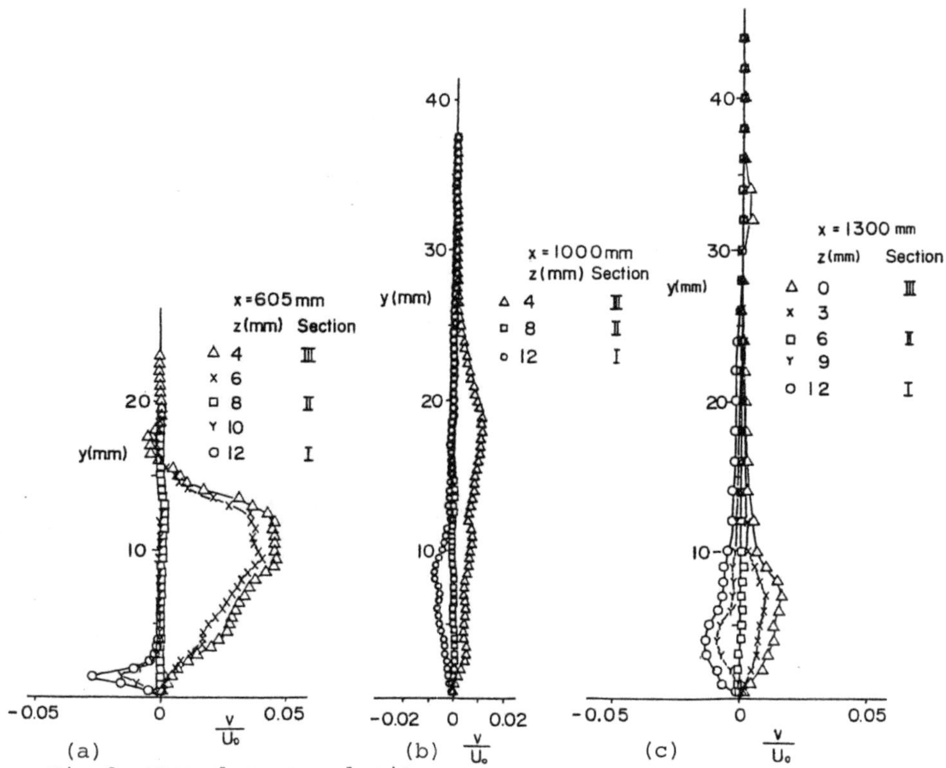

Fig.3. Normal mean velocity
(a) x=605mm (b) x=1000mm (c) x=1300mm

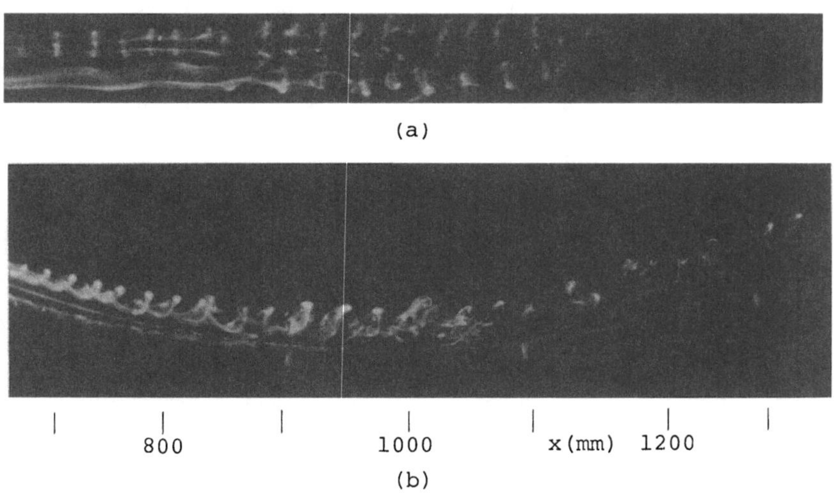

Fig.4. Flow visualization of horseshoe type vortices
(a) Plan view (b) Side view

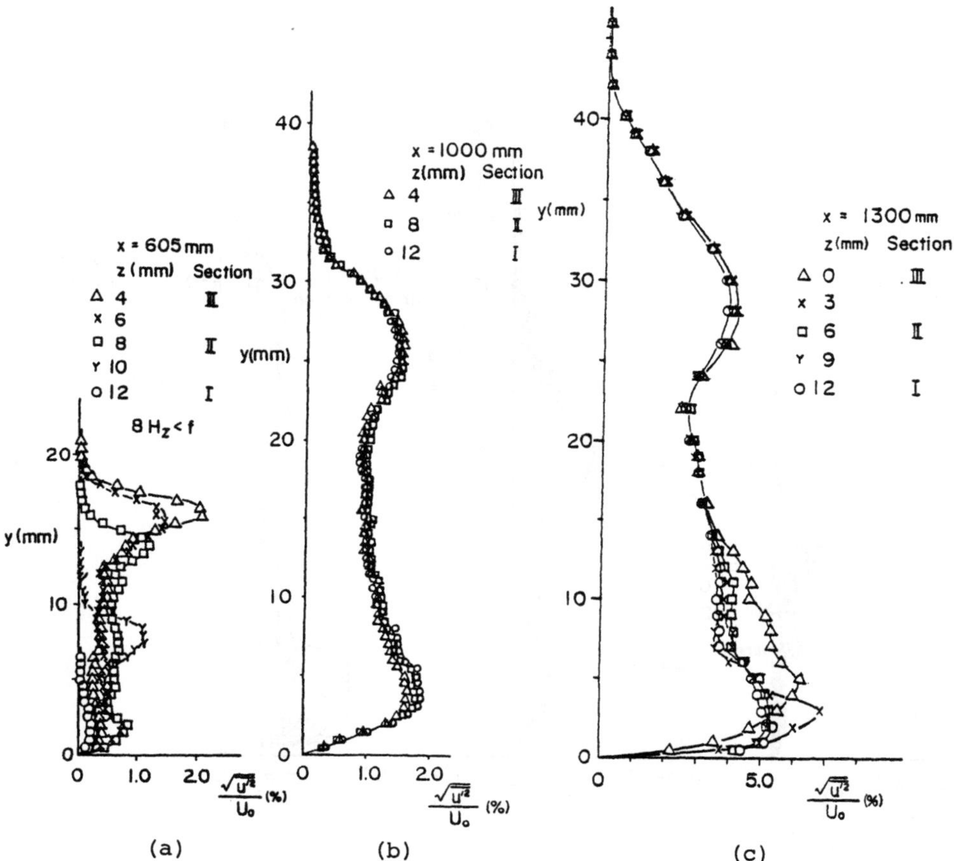

Fig.5. Streamwise turbulence, $\sqrt{\overline{u'^2}}/U_0$ (a) x=605mm (b) x=1000mm (c) x=1300mm

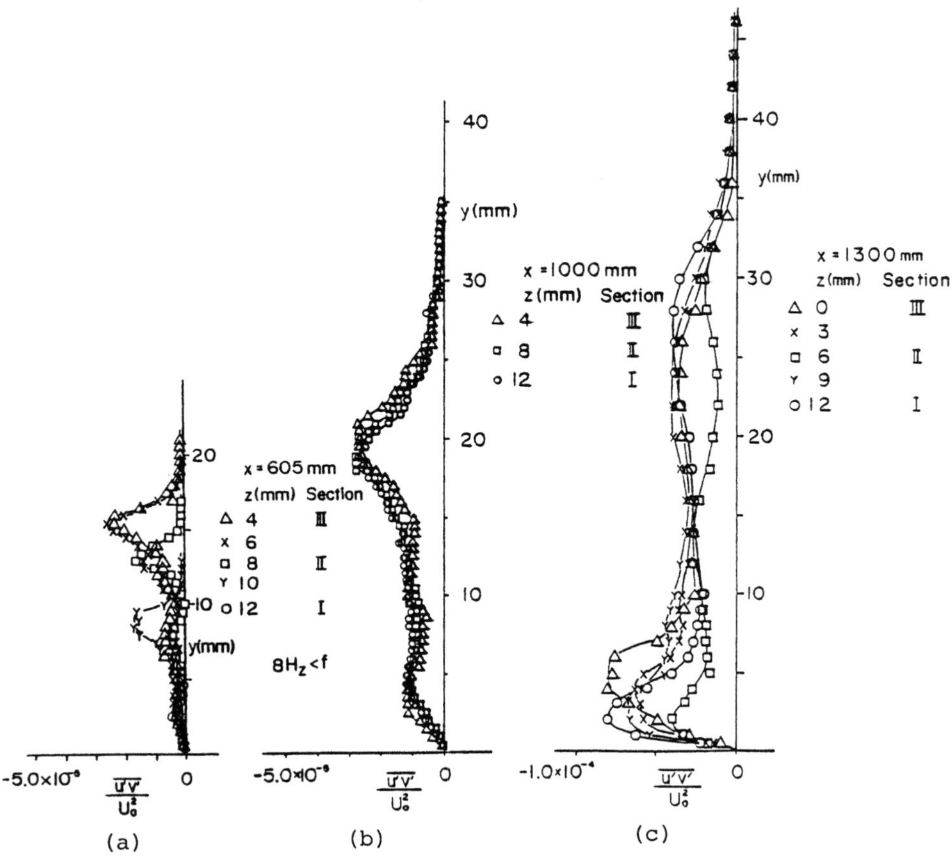

Fig.6. Reynolds stress, $\overline{u'v'}/U_0^2$ (a) x=605mm (b) x=1000mm (c) x=1300mm

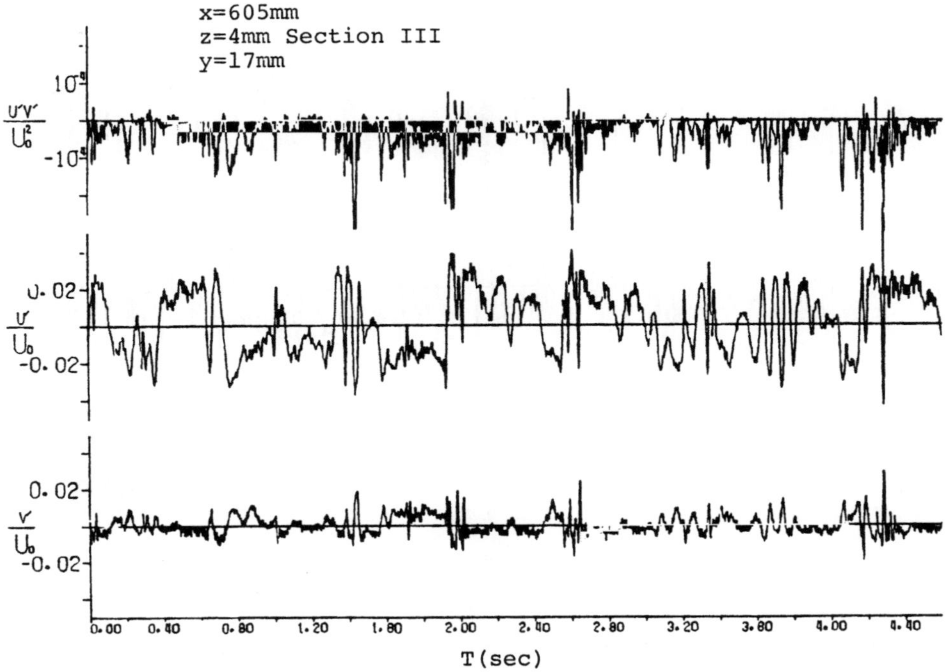

Fig.7. Typical traces of u',v' and u'v' at x=605mm and y=17mm in section III

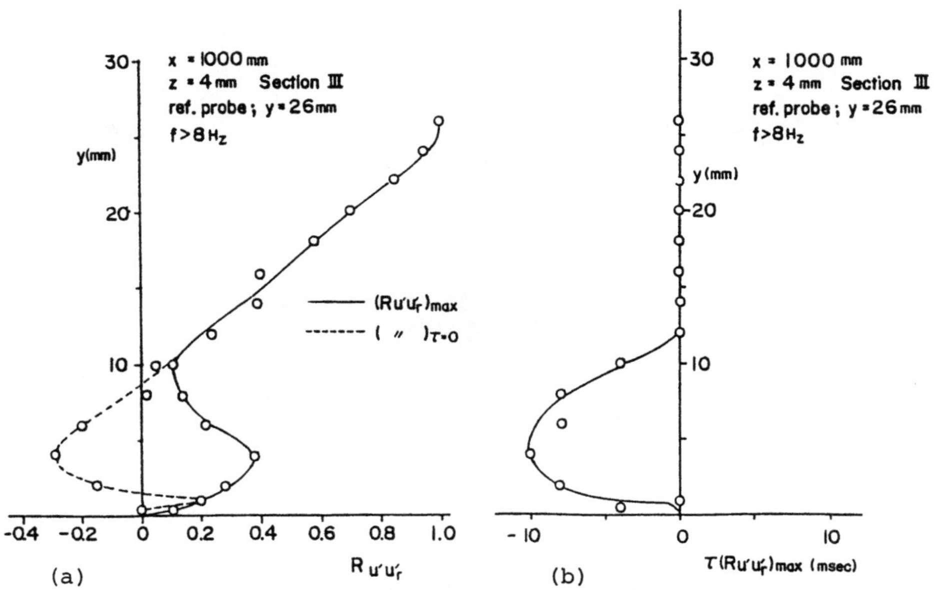

Fig.8. Two-point correlation of u', $Ru'u'_r$, at x=1000mm in section III for fluctuations higher than 8 Hz (a) Distributions of $Ru'u'_r$ (b) Time delay τ of maximum $Ru'u'_r$

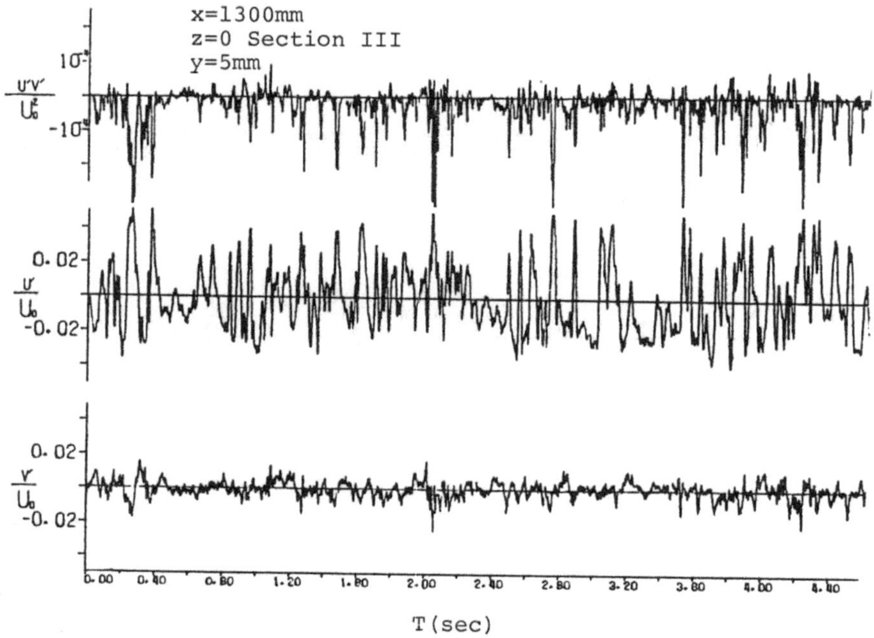

Fig.9. Typical traces of u', v' and u'v' at x=1300mm and y=5mm in section III

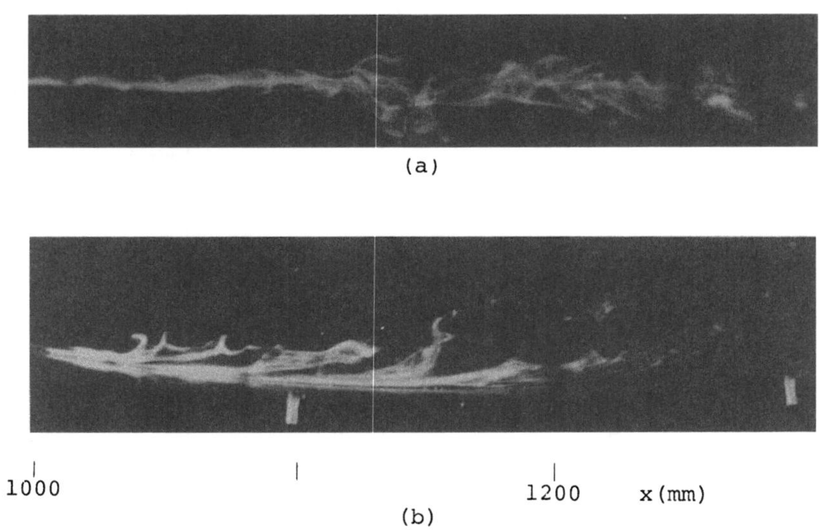

Fig.10. Flow visualization in developed turbulent region. Smoke is issued at x=1000mm and y=2mm in section III
(a) Plan view (b) Side view

Three-Dimensional Structure of the Turbulent Wake behind an Intersecting Cruciform Circular Cylinder

H. OSAKA and H. YAMADA

Department of Mechanical Engineering, Yamaguchi University, Ube, Japan

I. NAKAMURA

Department of Mechanical Engineering, Nagoya University, Nagoya, Japan

Summary

The three-dimensional structure of the turbulent wake produced by an intersecting cruciform circular cylinder has been investigated experimentally. The mechanism of secondary flows found in the near wake region was studied based on the budget of the mean streamwise vorticity equation, and seemed to be different from that in the streamwise corner of the fully developed turbulent wall flow. The turbulent kinetic energy budget was evaluated at the whole region of the y-z plane normal to the main flow in the far down stream region where the turbulent quantities were almost similar, and it was found that each term in the turbulent energy budget contributed significantly at the different parts in the y-z plane respectively.

Introduction

The experimental investigation has been carried out on the turbulent three-dimensional wake of the intersecting cruciform circular cylinder of which the length is effectively infinite. At first sight this configuration of the wake-producing obstacle seems to be peculiar having no generality, but it appears as a very fundamental element of a grid generating isotropic turbulence and the flow-developed region of the present wake is anticipated to have some properties connecting with the isotropic turbulence. Bradshaw's[1965], de Bray's[1967] and our experiment [1979] on the spatial instability of the flow passing through a screen having some range of open area ratio may be involved in this investigation. Also in the fluids engineering field, this problem has a practical significance with relation to the effect of the fluid dynamical force acting on off-shore structures and heat transfer coefficient of some types of heat exchanger.

In our previous papers[1982a,b], detailed measurements of

mean velocity and turbulent fields have been described and it is evident that both typical velocity and length scales are different from those of the two-dimensional and the axisymmetric wakes. Far downstream, the structure of the wake becomes similar, but doesn't agree with the one calculated, making a simple assumption that the present wake consists of two perpendicular two-dimensional wakes. The distortion of various contours suggests the occurrence of secondary currents.

Nomenclature (See Fig.1)

- d : diameter of the circular cylinder
- $b_{y'}$: half-width for the y=z plane
- L_x : integral length scale ($L_x=U\int_0^\infty R_{uu}(t)dt$, auto-correlation of u-fluctuating velocity)
- u,v,w : fluctuating velocity components in the x, y and z directions respectively
- U_1 : free stream velocity
- U_d : mean velocity defect ($U_d = U_1 - U$)
- $(1/2)\overline{q^2}$: turbulent kinetic energy ($\overline{q^2} = \overline{u^2} + \overline{v^2} + \overline{w^2}$)
- y',z' : distance along the y=z plane and normal to the y' direction respectively
- V' : mean velocity component in the y' direction
- v',w' : fluctuating velocity components in the y' and z' directions respectively
- $\Omega_x,\Omega_y,\Omega_z$: mean vorticity components in the x, y and z directions respectively ($\Omega_x=\partial W/\partial y-\partial V/\partial z$, $\Omega_y=\partial U/\partial z-\partial W/\partial x$, $\Omega_z=\partial V/\partial x-\partial U/\partial y$)
- ε : dissipation rate of turbulence

subscript
- $(*)_0$: value of (*) on the wake center axis

overbar
- $\overline{(*)}$: time average of (*)

Experimental Apparatus and Techniques

The wind tunnel used has a working section of 0.4 m square and about 4 m long, and is described in our previous paper[1982a] in detail. The intersecting cruciform circular cylinder consists of two sufficiently long brass cylinders of 5 mm in diameter, and is mounted in the uniform flow.

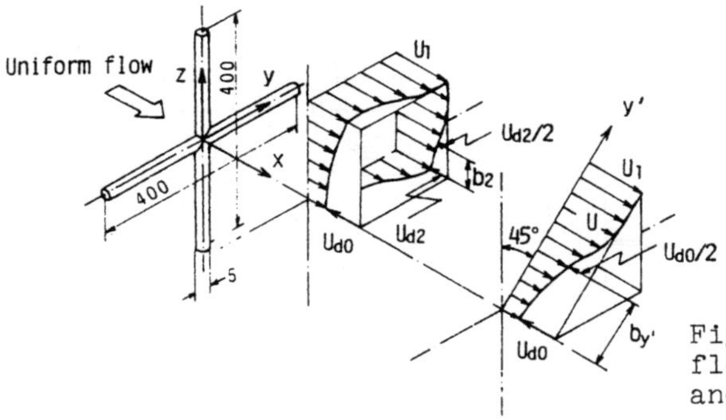

Fig.1. Configuration of flow field, nomenclature and coordinate system

Reynolds number $R_d=U_1 \cdot d/\nu$ was kept to be constant at 4×10^3 for all tests. Main measurements were conducted at 9 stations in the range of x/d=15 to 475 under a zero pressure gradient. A pitot tube, a constant temperature hot wire anemometer with X wire probe and a five-hole pressure probe were used for the measurement of mean velocity U, fluctuating velocities and secondary flows respectively. Uncertainty of the present experiment are ±1% for mean velocity U and ±7% for turbulent intensities and shear stresses, and ±15% for mean secondary velocity components. Most difficult is the error estimation of the terms in the mean streamwise vorticity equation.

Overall Flow Field

Fig.2 shows the mean velocity and the turbulent kinetic energy profiles in the y-z plane. The whole aspect of these profiles indicates the crisscross shape and development downstream. Near the corner region of the cruciform circular cylinder wake at x/d =30.5, the distortion due to secondary currents is clearly observed in both isovels and contours of the turbulent kinetic energy. Thus, these profiles on both cylinder axes are not monotonous and show a kind of saddle-back shape. In the far downstream section of x/d=298.5, isovels become gentle, and contours of the turbulent kinetic energy show the peak at symmetric locations about both the y=z plane and the cylinder axes. These locations are in connection with those of large production rate of the turbulent kinetic energy described below.

Fig.3 shows the streamwise variation of the characteristic scales for the cruciform cylinder wake. The maximum velocity de-

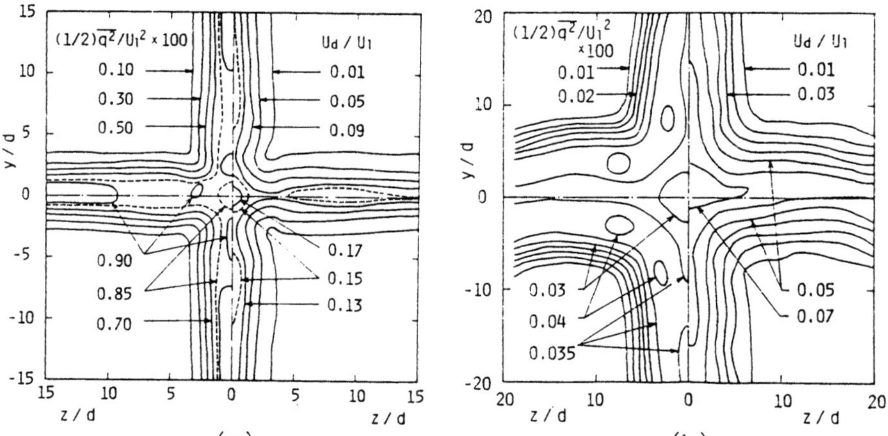

Fig.2. Mean velocity and turbulent kinetic energy profiles; (a) x/d=30.5, (b) x/d=298.5

fect U_{d0}/U_1 which exhibits the development of the wake significantly, varies as $x^{-0.37}$ beyond x/d=30. The rate of the decay is definitely slower than those of the two-dimensional and the axisymmetric wakes. The integral length scale closely related to the scale of shear layer has been calculated from the auto-correlation of u-fluctuating component on the wake center axis. As shown in the figure L_{x0}/d varies as $x^{0.37}$, so that Reynolds number $R_L=U_{d0}L_{x0}/\nu$ becomes constant. Therefore, the present wake steadily remains turbulent. On the other hand, the half-width $b_{y'}$ for the y=z plane which may be also regarded as a typical length scale of the present wake varies as $x^{0.48}$, and agrees quite well with that of the two-dimensional wake. With regard to

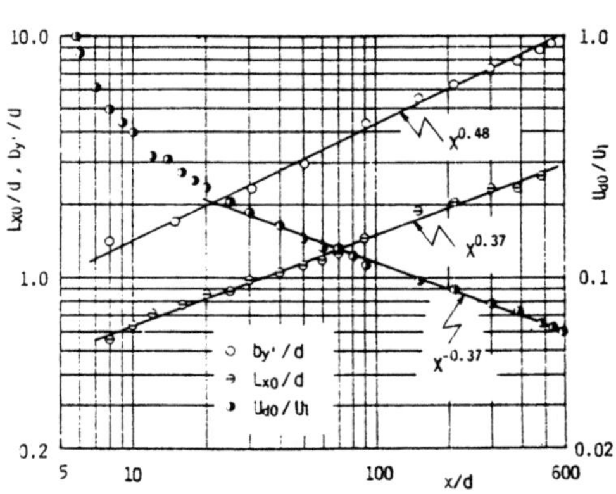

Fig.3. Velocity and length scales

the y=z plane, the mean velocity profile agrees precisely with that of the two-dimensional wake but profiles of turbulent quantities are significantly different. See Ref.[4] and [5] with respect to the details of mean velocity, turbulent intensities and turbulent shear stress.

Secondary Flow and Vorticity Budget

Fig.4 shows the secondary flow vectors and isovels at x/d=30.5. It is evident that there exist circulating flows which direct fluid towards on the y=z plane and propel it outwards on the cruciform cylinder axes. This secondary current distorts isovels towards the wake center along the y=z plane and the maximum value of one is about 4% of U_1. Fig.5 shows contours of the streamwise vorticity evaluated from components of secondary flow data at x/d=15 to 90.5. At all sections examined experimentally, the bisector (that is, the y=z plane) separates the independent secondary flow circulating zones which have the opposite sign vorticity of each other. Therefore, it reveals that there exist eight streamwise vortices which have clockwise or counterclockwise vorticities near the wake center. The streamwise vorticity normalized by U_{d0} and b_y, is almost constant within the range of x/d=100.

In order to investigate the mechanism of the secondary flow, each term of the following mean vorticity equation is examined experimentally and shown in Fig.6.

$$U\frac{\partial\Omega x}{\partial x}+(V\frac{\partial}{\partial y}+W\frac{\partial}{\partial z})\Omega x-\Omega x\frac{\partial U}{\partial x}-(\Omega y\frac{\partial}{\partial y}+\Omega z\frac{\partial}{\partial z})U-\frac{\partial^2}{\partial y\partial z}(\overline{v^2}-\overline{w^2})+(\frac{\partial^2}{\partial y^2}-\frac{\partial^2}{\partial z^2})\overline{vw}-\nu\nabla^2\Omega x = 0. \quad (1)$$

[P1] [P2] [P3] [P4] [P5] [P6] [P7]

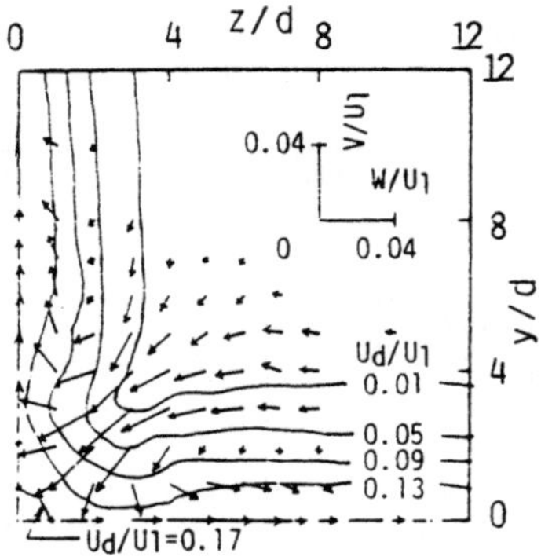

Fig.4. Secondary flow vectors at x/d=30.5

Here, P_7 is neglected because of very small value, and P_6 which couldn't be measured is determined as the residue in evaluating the equation(1). As can be seen from the figure, P_1, P_5 and P_6 have appreciable magnitudes, so that the equation (1) is reduced to

$$P_1 + P_5 + P_6 \simeq 0. \tag{2}$$

On the other hand, for the fully developed turbulent wall flow the following equation is proposed by Perkins[1970],

$$P_2 + P_5 + P_6 + P_7 \simeq 0. \tag{3}$$

Comparing the present flow field with the wall turbulence field, it is found that the contribution of P_1 and P_2 affecting the vorticity budget is quite different and P_5 which is called vorticity-production for the wall turbulence plays an opposite role for the occurrence of the secondary flow, because P_5 is positive and contributes to the vorticity budget as a loss. This fact indicates that in the region of streamwise vortices, the vorticities convected from the upstream region are compensated by the transverse heterogeneity of the turbulent normal stresses due to the three-dimensionality of the turbulence structure near the wake center.

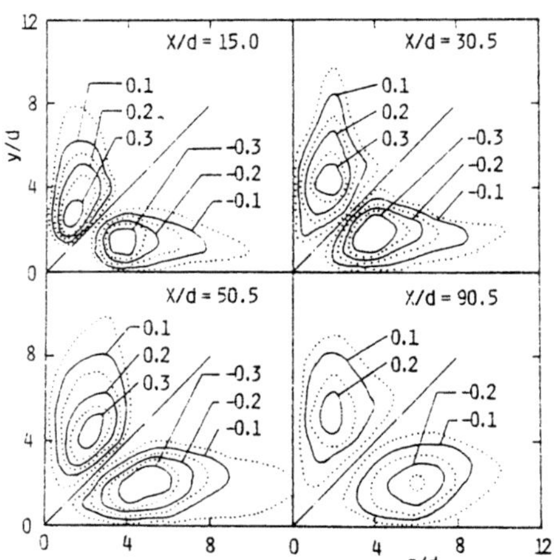

Fig.5. Contours of the streamwise vorticity

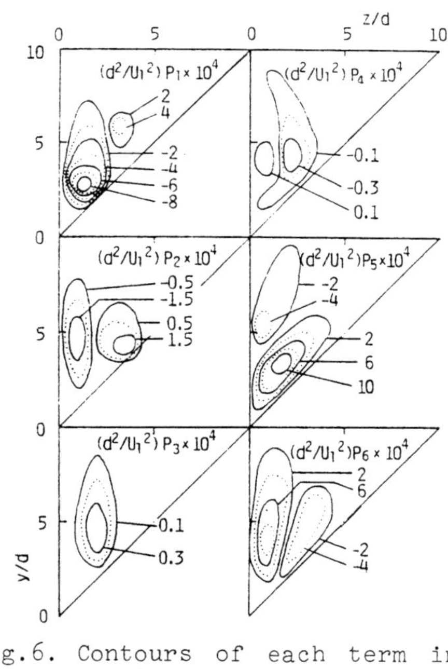

Fig.6. Contours of each term in the streamwise vorticity equation

Mean Flow and Turbulent Kinetic Energy Budget

The mean flow kinetic energy budget is examined for the far downstream region of x/d=386.5 where both the profiles of mean velocity and turbulent quantities are almost similar. Firstly, each term of the following mean flow kinetic energy equation is estimated for the y=z plane, and shown in Fig.7.

$$\underset{\text{Advection ①}}{U\frac{\partial}{\partial x}(\frac{1}{2}U^2)}+\underset{\text{Advection ②}}{V'\frac{\partial}{\partial y'}(\frac{1}{2}U^2)}+\underset{\text{Work ①}}{\frac{\partial}{\partial y'}(U\overline{uv'})}+\underset{\text{Work ②}}{\frac{\partial}{\partial z'}(U\overline{uw'})}\underset{\text{Turbulent-production}}{-\overline{uv'}\frac{\partial U}{\partial y'}} = 0. \qquad (4)$$

As can be seen from the figure, the advection and the work terms contribute significantly to the mean flow kinetic energy budget. Here, the advection term ② could not be estimated. The turbulent-production term is also negligible compared with other main terms. The contribution of the work term ② which is usually negligible in the two-dimensional wake is markedly large due to the three-dimensionality of the cruciform wake.

The statement mentioned above suggests that the mean flow kinetic energy equation in the entire region of the y-z plane is approximately reduced to

$$\underset{\text{Advection}}{U\frac{\partial}{\partial x}(\frac{1}{2}U^2)} + \underset{\text{Work}}{\frac{\partial}{\partial y}(U\overline{uv}) + \frac{\partial}{\partial z}(U\overline{uw})} \simeq 0. \qquad (5)$$

Fig.8 shows the profile of each term of equation (5) estimated experimentally. Over all the y-z plane contours of the advection and the work terms which make an opposite contribution as a gain

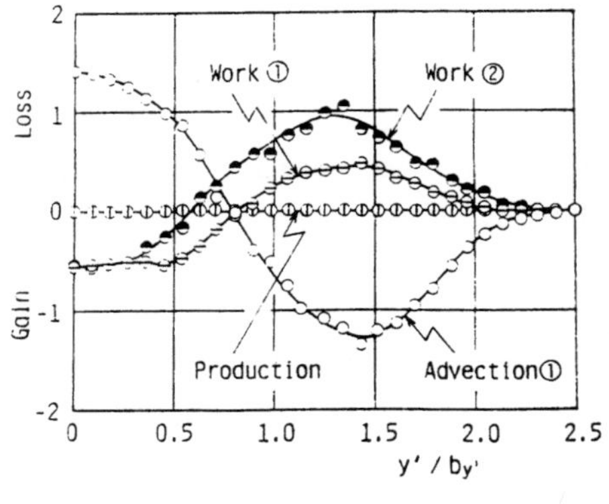

Fig.7. Mean flow kinetic energy budget for the y=z plane at x/d=386.5

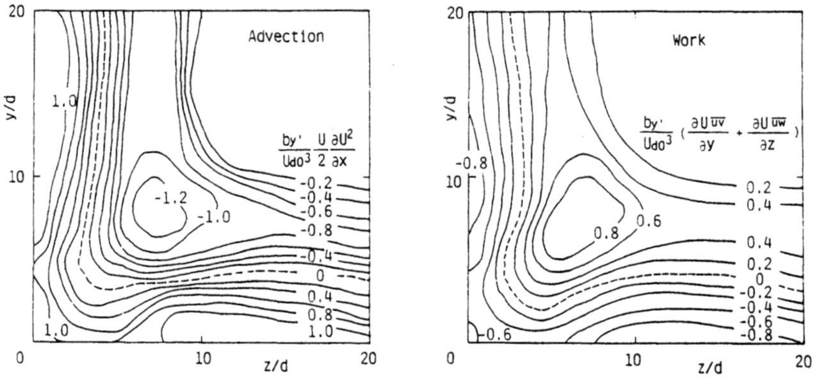

(a) (b)

Fig.8. Contours of the main terms of the mean flow kinetic energy equation at x/d=386.5 ; (a) Advection term, (b) Work term

or a loss respectively, agree quite well with each other, so that the mean flow kinetic energy budget over the whole region of the y-z plane seems to be almost satisfied by both terms mentioned above.

In the following, each term of the turbulent kinetic energy equation described below is examined for the far downstream region of x/d=386.5.

$$U\frac{\partial}{\partial x}(\frac{1}{2}\overline{q^2}) + \overline{uv}\frac{\partial U}{\partial y} + \overline{uw}\frac{\partial U}{\partial z} + \frac{\partial}{\partial y}(\frac{1}{2}\overline{vq^2}) + \frac{\partial}{\partial z}(\frac{1}{2}\overline{wq^2}) + \varepsilon = 0. \tag{6}$$

Advection Production Diffusion Dissipation

The pressure-velocity diffusion term could not be measured, and not involved in equation (6) in advance. The turbulent velocity diffusion is inferred assuming isotropic structure as follows;

$$\overline{vq^2} = \frac{3}{2}\overline{v(u^2+v^2)}, \qquad \overline{wq^2} = \frac{3}{2}\overline{w(u^2+w^2)}.$$

The dissipation term is calculated from the measurements of $\overline{(\partial u/\partial t)^2}$ by use of Taylor's hypothesis. Fig.9 shows the turbulent kinetic energy budget for the y=z plane. Each term of equation (6) is normalized by U_{d0} and $b_{y'}$. The magnitude of the production term is fairly small as compared with that of the two-dimensional wake[1980], and the magnitudes of the diffusion and the advection terms are very high in the region of $y'/b_{y'} \gtrsim 1$.

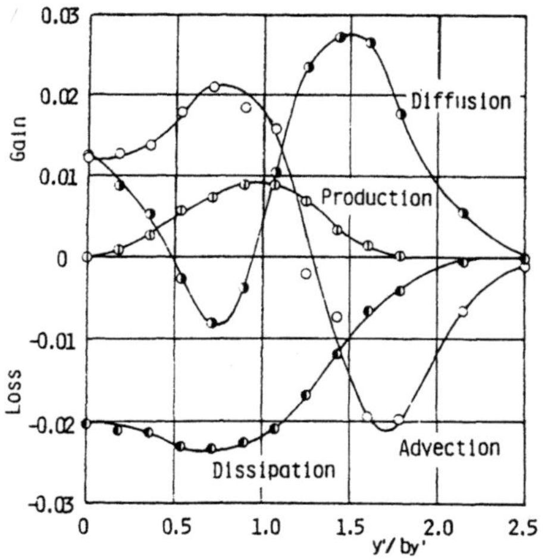

Fig.9. Turbulent kinetic energy budget for the y=z plane at x/d=386.5

Integration of the diffusion term calculated over the entire region of the y=z plane contributes as a gain, but the turbulent kinetic energy budget is approximately balanced by the main four terms described in equation (6).

Fig.10 shows the profiles of the main four terms of equation (6) in the y-z plane. As can be seen from the figure, each term in the turbulent kinetic energy equation becomes most important at the different parts in the y-z plane respectively. The integrated value of the diffusion term calculated over all the y-z plane is about zero, so that the diffusion term doesn't contribute to the turbulent kinetic energy balance as a whole. For the sake of convenience of explanation, the region close to the wake center where the advection term has a positive value is termed as inner layer side, whereas the region near the main flow where the advection term has a negative value is termed as outer layer side. In the inner layer side, the positive peaks of the advection and the production terms take place at a pair of symmetric locations about the y=z plane, while at the same locations as above the negative peaks of the diffusion term occur. Then, the sum of the advection and the production terms which contribute to the energy budget as a gain are roughly balanced by those of the diffusion and the dissipation terms which contribute as a loss except for the nearby wake center where the diffusion term is significant. In the outer layer side, the advection term as

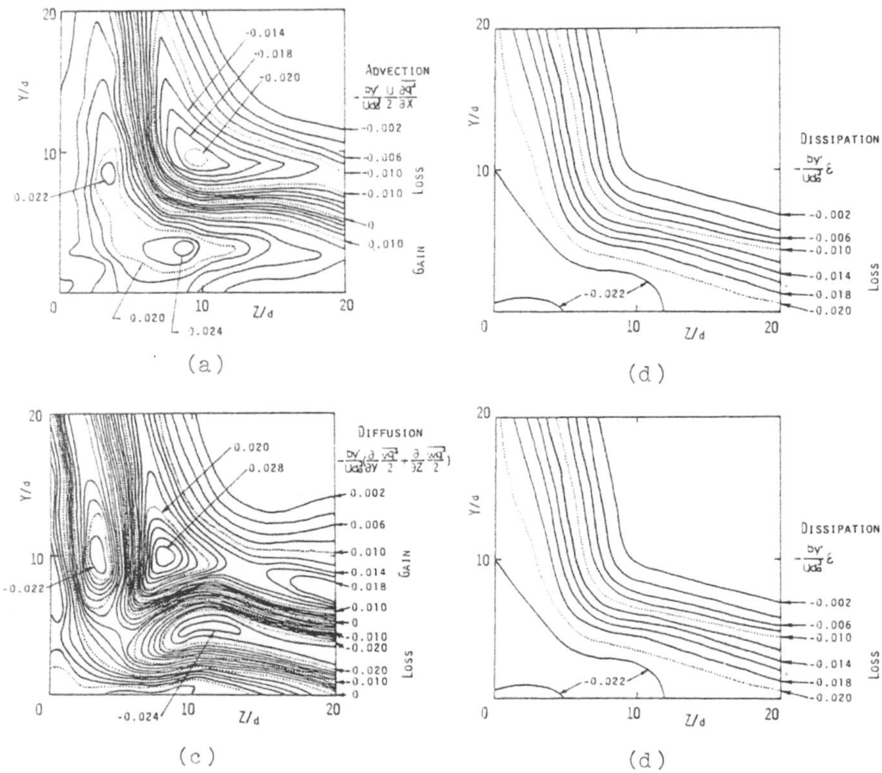

Fig.10. Contours of each term of the turbulent kinetic energy equation at x/d = 386.5 ; (a) Advection term, (b) Production term, (c) Diffusion term, (d) Dissipation term

a loss to the energy budget is almost balanced by the diffusion term as a gain. Therefore, the turbulent kinetic energy balance over all the y-z plane seems to be approximately satisfied by each term of equation (6).

Conclusion

Experimental investigation of the turbulent wake behind an intersecting cruciform circular cylinder showed the following results.

(1) The maximum velocity defect and the integral length scale on the wake center axis regarded as the characteristic scales of the present wake vary as $x^{-0.37}$ and $x^{0.37}$ respectively. These powers are different from those of the two-dimensional and the axisymmetric wakes significantly.

(2) The secondary currents found in the relatively upstream region of the present wake form eight streamwise vortices which are symmetric with respect to the geometrically symmetric planes of the flow field. The behavior of the streamwise vortices is described by using equation (2) and is different from that well known in the turbulent wall flow.

(3) The mean flow kinetic energy balance is almost satisfied by both the advection and the work terms which make an opposite contribution to each other over all the y-z plane.

(4) Each term in the turbulent kinetic energy budget contributes significantly at the different parts in the y-z plane respectively. In particular, the contribution of the diffusion term nearby the wake center is markedly different from that of the two-dimensional wake. At a pair of symmetric locations about the y=z plane where both the advection and the production terms have the maximum value as a gain, the turbulent kinetic energy also takes a maximum value.

A part of this investigation was financially supported by Grant-in-Aid for Special Project Research 1 (No. 56116001).

References

1. Bradshaw, P.: The effect of wind-tunnel screens on nominally two-dimensional boundary layer, J. Fluid Mech., 22 (1965) 679-687.
2. de Bray, B.G.: Some investigations into the spanwise non-uniformity of nominally two-dimensional incompressible boundary layers downstream of gauze screens, Aeron. Rep. Council, R & M, No.3578 (1967).
3. Furuya, Y., Nakanura, I. & Osaka, H. : Three-dimensional structure of a nominally planar turbulent boundary layer, ASME J. Fluids Eng., 101 (1979) 326-330.
4. Osaka, H.,et al.: The structure of turbulent wake behind a cruciform circular cylinder (1st Rep.), Trans. JSME(B), 48-431 (1982), (to be published).
5. Osaka, H., et al.: The structure of turbulent wake behind a cruciform circular cylinder (2nd Rep.), Trans. JSME(B),(to be published in 1982).
6. Perkins, H.J.: The formation of streamwise vorticity in turbulent flow, J. Fluid Mech., 44 (1970) 721-740.
7. Yamada, H., et al.: Turbulence measurements in a two-dimensional turbulent wake, Tech. Rep. Yamaguchi Univ., 2-4 (1980) 329-339.

Three-Dimensional Shock Wave/Turbulent Boundary Layer Interactions: Some Observations and Conjectures

S. M. Bogdonoff

Department of Mechanical and Aerospace Engineering
Gas Dynamics Laboratory
Princeton University
Princeton, New Jersey

ABSTRACT

The subject paper briefly reviews the current studies of four three-dimensional configurations of shock wave turbulent boundary layer interactions at a Mach number of 3. These configurations are a sharp fin, a sharp swept fin, a swept corner, and a blunt fin. Surface pressure distributions and surface flow visualization were obtained for all configurations. Some flow field surveys and optical studies were carried out and some high frequency static pressure measurements were made primarily for one configuration. These measurements, carried out for a range of test conditions, provide the basis for a series of observations, the development of an upstream influence correlation, and some conjectures of possible future directions in research in this complex flow field.

INTRODUCTION

For the past ten years, the Gas Dynamics Laboratory of Princeton University has been involved in a study of three-dimensional shock wave turbulent boundary layer interactions. Initiated as an extension of a long range program of two-dimensional studies, much of the early work was based on the observations of McCabe [1] and Lowrie [2]. The major emphasis of the still incomplete studies has been on a limited class of geometries, fully turbulent boundary layers, and at Mach 3, with an evaluation of the effect of Reynolds number and boundary layer thickness a primary goal. The research program was directed by the author and his colleagues, Drs. G. Settles and D. Dolling.

EXPERIMENTAL FACILITIES AND TEST GEOMETRIES

All of the studies have been carried out in the Princeton University High Reynolds Number Supersonic Tunnel. This tunnel has a test cross section 20 cm x 20 cm, a nominal free stream Mach number of 2.95, and a stagnation presssure range of 4×10^5 to 3.4×10^6 N_m^{-2} (60-500 psia). In all cases, the tests were made under approximately adiabatic wall conditions. Four basic geometries were studied, Fig. 1. Each configuration was designed to focus on a specific aspect of the three-dimensional shock wave boundary layer interaction. Classical surface pressure distributions were obtained for all configurations, as well as surface flow traces by a kerosene-graphite technique. In addition, flow field surveys, high frequency wall static pressure distributions, and special optical studies were carried out for specific conditions and test configurations.

SHARP FIN - This was the first configuration that was studied. Detailed flow field measurements were made for two cases. Surface pressures and flow traces, over a wide range of fin angles of attack, were obtained for several initial boundary layer conditions. The full results are reported in References 3 and 4.

SWEPT SHARP FIN - This configuration is a recent addition to the test group. No data on this configuration is presented herein but a report is soon to be published. The results are part of the correlation studies of upstream influence noted later.

SWEPT CORNER - This configuration was designed to examine the transition between two-dimensional flows, where extensive data existed, to a highly swept configuration. The variable was the sweep angle, and the test was carried out for two corner angles, (measured with respect to the free stream), which corresponded to specific flows examined in detail in two-dimensions. There has been extensive surface observations, and exploratory flow field measurements. The full results are presented in References 5-7.

BLUNT-FIN - This is a "geometry dominated" flow, as compared to the previous three configurations. A part of the flow has been found to be non-steady, and much of the work has been involved with high frequency wall static pressure measurements and optical studies. The flow field close to the fin is quite different than those noted for the previous configurations. The results obtained thus far are reported in References 8 and 9.

RESULTS AND OBSERVATIONS

SHARP FIN - A typical pressure distribution on the wall, measured in the initial stream direction and at some distance from the fin, is shown in Fig. 2, for a range of fin angles of attack from 2-13°. The scale of the interaction is almost independent of shock wave strength, and is very large as compared to comparable two-dimensional interactions. All interactions start at about the same position. As the shock strength increases, the monotomic pressure rise at low angles changes to an inflected distribution and an overshoot of the theoretically calculated final pressure. Surface flow observations show what appears to be converging traces starting at about 8°, which corresponds to the beginnings of a dip in the pressure distribution curves. Detailed flow field surveys were made at 4° and 10°. Contours of static pressure, Mach number, and yaw angle were determined and, from these, pitch angles were calculated. The yaw angle contour plot for a significant part of the interaction for a 10° fin angle is presented in Fig. 3. The survey show that large angles are limited to a thin region close to the wall. The interaction extends 3-4 boundary layer thicknesses into the free stream. The specific details shown on the surface flow visualization at approximately 3 boundary layer thicknesses upstream of the shock are not evident in either the presented yaw angle contours nor any of the other flow field information.

Using the coordinate system of Fig. 4, the effect of shock wave strength and boundary layer thickness can be seen in Fig. 5. It is clear that the results cannot be scaled by boundary layer thickness and, although the effect of shock wave strength is measureable, it is not large considering the range of shock strengths which were examined. If this data is replotted with the vertical axis scaled by the ratio of the incoming normal Mach number to a reference normal Mach number, the effect of shock strength disappears. A single check with the results of Law [10] at Mach 6 seems to support this correlation. The normal upstream influence as a function of shock wave strength and fin angle of attack is shown in Fig. 6. A clear difference in behavior between two-and three-dimensional interactions is obvious. Although there is an influence of shock strength, there is an important anomaly in that weak shock waves have almost as much upstream influence as very strong ones.

At a fixed angle of attack, the upstream influence occurring for different Reynolds numbers and boundary layer thicknesses can be correlated by scaling by the ratio $Re_u^{1/3}/\delta^{2/3}$. With the inclusion of the dependence on normal Mach number, data at different angles of attack can also be correlated.

CORNER FLOW - This geometry provides the only link between the well examined two-dimensional flows and three-dimensional configurations with highly swept shock waves. As soon as a small amount of sweep was introduced, the surface flow patterns showed strong lateral effects. For small sweep angles, of the order of 10°, no effect on the pressure distribution or upstream influence was noted. A typical surface flow pattern and the nomenclature used for this configuration is shown in Fig. 7. In Fig. 8, the pressure distributions obtained along the free stream direction is presented, with the origin of the horizontal axis corresponding to the corner line. The significant upstream spreading of the interaction with sweep is clearly seen. It should be noted that the wedge angle in the stream direction has been held constant. An exploratory yaw angle survey is presented in Fig. 9. Large yaw angles are restricted to a relatively thin region close to the wall, similar in many respects to the sharp fin studies noted earlier.

A detailed examination of the variable sweep information has resulted in a new concept of spanwise scaling. Since there are no characteristic physical dimensions, the flow characteristics, corner angle, and sweep angle must dictate the full scale. Based on dimensional analysis and two empirically determined constants, an overall correlation for both Reynolds number and boundary layer thickness has been derived. Figure 10 presents the experimental validation of this approach.

The correlation of this new experimental data reveals that the three-dimesional scaling with boundary layer thickness and Reynolds number, in both the streamwise and spanwise direction, is quantitatively similar to the streamwise scaling of two-dimensional interactions. The observed three-dimensional scaling law reveals that an increase in unit Reynolds number or a decrease in boundary layer thickness compresses the "footprint" of the interaction.

BLUNT FIN - This configuration is quite different than those noted previously. The geometry has a dominant dimension set by the diameter of the nose of the fin. The flow close to the nose was dominated by this dimension, with only small effects of boundary layer thickness and Mach number. Far from the nose, the shock interaction structure became similar to that for the sharp fin and scaled on the same parameters as for that configuration. In Fig. 11, a typical kerosine-graphite surface flow trace is presented. Figure 12 presents four microsecond Schlieren photograhs of the nose region. Many such photographs were taken and they showed that there was a wide range of shock configurations and positions. This data lead to an extension of the pressure distribution studies using high frequency pressure transducers. The pressure distribution on the centerline is shown in Fig. 13. The mean pressure distribution shows the characteristic rise/fall and then a sharp increase close to the leading edge of the fin. This pressure distribution shows little effect of boundary layer thickness and is almost invariant with Mach number. The shape of this pressure distribution has lead to conjectures that the flow structure was like that seen at low speed, a horseshoe vortex wrapped around the base of the fin. The Schlieren photographs and high frequency pressure measurements show that this flow is violently unsteady. Pressure variations 15-20 times the original pressure are measured in various areas of the interaction. Optical studies have shown that the motion of the shock structure is not simply longitudinal. The frequency analysis of the data shows no unique characteristic frequency, but rather a spectrum similar to that found in the turbulent boundary layer. The magnitude of the nonsteadiness is shown in the pressure time histories of Fig. 14. At an X/D of -2.5, where the mean pressure has already risen 50%, it is clear that there is a significant amount of time when the interaction is downstream of this station. The spanwise extent of these regions are shown in Figs. 15 and 16. There is a considerable region around the nose where the concept of steady flow appears to be inappropriate.

In light of the high frequency data, the dependence of flow analysis and upstream influence on surface flow techniques (with zero frequency response) is clearly inappropriate. Considerable further work will be required to understand the basic mechanisms of this area of the interaction.

GENERAL OBSERVATIONS

In examining the characteristics of the four geometries studied, several overall comments can be made: the size of the three-dimensional interaction is, in many cases, much larger than that of the classical two-dimensional case. The relative insensitivity of the phenomena to shock wave strength is unique, and the physical explanation of this is, as yet, unknown. At least part of the scaling, the upstream influence of sharp fins, sharp swept fins, corners, and the far flow field around blunt fins, can be reasonably well correlated, giving some support to the idea that the underlying physics is similar. Although the upstream influences are quantitatively different, the functional dependence on Reynolds number and δ is the same for both unswept and swept compression corners. In the sharp fin case, the upstream influences are typically order of magnitude or more greater than those seen ahead of unswept compression corners having free stream shocks of the same strength. When the compression corners are swept to angles close to those of the sharp fin shock waves, the upstream influences are similar in magnitude. This suggests that three-dimensional flow fields have a scale which is critically dependent on shock orientation.

COMPUTATIONS

Large scale computation has only been applied with any success to the sharp fin case. The work of Horstman and Hung [11] present, to the authors knowledge, the only specific detailed comparisons. The author believes that the reasonable comparison of computation and experiment is due to the fact that the flow is not separated.

CONJECTURES

Although the test program described herein provided an extended data base, many new questions have been brought up. With a lack of physical understanding of the fundamental mechanisms, future directions are not easy to project. It seems clear that many of the two-dimensional experiments and computations may not extrapolate to three-dimensional flows in all cases. The relatively large scale of three-dimensional interactions may make them considerably easier to compute. In many cases, the gradients are much smaller and, therefore, the major concentration on two-dimensional configurations may be misplaced. There is a major question about turbulence in three-dimensional highly sheared flows of the present experiments. The observation of surface flows, which characterize the low speed region close to the surface, gives, in many cases, little or no information on what is going on in the high speed exterior flow. Very large flow deflections at the surface are the result of rather small lateral gradients, and small deflections of the external flow may result in very large surface flow deflections. The very strong lateral flows caused by small lateral gradients provides the possibility of boundary layer control techniques not heretofore considered. The surface flow visualization results seem to indicate that, at least in some three-dimensional cases, boundary layer control is not being applied in the most efficient manner. The discovery of significant areas of non-steady flow, and preliminary results which indicate the same effects in two-dimensions, open the possibility that a primary parameter of the flow has not been measured, predicted, or calculated. This may be particulary true for flows that are separated, and may be the reason why calculations of such flows have not been particularly successful. It is the authors belief that flow field information, either by surveys or visualization, and high frequency instrumentation will be critical to the understanding of three-dimensional flows.

REFERENCES

1. McCabe, A., 1966, The Three-Dimensional Interactions of a Shock Wave with a Turbulent Boundary Layer, Aeronautical Quarterly, pp. 231-252, August.

2. Lowrie, B. W., 1965, Cross-Flow Produced by the Interaction of a Swept Shock Wave with a Turbulent Boundary Layer, Ph.D. Thesis, Cambridge University.

3. Oskam, B., Vas, I.E., and Bogdonoff, S. M., 1976, Mach 3 Oblique Shock Wave/Turbulent Boundary Layer Interactions in Three Dimensions, AIAA Paper No. 76-336, July.

4. Dolling, D. S., and Bogdonoff, S. M., 1981, Upstream Influence Scaling of Sharp Fin-Induced Shock Wave Turbulent Boundary Layer Interactions, AIAA Paper No. 81-0336, January.

5. Settles, G. S., and Perkins, J. J., 1979, Investigation of Three-Dimensional Shock/Boundary Layer Interactions at Swept Compression Corners, AIAA Paper No. 79-1498, July.

6. Settles, G. S., Perkins, J. J. and Bogdonoff, S. M., 1980, Investigation of Three-Dimensional Shock/Boundary-Layer Interactions at Swept Compression Corners, AIAA Journal, Vol. 18, No. 7, pg. 779, July.

7. Settles, G. S., Perkins, J. J., and Bogdonoff, S. M., 1981, Upstream Influence Scaling of 2D & 3D Shock/Turbulent Boundary Layer Interactions at Compression Corners, AIAA Paper No. 81-0334, January.

8. Dolling, D. S., and Bogdonoff, S. M., 1981, An Experimental Investigation of the Unsteady Behavior of Blunt Fin-Induced Shock Wave Turbulent Boundary Layer Interactions, AIAA Paper No. 81-1287, June.

9. Dolling, D. S. and Bogdonoff, S. M., 1981, Scaling of Interactions of Cylinders with Supersonic Turbulent Boundary Layers, AIAA Journal, Vol. 19, No. 5, pg. 655, May.

10. Law, H. C., 1975, Three-Dimensional Shock Wave Turbulent Boundary Layer Interactions at Mach 6, ARL TR-75-0191, June.

11. Horstman, C. C. and Hung, C. M., 1979, Computation of Three-Dimensional Turbulent Separated Flows at Supersonic Speed, AIAA Journal, Vol. 17, No. 11, pp. 1155-56, November.

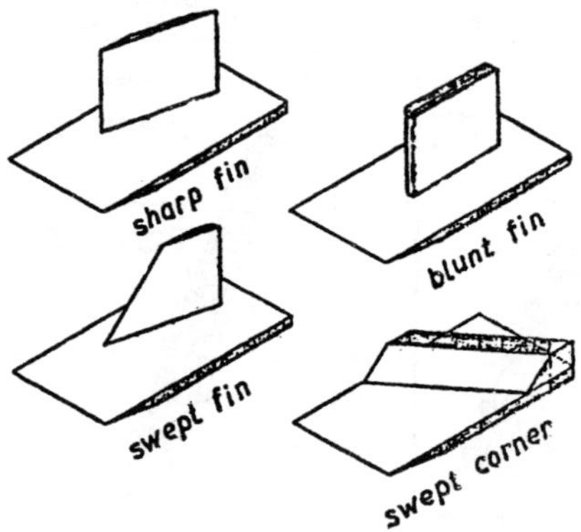

Fig. 1 Test Geometries

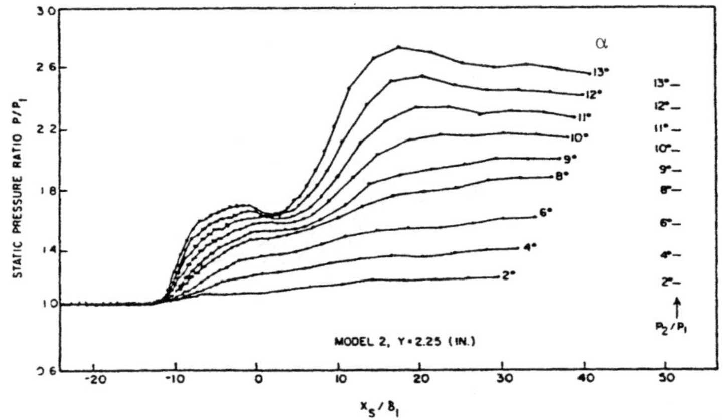

Fig. 2 Static Pressure distributions, variable angle Sharp Fin [Model 2, Y = 2.25 inch]

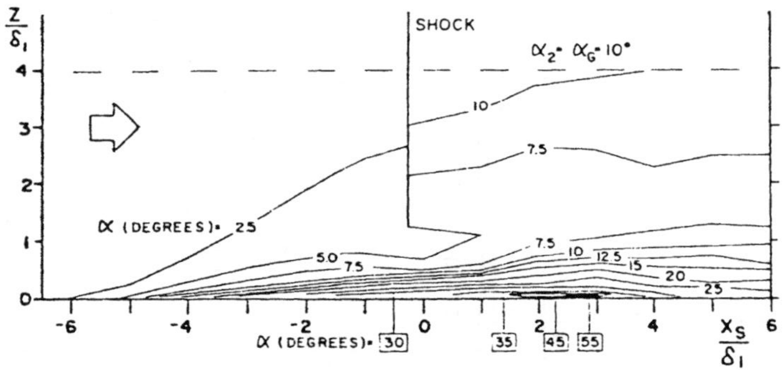

Fig. 3 Yaw Angle Contour Plot, Sharp Fin at 10° [Model 1, Y = 4.0 inch]

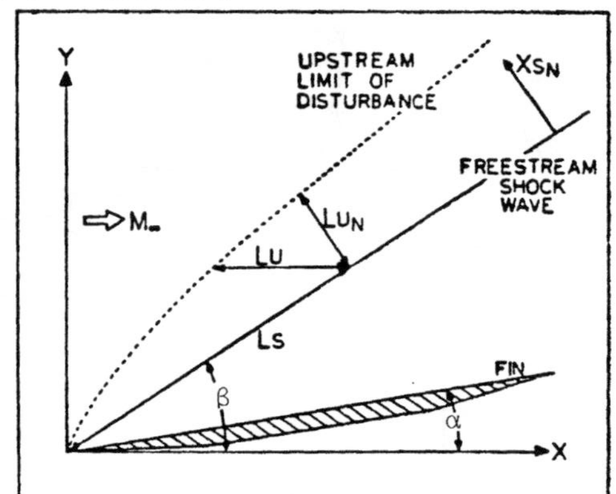

Fig. 4 Coordinate System, Sharp Fin

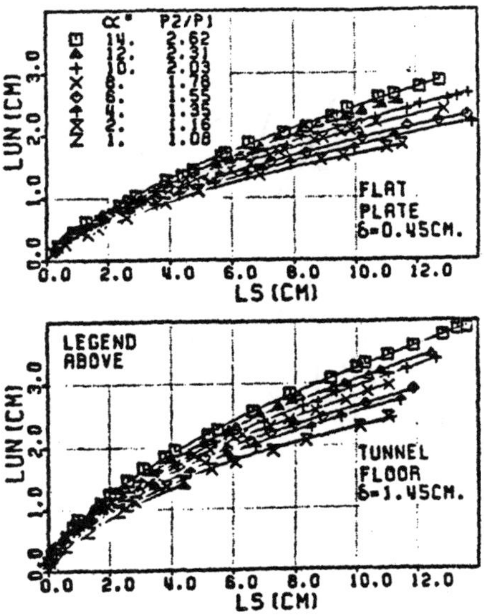

Fig. 5 Effect of Shock Wave Strength on Normal Upstream Influence, Sharp Fin

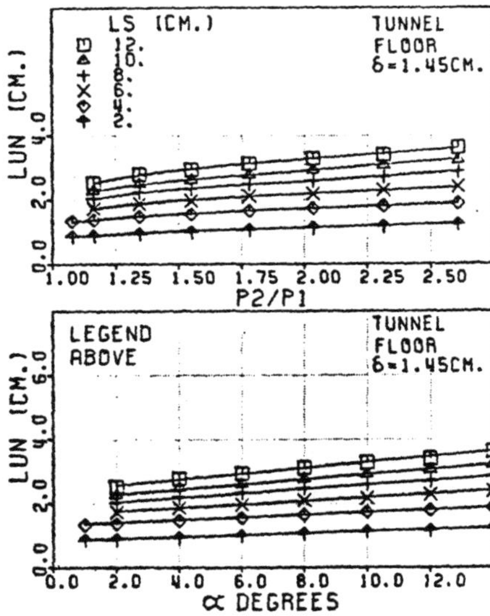

Fig. 6 Normal Upstream Influence, Sharp Fin

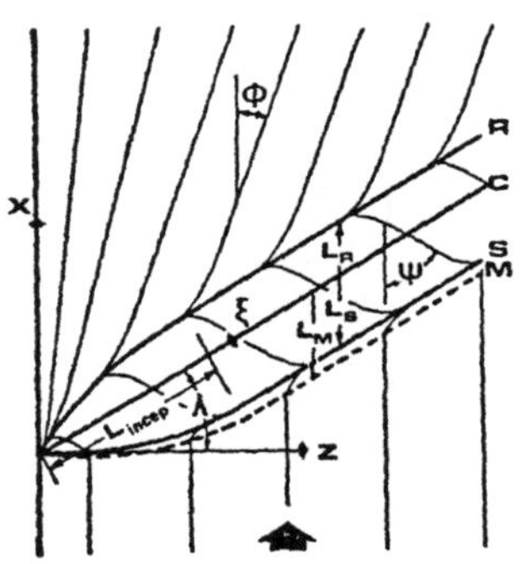

Fig. 7 General Sketch of Surface Flow Pattern on Swept Compression Corner with Definition of Parameters

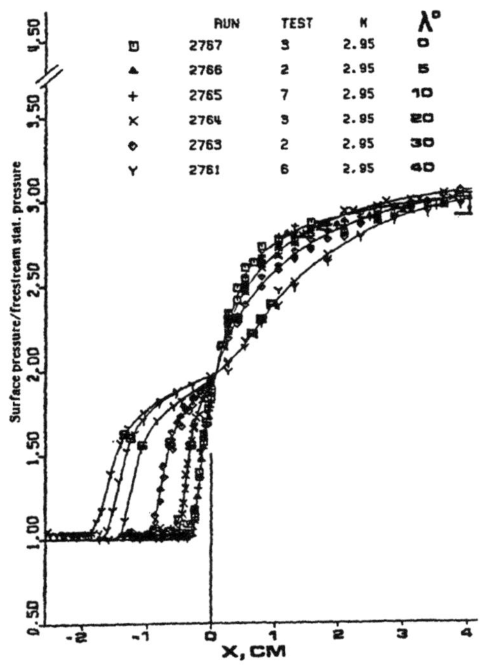

Fig. 8 Surface Pressure Distribution for α = 16 deg Corner at Various Sweep Angles, $Re_x = 1.87 \times 10^7$ (x = 0 is Corner Line in each case)

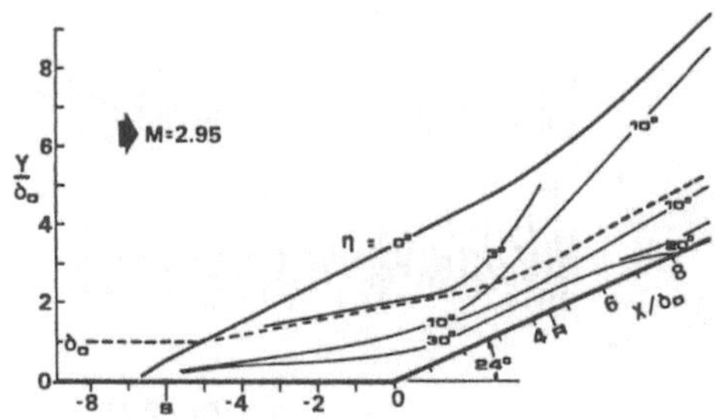

Fig. 9 Yaw Angle Contour Plot, 24° Corner Swept 40°

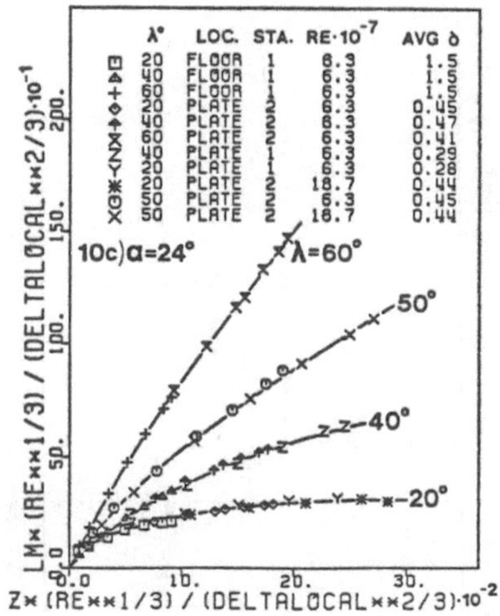

Fig. 10 Scaling of Upstream Influence Lengths of Swept Compression Corners

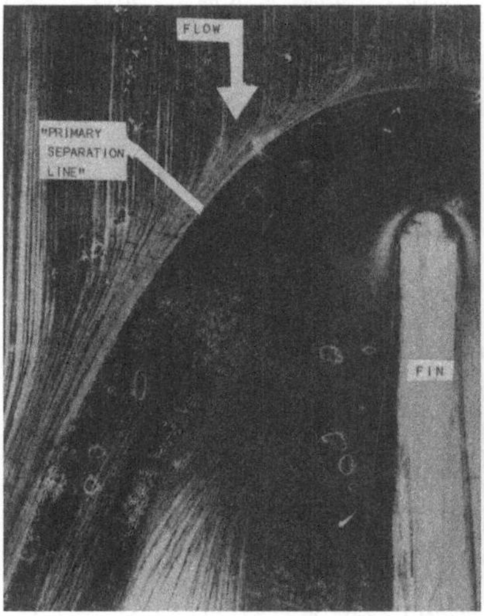

Fig. 11 Typical Kerosine-Graphite Surface Flow Pattern around a Hemi-Cylindrical Nosed Fin

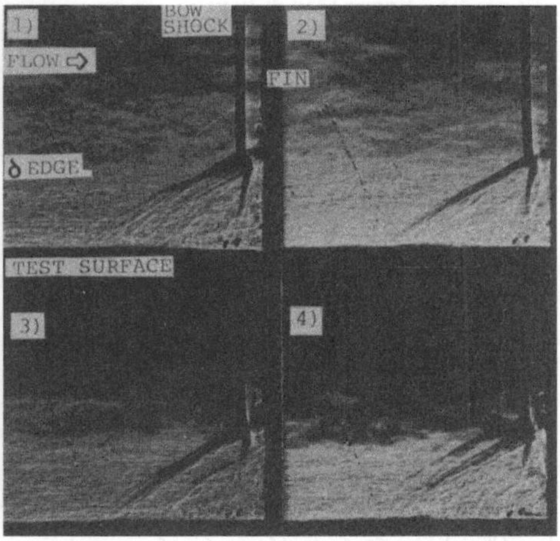

Fig. 12 Typical Schlieren Picture of Shock Wave Structures Ahead of the Blunt Nosed Fin

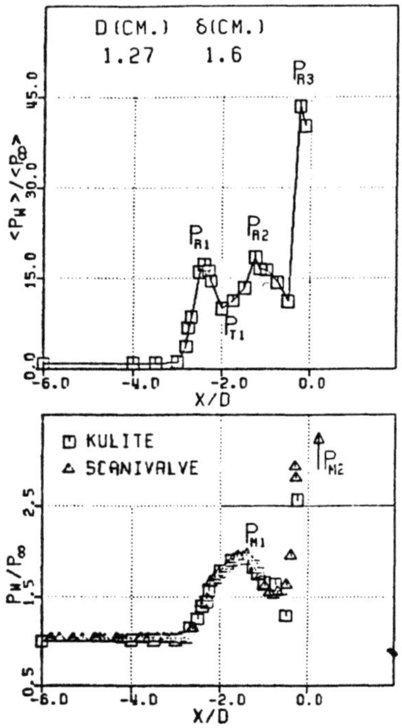

Fig. 13 Normalized r.m.s. and Mean Pressure Distributions on Blunt Fin Centerline

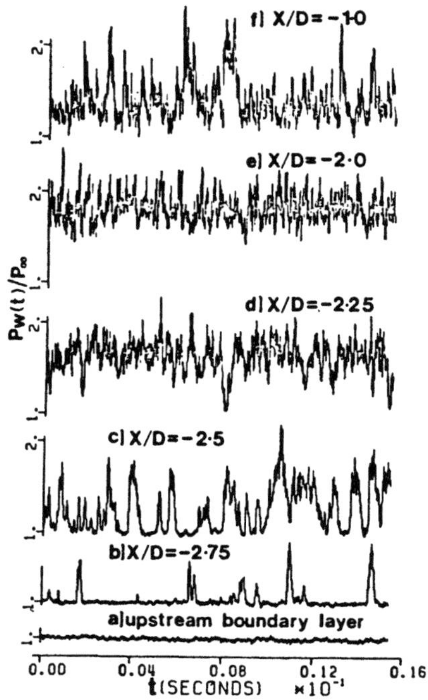

Fig. 14 Samples of Pressure-Time Histories on Flunt Fin Centerline

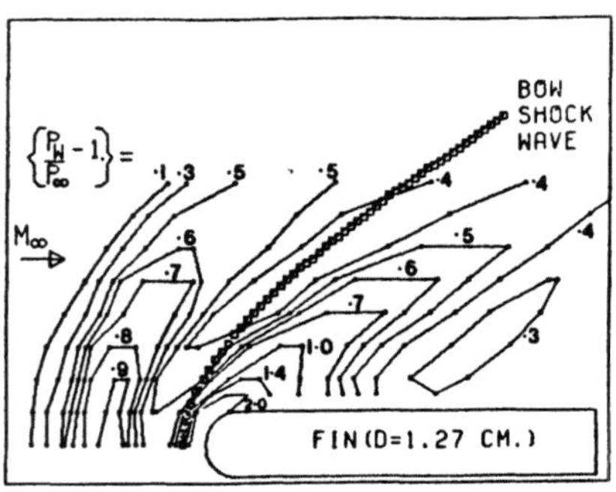

Fig. 15 Normalized Mean Pressure Contours on the Wall

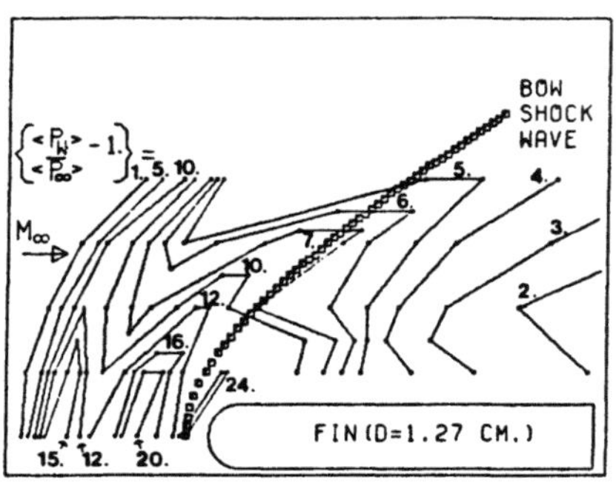

Fig. 16 Normalized r.m.s. Pressure Contours on the Wall

An Investigation of Surface Flow Pattern and Pressure Distribution for Viscous, Sonic Flow over Hemisphere-Cylinder at Incidence

TSUYING HSIEH
Naval Surface Weapons Center
White Oak, Silver Spring, Maryland

Summary

An experimental and numerical investigation of surface flow pattern and pressure distribution about a hemisphere-cylinder at 19 degree incidence in a sonic freestream velocity is reported. The appearance of nose separation line and standing vortex in addition to the well-known primary and secondary separation lines on the body can be identified from the oil flow pictures. Numerical simulation of the observed flowfield by solving the Navier-Stokes equations with thin-layer approximation is shown to produce similar separation features. Comparison between the measured and the calculated surface pressure distribution are presented and discussed.

Introduction

An extensive experimental investigation of surface flow pattern and pressure distribution for hemisphere-cylinder at 0 to 19 degree incidence and freestream Mach number between 0.6 to 1.5 range was conducted several years ago [1]. Among the many three dimensional separation phenomena, of great interest is the appearance of nose separation and standing vortex on the leeside of the forebody at α = 15 and 19 degrees as indicated by the oil flow pictures taken from wind tunnel testing. The nose vortex also appears to be most pronounced at α = 19 and M_∞ = 1.0. The mechanism for the formation of nose vortex was later proposed by Hsieh and Wang [2] within the context of the open-and-closed separation idea. Under the proposed mechanism, the direction of the nose vortex is counterclockwise (viewed from the left side of the pitching plane facing upstream). This finding also provides an explanation to the flowfield observations made by Jorgensen [3] using vapor screen techniques for body of revolution at large incidence, where a pair of small counterclockwise vortices were found on top of the well-known body vortices.

In Ref. [4], calculation was performed for the case of viscous sonic flow over a hemisphere-cylinder at 19 degree incidence based on the work of Pulliam and Steger [5] who solved the three dimensional unsteady Navier-Stokes equations with thin-layer approximation and use an algebraic turbulence model of Baldwin and Lomax [6]. Because of the limitation of available grid points to resolve the viscous effects and the simplicity of the turbulence model, the calculated results are only qualitative in nature. Nevertheless, all the important separated flow phenomena, including the nose standing vortex are given by the numerical results. In this paper, the experimental work will first be described and followed by an outline of the calculation procedure. Finally comparisons between the measured and calculated surface flow pattern and pressure distribution will be presented and discussed.

Experiments

The experiments were conducted in the AEDC Aerodynamic Wind Tunnel (1T). The test section is 1 ft. square and 37.5 in. long with six-percent porous walls in the top and bottom and two plexiglass side walls for flow visualization. The operating Mach number range is from 0.2 to 1.5 with accuracy of Mach number in the test section about $\pm$ 0.003. The model used is a stainless steel hemisphere-cylinder 1 in. in diameter and 10 in. long. The model is sting mount with a sting diameter of 0.75 in. The model incidence can vary from 0 to 19 degrees with accuracy of $\pm$ 0.05 degree. Eighteen pressure orifices are located along a single meridianal plane. The model can be rolled every 15 degrees to obtain pressure data at different meridianal planes. A system of Scanivalves was used for mean flow pressure measurement with accuracy of $\Delta C_p = \pm 0.01$. The free stream Reynolds number at sonic speed is 5.2×10^6 per foot.

The surface flow pattern is revealed by the technique of injecting oil during testing. The pressure tubes from the model were connected into an oil tank which was pressurized outside the tunnel. Six pressure ports were selected for oil

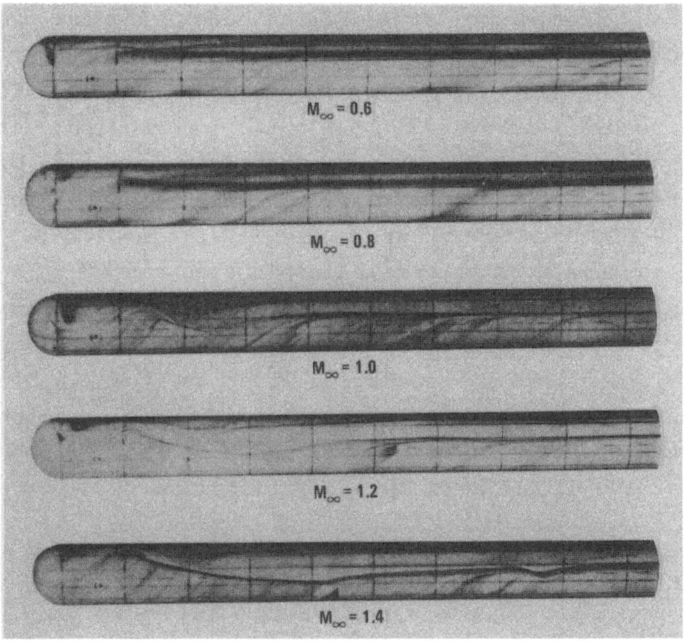

a. At different M_∞

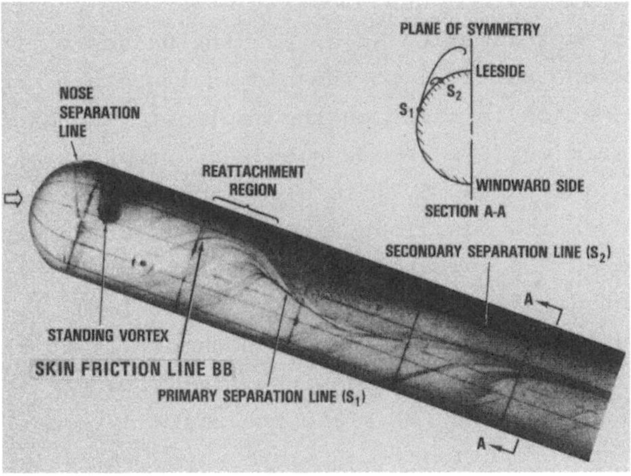

b. Features of separation at $M_\infty = 1$

Fig. 1. Oil flow pictures showing surface flow patterns about a hemisphere-cylinder at 19 deg incidence injection. Black dye was added to the oil and the model was sprayed with a light coat of white paint to increase the contrast. Grid at intervals of 30 degrees in the circumferential direction and every 1 in. in the longitudinal direction are marked on the model surface to provide readings of separation lines. Typical surface flow patterns as revealed by the oil flow pictures are shown in Fig. 1. From Fig. 1a, one finds that the nose vortex is most pronounced at $M_\infty = 1$. One can also identify the nose separation line, the standing counterclockwise nose vortex, the reattachment region, the primary and secondary separation lines as shown in Fig. 1b. It should be pointed out that the open-type separation described by Wang [9] is vividly shown in Fig. 1 around the fore portion of the primary separation line.

Numerical Calculation

Numerical calculations of viscous and inviscid sonic flow over hemisphere-cylinder at 19 degree incidence were performed using the computer code of Pulliam and Steger [5] who solve the three dimensional unsteady Navier-Stokes equations with thin-layer approximation, i.e. all the diffusion terms in both the directions parallel to the body surface are neglected. The numerical method is based on the implicit, factored scheme of Beam and Warming [8]. A turbulence and transition model due to Baldwin and Lomax [6,7] is also included in the code. The lengthy governing equations, numerical algorithm and turbulence model will not be repeated herein because of the limitation of space. For detail, please refer to Refs. [4-8]. It should be pointed out that: (1) unlike the boundary layer theory, the pressure is calculated along with other flow variables in the thin-layer theory and (2) since the unsteady equations are solved (steady solution is obtained asymptotically), it can compute flowfield with simultaneous streamwise and crossflow separation.

For hemisphere-cylinder calculation, a warped spherical coordinate system was used as shown in Fig. 2. The boundary conditions are implemented as follows: (1) On the surface, the tangency condition is imposed for inviscid flow and zero velocity components for viscous flow. (2) At the far field boundaries (RF/R = 3, $R_{min}/R = 12$ and $L_{min}/R = 20$), the constant free stream values are specified. (3) On the downstream boundary, a simple first order extrapolation is used for $M_\infty > 1$, while extrapolation and the condition that pressure is fixed at free stream value is used for $M_\infty < 1$. (4) On the singular line, S or AB, extrapolate flow variables and take average value. (5) On the plane of symmetry (pitching plane), the symmetry boundary condition is imposed using the reflection principle.

To initiate the calculation, the free stream condition is imposed everywhere (a sudden introduction of the body). A grid distribution of 48 (streamwise direction) x 15 (circumferential direction) x 30 (normal direction) was used for viscous flow solution with minimum spacing 0.00005R above the body surface. The starting time step is $\Delta T = 0.001$ which

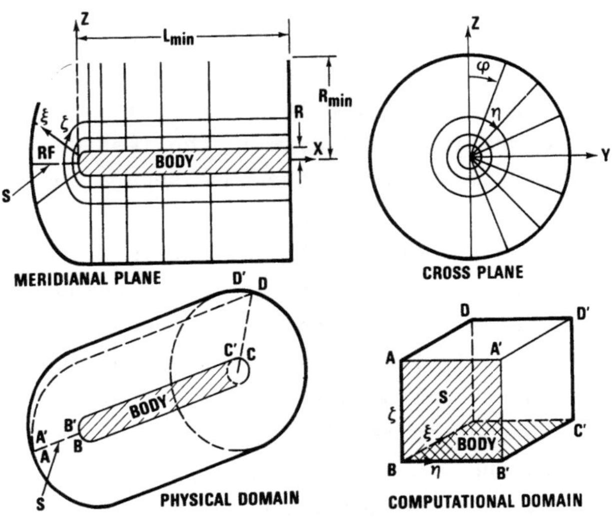

Fig. 2. Warped spherical coordinate system

corresponds to a Courant number CN = 15 and gradually increase to $\Delta T = 0.1$ (CN = 1500) and hold constant from there on. The steady solution is obtained when the flow variables do not show discernible change in between 50 time steps and the residue of the solution is in the order of 10^{-4}. A steady state solution may be obtained in about 800 time steps for viscous laminar flow calculation. Solutions for transition and turbulent flows may be obtained using the laminar flow solution as the initial flowfield and turn on the transition or turbulence model and continue the calculation for about 200 to 400 time steps. Inviscid calculations are obtained separately with less grid points in the normal direction, i.e. 48 x 15 x 12 and fewer time steps are required to reach steady state solution, about 400 to 600 steps. Calculations were performed for hemisphere-cylinder at sonic freestream Mach number and 19 degree incidence for inviscid, laminar, transition and turbulent flows and results are described in the following paragraphs.

Results and Discussions

Surface Pressure Distribution

A comparison of surface pressure distribution between the measured data and the numerical solutions for inviscid, laminar, transition and turbulent flows is given in Fig. 3. The agreement is generally good except in the leeside of the cylinder in the region between the nose separation line and the flow reattachment. In this region, a constant pressure area was measured but were not calculated. It is noted that the Reynolds number of the experiment is in the transition flow regime. Thus, it is possible that the turbulence or transition model used in the calculation is not sufficient to describe the actual three dimensional separated flow there. It is interesting to point out that inviscid solution agrees better with the measured data in the windwardside and, in general, the laminar solution gives overall better agreement with the measured data, particularly in the downstream portion (X/R > 10) which is not shown in Fig. 3 but can be found in Ref. [4].

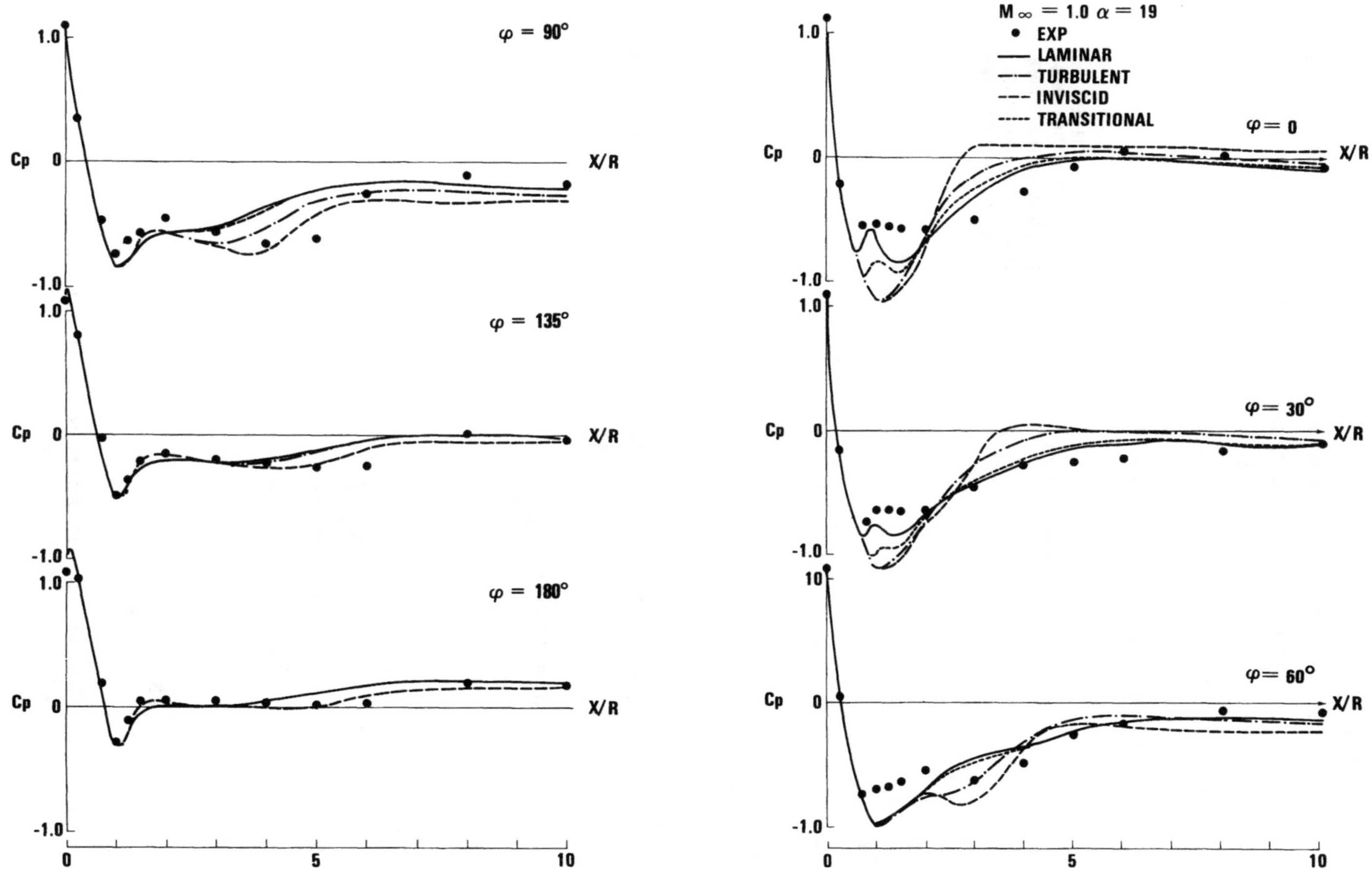

Fig. 3. Comparison of surface pressure distribution

Surface Flow Pattern

In Fig. 4, the direction and the relative magnitude of the skin friction at each grid point on the surface of the hemisphere-cylinder are plotted. From the skin friction vectors, it is possible to determine the separation lines and the surface flow pattern. It is interesting to see that the nose separation line, the nose standing vortex and the primary

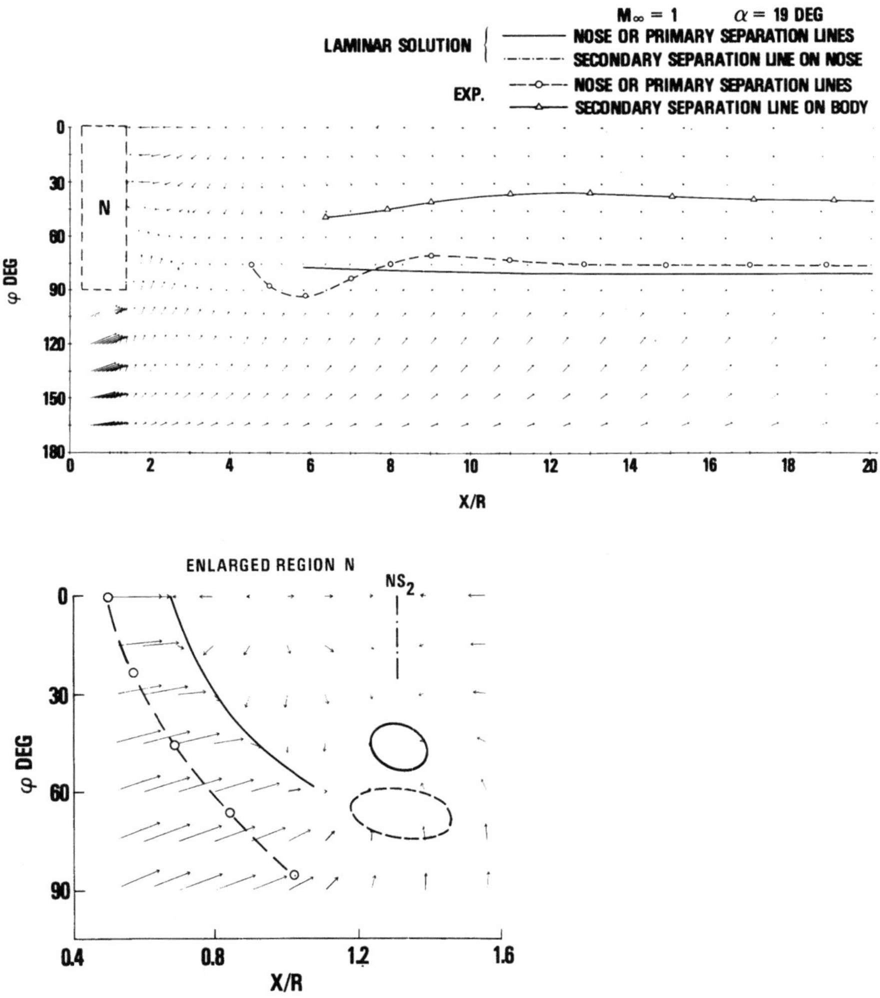

Fig. 4. Comparison of surface flow pattern between laminar solution and measured data

separation line on the body can easily be drawn and match qualitatively with the measured locations of the separation lines and nose vortex. The direction of the nose vortex is indeed counterclockwise. The missing feature in the numerical simulation is the secondary separation line on the body. This is believed to be due to insufficient grid points in the circumferential direction ($\Delta\phi$ = 15 degrees is too sparse. At this time of writing, results of an improved calculation with additional grid points in the circumferential direction have not been completed yet). The laminar solution also shows a small secondary separation bubble in the nose separation region with separation line NS_2. However, this is not observed experimentally.

Another important feature observed experimentally is the open-type separation where skin friction lines initiated from the windwardside enter the separated region created by the primary separation line as shown in Fig. 1 by the line BB. It is interesting to see that this feature is indicated in the plot of the skin-friction vectors for the turbulent solution as shown in Fig. 5.

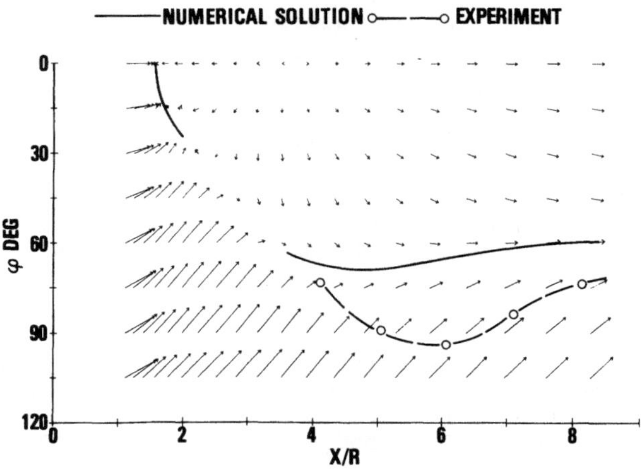

Fig. 5. Comparison of surface flow pattern between turbulent solution and measured data

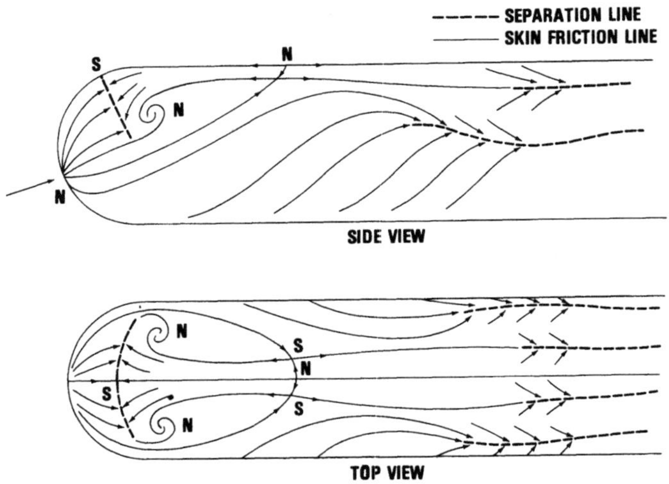

Fig. 6. Sketch of skin friction lines for hemisphere-cylinder at $M_\infty = 1.0$ and $\alpha = 19$ deg

The turbulent solution gives a smaller nose and body separation regions as expected, in the meantime, the nose vortex disappears.

Surface flow pattern can best be shown with the nodal (N) and saddle (S) singular points [9,10]. With the help of oil flow picture and the numerical calculation, the surface flow pattern or the topograph of skin friction lines can be sketched as shown in Fig. 6. It should be noted that Fig. 6 differs slightly from Fig. 26 of Ref. [10].

Acknowledgement

The work reported herein is partly funded by the Fluid Mechanics Program of Naval Air Systems Command at Washington, DC monitored by Mr. D. E. Hutchins.

References

1. Hsieh, T.: An investigation of separated flow about a hemisphere-cylinder at 0- to 19-degree incidence in the Mach number range from 0.6 to 1.5. AEDC-TR-76-112, Arnold Engineering Development Center, Arnold AFS, TN, Nov. 1976.

2. Hsieh, T.; Wang, K. C.: Concentrated vortex on the nose of an inclined body of revolution. AIAA J. 14 (1976), pp. 698-700.

3. Jorgensen, L. H.: Prediction of static aerodynamic characteristics for slender body alone and with lifting surfaces to very high angles of attack. NASA TR-R-474. Ames Research Center, Moffett Field, CA. Sept. 1977.

4. Hsieh, T.: Calculation of viscous, sonic flow over hemisphere-cylinder at 19 degree incidence: the capturing of nose vortices. AIAA Paper 81-0189. Jan. 1981.

5. Pulliam, T. H.; Steger, J. L.: On implicit finite difference simulations of three dimensional flow. AIAA J. Vol. 18, No. 2. Feb. 1980.

6. Baldwin, B. S.; Lomax, H.: Thin layer approximation and algebraic model for separated turbulent flow. AIAA paper 73-257. 1978.

7. Pulliam. T. H.; Lomax, H.: Simulation of three-dimensional compressible viscous flow on the illiac IV computer. AIAA Paper 79-0206. Jan. 1979.

8. Warming, K. F.; Beam, R.: On the construction and application of implicit factored schemes for conservation laws. SIAM, ATNS Proceedings, 11 (1978) pp. 85-129.

9. Wang, K. C.: Separation of three dimensional flow. MML TR-76-54-C, Martin Marietta Laboratories, Baltimore, MD. 1976.

10. Peak, D. J.; Tobak, M.: Three dimensional interactions and vortical flows with emphasis on high speeds. AGARD-AG-252, North Atlantic Treaty Organization, July 1980.

Cross-Hatching – An Interaction between Shock and Turbulent Boundary Layer

Pu Qun

Institute of Mechanics, Academia Sinica,
Beijing, P.R. China

Abstract

On the basis of experimental studies and analysis of the cross-hatching pattern, the present paper shows how the streamwise vortices, the turbulent boundary layer with ablation, and the external supersonic flow play an important role in the formation of the pattern. A flow model is proposed to explain the formation and development of the pattern. A correlation was obtained for some ablation materials between the wavelength of the cross-hatching and the pressure of the external supersonic flow and the recovering temperature of the boundary layer.

Introduction

On the ablation surfaces in a supersonic turbulent boundary layer, an orderly arranged diamond-shaped pattern has often been observed, which is known as cross-hatching pattern (Fig.1). This type of pattern occurs not only on ablation surfaces of different geometries, such as flat plates, pointed cones and sphere-cones of different sizes, but also on those made of different materials, such as wax, phenolic silica, teflon and camphor (Plate 1). These materials are of widely different physical properties: some of them will melt in ablation, while others will sublime or become carbonized (Swigart (1974)). Besides, the cross-hatching occurs on the natural fusion crust of meteorites, such as Jilin Meteorites (Ref.3) as well as on man-made model surface including those of recovered re-entry vehicles. It occurs not only in the external flow, but also in the internal flow (Plate 1). Furthermore, it can be found not only on ablation surfaces but also on painted model surfaces in a supersonic turbulent boundary layer. All this indicates that cross-hatching is a very stubborn phenomenon and an inherent physical property in supersonic turbulent boun-

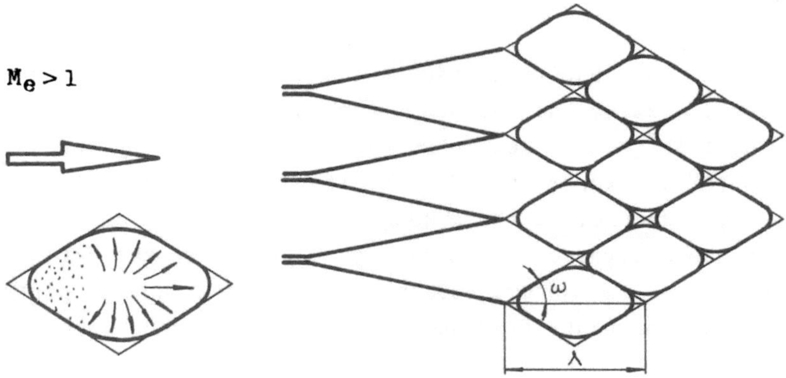

Figure 1. Sketch of cross-hatching pattern

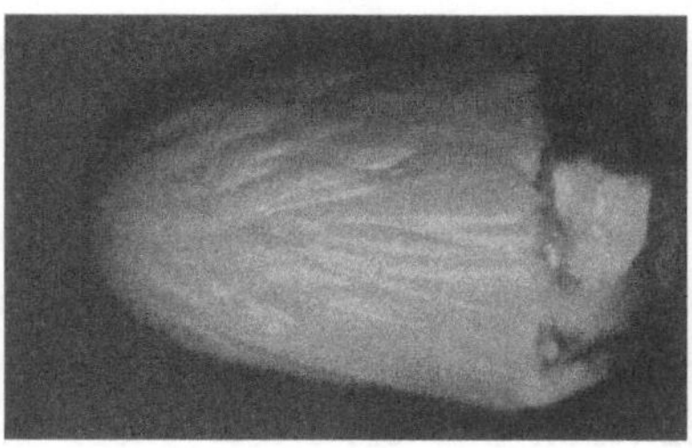

(1)

(2)

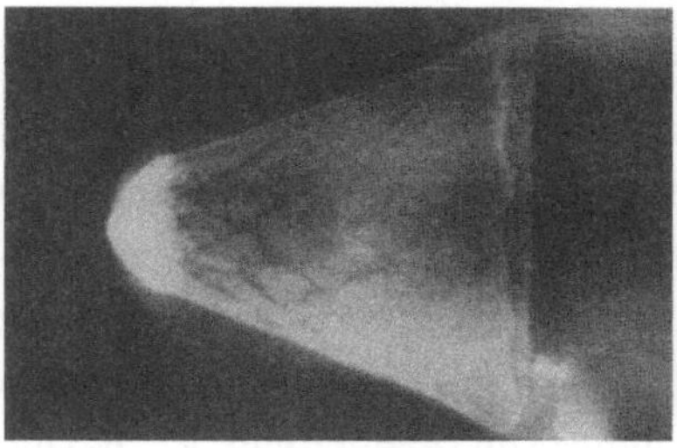

(3)

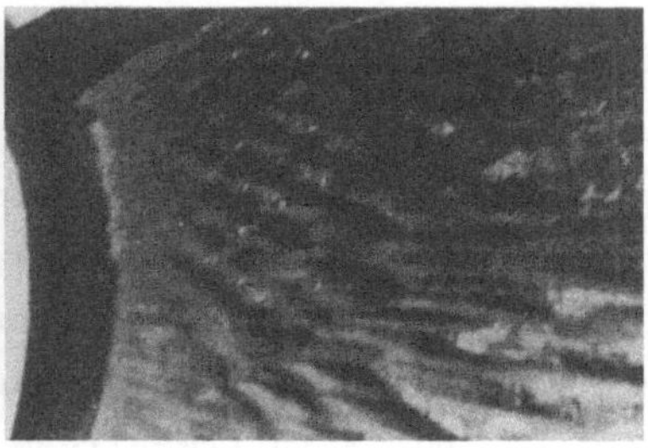

(4)

Plate 1. Cross-hatching on various ablation surfaces: (1) wax model, (2) camphor model, (3) phenolic silica model, (4) internal Laval nozzle liner dary layer flow with surface ablation or deformation.

Although much study has been made on the cross-hatching pattern since the mid-sixties and various hypotheses proposed (Swigart (1974)), no satisfactory mechanism has been offered, by which the inherent physical properties of the pattern can be adequately explained. For example, is the presence of streamwise vortices essential to the occurence of the pattern ? What are the triggering factors of its occurence? What are the respective parts played by the supersonic flow and the turbulent boundary layer in the formation of the pattern? Why in some cases it does not occur on the ablation surface, although the conditions of supersonic external flow and turbulent boundary layer are provided? What are the major physical parameters that affect the wavelength of the pattern? How can the surface pattern induce rolling torque on test models, such as sharpnosed and self-blunting right circular cones? All these problems remain to be solved.

This paper is an attempt to answer some of the problems mentioned above, with reference to a unified underlying physical mechanism based on close observation and comparative analysis of the various detail physical phenomena of the ablation patterns and their interelations as obtained from our ground and flight tests as well as from the fusion crust of the Jilin Meteorites.

The interelation between ablation patterns

A series of tests on sphere-cone models made of wax were performed in hypersonic tunnels under various stagnation pressures and temperatures, with an attempt to expose the relations between different types of ablation patterns. For a given sphere-cone model tested under fixed freestream Mach number M_∞ and stagnation temperature T_0, but with freestream stagnation pressures P_0 being increased step by step, a series of different types of ablation patterns appeared successively on the sphere-cone surface of the model as follows:

1. Fine striations: They are a great number of very fine and shallow streamwise grooves on the sphere-nose surface.
2. Smooth surface.

3. Narrow grooves: They are also streamwise grooves similar to the fine striations but deeper and wider and smaller in number.
4. Pits: They are in general rather large and very few, distributed unevenly around the sonic line of the sphere-cone surface.
5. Ablation wedges: They are wedge-shaped grooves similar to the turbulent wedges appearing in the transition region of boundary layer without ablation. These ablation wedges distribute circumferentially in an orderly way on the sphere-nose surface and have a difinite configuration (Ref.4).
6. The cross-hatching pattern: The pattern follows closely the ablation wedges and starts at the intersection points of the adjacent side grooves of ablation wedges. It is noted that the angle of the cross-hatching is very different from that of the ablation wedge (Plate 1 and Fig.1).
7. Regmaglypts: With P_0 stepping up, ablation wedges move forward and become shorter and regmaglypts appear on the ablation surface instead of the cross-hatching.

By repeating the tests at different oncoming Mach numbers with T_0 fixed and P_0 varying as before, we found the order of the occurence of the different types of patterns remained the same, as is shown in the log P_0 against log M_∞ diagram in Fig.2. It is interesting to note that the lines joining the same types of the patterns coinside with the iso-Raynold's lines of the boundary layer. It shows that the type of the ablation pattern appearing on the surface is governed by the Raynold's number of the boundary layer.

The occurrence of the cross-hatching pattern is possible only when the ablation rate is high enough to induce a disturbance in the supersonic external flow. In our experiments we found the minimum order of magnitude of the ablation rate for the occurrence of the cross-hatching pattern for several materials.

It is seen from Fig.2 that the appearance of a cross-hatching requires a larger Raynold's number than that of an ablation wedge. Since the ablation wedge is a characteristic feature in the transition region of the boundary layer (Ref.4), it follows that a

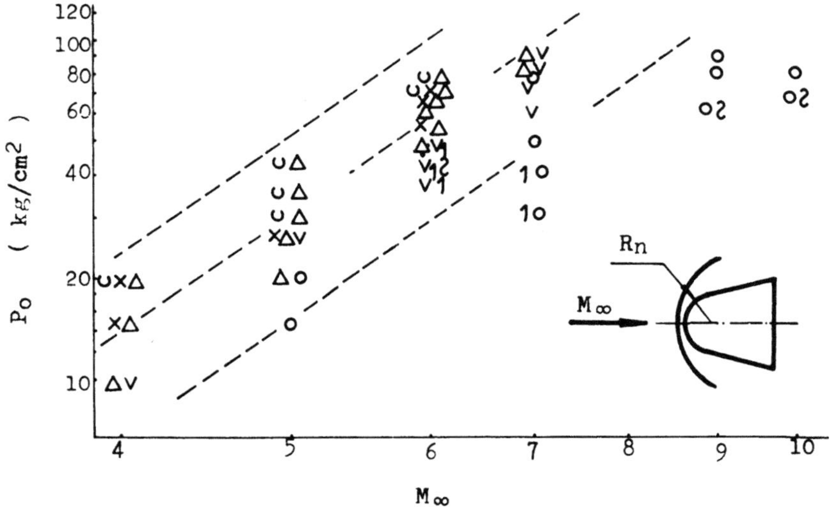

Figure 2. Ablation patterns and the related Raynold's numbers. Here the symbles ʔ, o, 1, v, Δ, x and ᴗ denote respectively fine striation, smooth surface, narrow groove, pit, ablation wedge, cross-hatching and regmaglypt. R_n = 25 mm, T_o = 130°C.

turbulent boundary layer is required for the appearance of a cross-hatching.

Streamwise vortex —— the triggering role

In a normally developed boundary layer, such as that on a smooth surface of a flat plate, a pointed cone, or a sphere-cone, cross-hatching usually appears immediately after the ablation wedges. On close examination of the detailed surface structure of the ablation patterns, we found that the starting points of the foremost row of the cross-hatchings usually lie at the intersection points of the adjacent side grooves of ablation wedges. As these grooves are the traces of the streamwise vortices generated along the edges of the ablation wedges (Ref.4), it is evident that these streamwise vortices play an important role in the formation of the cross-hatching. In the case that the boundary layer is disturbed by a surface irregularity, such as a small back-facing step, a series of streamwise grooves sometimes appear on the surface downstream of the step and the cross-hatching follows. These streamwise grooves are the traces of streamwise vortices generated from the disturbance of the back-facing step. When the Raynold's number of the boundary layer is high enough, the disturbance will grow and force the boundary layer to turn to turbulent and provides a condition for the appearance of the cross-hatching. If the Raynold's number of the boundary layer is not high enough, the disturbance dies off at a short distance downstream of the back-step and no cross-hatching can be formed. From the cases mentioned above, it is evident that the streamwise vortices are essential to the formation of the cross-hatching. But this does not mean that they trigger off the cross-hatching directly. In fact, it is only through the introduction of a conical shock wave in the external supersonic flow outside the boundary layer that the streamwise vortex precipitates the formation of the cross-hatching. This is the reason why for the appearance of a cross-hatching the external flow has to be supersonic.

In order to introduce a conical shock, the streamwise vortex has to be strong enough to create a disturbance outside the boundary layer. It takes a certain time or a certain distance along the ablation surface for the vortices to grow or to merge into stronger ones. Therefore for the appearance of a cross-hatching, the thicker the boundary layer, the larger the streamwise vortex required and therefore the larger the ablation wedge that evolves. This is indeed the case that we have found in our experiments.

Detailed flow pattern on the ablation surface

Once a conical shock outside the boundary layer is induced, along the intersection line of the conical shock and the boundary layer there is a sharp pressure rise in a direction normal to the intersection line underneath the shock and inside the turbulent boundary layer (Barry (1951), Kussoy (1980)). As the wall friction is reduced by the injection of ablation products, this sharp pressure rise forces the melted materials to flow backward in a local region, where the shock wave and the turbulent boundary layer intersect. From the ablation tests in exhaust streams of jet engines, we found that drops of melted material flew backward in the local region of the ablation surface in a turbulent boundary layer. This backward flow togather with the external downstream flow forms a new pair of vortices along the intersection line of the shock and the boundary layer. These vortices and the sharp pressure rise produce higher ablation rate in the local region and form the leading edges of a cross-hatching. The conical shocks induced by the streamwise vortices are in general weak shocks. Therefore one half of the angle made by the two leading edges of a cross-hatching is equal or very close to the Mach angle of the external supersonic flow (Fig.1). In the case of a laminar boundary layer, the pressure rise underneath the shock inside the boundary layer is in general an order of magnitude smaller than that in a turbulent boundary layer. This is one of the main reasons that in a laminar boundary layer the cross-hatching pattern does not appear.

Furthermore, the vortex along the leading edge of cross-hatching grows and introduces a new disturbance in the external supersonic flow and causes a new conical shock wave. For the same reason mentioned above, a second row of the leading edges of cross-hatchings are formed, and they are also the trailing edges of the first row of the cross-hatching. Therefore, a diamond-shaped pattern is formed on the ablation surface. The flow pattern on the ablation surface of the cross-hatching is clearly seen on the fusion crust of Jilin Meteorite (Plate 2). The detailed striation on the regmaglypt surface of Jilin Meteorites reveals that on the leeward surface of the pits, there are fine grains of melted materials, and on the windward surface there are striations radiating from the central part of the pit in the downstream and side directions (Ref.3). These radiating striations are also found on the cross-hatching surface of Nandan Iron Meteorites, collected by the Guiyang Institute of Geochemistry, Academia Sinica, Guiyang, China. From the surface pattern of the cross-hatching it is obvious that a three-dimensional boundary layer is developed on the ablation surface of the cross-hatching. Besides, the effect of the temperature ratio, $(T_r-T_c)/T_c$, on pattern wavelength is appearent not only for wax, but also for other materials, such as camphor. For some high ablation tem-

Plate 2. The detailed striations on cross-hatching surface

perature materials, as a result of the small variation in the temperature retio in the experiments, the effects of the temperature are usually not obvious.

Flow Model

On the basis of the above mentioned observation and analysis, a preliminary flow model is proposed to explain various phenomena in regard to cross-hatching pattern.

To begin with (Fig.3), with the accumulation of ablation products, the vortices on both side of the ablation wedges gradually build up. They grow as they extend until they induce disturbances in the external supersonic flow and cause the conical shock waves. This is the trigger factor in the formation of cross-hatching. At the place where the shock wave and the turbulent boundary layer intersect (Fig.3), a new pair of vortices are formed on the ablation surface by the accumulation of the backward flow of the melted material. They, in tern, grow and induce new disturbances and conical shocks in the external flow. This process repeats itself, resulting in a series of intersections between shock waves and the turbulent boundary layer. The sharp pressure rise underneath the shock and inside the turbulent boundary layer pro-

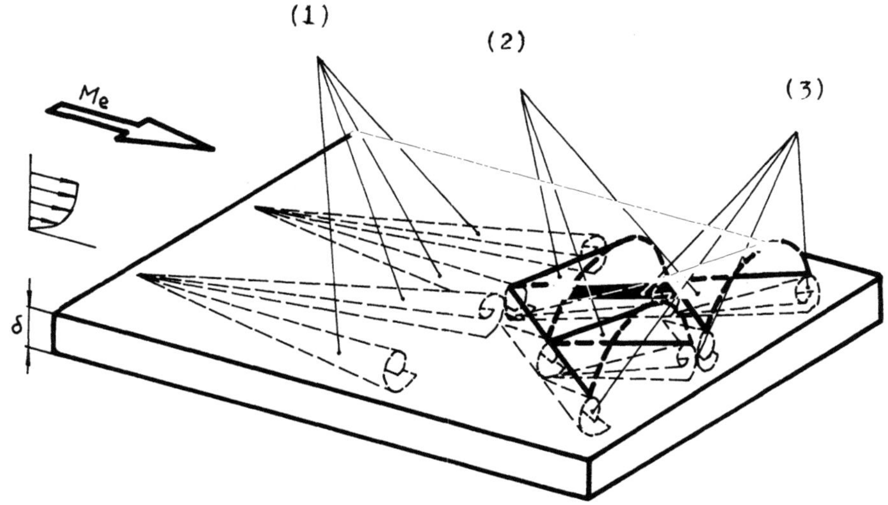

Figure 3. Sketch of flow model of the cross-hatching. (1) Vortices along edges of ablation wedges, (2) Conical shock waves caused by growth of vortices, (3) Vortices along edges of cross-hatching.

duces higher ablation rate in the local region and a cross-hatching pattern is formed on the ablation surface.

From the above flow model it is obvious that the higher the pressure, the larger Raynold's number of the boundary layer, and the thinner the boundary layer, which in turn gives a shorter wavelength of the cross-hatching. Similary, the higher the stagnation temperature, the smaller Raynold's number and the thicker the boundary layer, which leads to a longer wavelength. For some ablation materials, a correlation between the wavelength λ of the cross-hatching, the external pressure P_0, and the recovered temperature T_r of the boundary layer is obtained as follows:

$$\frac{\lambda}{\lambda_c} = K \left(\frac{P_e}{P_c} \right)^a \left(\frac{T_r}{T_c} - 1 \right)^b \quad .$$

Here the subscript c denotes the reference condition; k, a and b are dimensionless material constants. In our experiments, a=-0.77 and b=0.5 for wax and camphor.

The fact that the surface patterns induce a rolling torque on test model can be explained by the model proposed above. Under a high speed flow, an axisymmetric model, such as a sharpnosed or a self-blunting right circular cone, with any small surface irregularity or asymetry, may start to rotate on its axis of symmetry at a small angular speed in either direction, say the positive direction. Because of this rotation, the relative speed of the oncoming flow with respect to the surface of the model will twist in the oppsite, i.e. the negative direction, and so does the cross-hatching. Owing to the oriented surface roughness, the flow over an axisymmetric model with cross-hatching twisted in the negative direction, results in a larger rolling torque in the same positive direction as that due to the initial irregularity. This process will magnify by itself, until an equilibrium condition between the rolling and the frictional torque of the model is established. A similar result will be obtained if the model starts to turn in the negative direction.

Conclusions

1. The vortices along the edges of the ablation wedges in the transition region of the boundary layer or the streamwise vortices on the ablation surface behind a back-facing step and their inevitable growth play a triggering role in the occurence of the cross-hatching pattern.
2. The final formation of the cross-hatching pattern is a result of the intersection of the turbulent boundary layer with the conical shock waves. The sharp pressure rise inside the turbulent boundary layer underneath the shock wave produces higher ablation rate in the local region and forms a cross-hatching pattern on the ablation surface.
3. The interelations between various ablation patterns and the detailed striations on the surfaces of both the ablation wedge and the cross-hatching provide strong evidence for the establishment of a flow model for the formation of the cross-hatching pattern.

Acknowledgment

The author is grateful to Prof. T. C. Lin for his valuable discussions and suggestions on this paper and thank to Z.Y.Chen, M.S.Yan, Z.J.Zhou, R.Y.Wang for their help in carrying out part of the experiment.

References

1. Barry, F.W.; Shapiro, A.H.; Neumann, E.P.: The interaction of shock waves with boundary layer on a flat surface. Jour. Aero. Sci. v.18 No.4 (1951) 229

2. Kussoy, M.I.; Viegas, J.R.; Horstman, C.C.: Investigation of a three-dimensional shock wave separated turbulent boundary layer. AIAA J. v.18 (1980) 1477

3. Lin Tongji; Chen Zhongying; Wang Ruyong; Yan Mingshan; Zhou Zhengjin; Analysis of the ablation patterns on Jilin Meteorite. Symposium on the Jilin Meteorites. The Publishing House of Sciences, Beijing, China. (1979) 270-279

4. Lin Tongji; Chen Zhongying; Wang Ruyong; Pu Qun; Zhou Zhengzin; Yan Mingshan: Ablation Wedges and boundary layer transition. ACTA Mechanica Sinica, Fourth Issue, 1981, Beijing, China 326-332

5. Swigart, R.J.: Cross-hatching studies —— a critical review. AIAA J. v.12 (1974) 1301-1318

The Numerical Computation of Three-Dimensional Boundary Layers

P. D. SMITH

Royal Aircraft Establishment
Bedford, UK

Abstract

The governing equations for compressible three-dimensional turbulent boundary layers in a general coordinate system are given. Methods for the solution of these equations and their integral counterparts are described and compared. The available finite difference techniques are discussed in detail.

Introduction

I would like to begin by saying what an honour and a pleasure it is for me to give this review lecture. To some extent my pleasure is enhanced by a sense of déja vu as it is almost exactly ten years since I spoke on a similar topic at this very institute. The occasion then was EUROMECH 33 on three-dimensional turbulent boundary layers. At that meeting I described the integral calculation method[2] I had developed for compressible turbulent boundary layers and laid particular stress upon the generality of the coordinate system I had adopted. At that time the method was probably the most advanced (at least in its range of applicability) available and will serve as a useful datum against which to measure progress during the past decade. Of course, as I hope to show, real progress has been made. Other integral methods have adopted the general coordinate system and have used more sophisticated assumptions about the velocity profiles and entrainment. The major advance has been, however, in the development of the finite-difference methods and I hope to spend some time describing these in detail. As a guard against complacency let me state here however, that progress has not been achieved in the types of turbulence modeling we use in these methods. In 1972 Wesseling and Lindhout[3] described their finite-difference method which although confined to incompressible flow and Cartesian coordinates used Bradshaw's[4] vector equation for the turbulent shear stress. Ten years later almost all the finite-difference methods appear to have reverted to the more primitive eddy viscosity concept of turbulence modelling.

EUROMECH 33 also marked the birth of the other significant development in this field, the organisation of a series of meetings to compare results from the various calculation methods with each other and where possible against experiment. The first of these meetings, EUROMECH 60, became known as the Trondheim Trials[5] and prompted the EUROVISC Committee of national representatives to form a working party which has since organised two further meetings, at Stockholm[6] in 1978 and Amsterdam[7] in 1979. Finally, as you will all know a third meeting will immediately follow this Symposium so that any predictions I may be foolish enough to make concerning the capabilities of the various calculation methods will rapidly be put to the test and almost certainly shown to be false!

As with any review paper little of what follows is original. In particular I have drawn heavily upon the works of J P F Lindhout[7] of the NLR and my colleague at the RAE, Caroline Boyd[8]. To them and also to all those whose names appear in the list of references I offer my sincere thanks.

2 THE THREE DIMENSIONAL BOUNDARY LAYER EQUATIONS[9,10,11]

For general applicability most three-dimensional boundary layer calculation methods use a non orthogonal curvilinear coordinate system; x, y in the surface, z normal to the surface with an angle λ (x, y) between the x and y directions. In this coordinate system the continuity, momentum and energy equations may be written as:

Continuity

$$\frac{\partial}{\partial x}(\rho u h_2 \sin\lambda) + \frac{\partial}{\partial y}(\rho v h_1 \sin\lambda) + \frac{\partial}{\partial z}(\overline{\rho w} h_1 h_2 \sin\lambda) = 0 \qquad (1)$$

x momentum

$$\frac{\rho u}{h_1}\frac{\partial u}{\partial x} + \frac{\rho v}{h_2}\frac{\partial u}{\partial y} + \frac{\overline{\rho w}\partial u}{\partial z} - \rho\cot\lambda k_1 u^2 + \rho\csc\lambda k_2 v^2 + \rho k_{12} uv$$

$$= - \csc^2 \frac{\lambda}{h_1}\frac{\partial p}{\partial x} + \cot\lambda \frac{\csc\lambda}{h_2}\frac{\partial p}{\partial y} + \frac{\partial}{\partial z}(\mu\frac{\partial u}{\partial z} - \rho\overline{u'w'}) \quad (2)$$

y momentum

$$\frac{\rho u}{h_1}\frac{\partial v}{\partial x} + \frac{\rho v}{h_2}\frac{\partial v}{\partial y} + \overline{\rho w}\frac{\partial v}{\partial z} - \rho\cot\lambda k_2 v^2 + \rho\csc\lambda k_1 u^2 + \rho k_{21} uv$$

$$= \cot\lambda \frac{\csc\lambda}{h_1}\frac{\partial p}{\partial x} - \frac{\csc^2\lambda}{h_2}\frac{\partial p}{\partial y} + \frac{\partial}{\partial z}(\mu\frac{\partial v}{\partial z} - \rho\overline{v'w'}) \quad (3)$$

Energy equation

$$\frac{\rho u}{h_1}\frac{\partial H}{\partial x} + \frac{\rho v}{h_2}\frac{\partial H}{\partial y} + \overline{\rho w}\frac{\partial H}{\partial z} = \frac{\partial}{\partial z}\left[\frac{\mu}{P_r}\frac{\partial H}{\partial z} + \mu(1 - \frac{1}{P_r})\frac{\partial}{\partial z}(\frac{U^2}{2}) - \rho\overline{w'H'}\right] \quad (4)$$

Here

$$\overline{\rho w} = \rho w + \overline{\rho' w'}$$

An element of length on the surface ds is given by

$$ds^2 = h_1^2 dx^2 + h_2^2 dy^2 + 2h_1 h_2 \cos\lambda dxdy \quad (5)$$

and the total velocity U is given by

$$U^2 = u^2 + v^2 + 2uv\cos\lambda \quad (6)$$

Where the metric coefficients h_1 and h_2 are to the boundary layer approximation functions of x and y but not z.

$$\text{ie} \quad h_1 = h_1(x, y), \; h_2 = h_2(x, y)$$

The boundary conditions are

$$z = 0, \; u = v = 0 \; w = w_w \; (\frac{\partial H}{\partial y})_w = \text{given}$$

$$z = \delta \; u = u_e(x,y) \; v = v_e(x,y) \; H = H_e(x,y) \quad (7)$$

The curvature parameters k_1, k_2, k_{12} and k_{21} are given by

$$k_1 = \frac{1}{h_1 h_2 \sin\lambda}\left[\frac{\partial}{\partial x}(h_2\cos\lambda) - \frac{\partial h_1}{\partial y}\right] \quad (8)$$

$$k_2 = \frac{1}{h_1 h_2 \sin\lambda}\left[\frac{\partial}{\partial y}(h_1\cos\lambda) - \frac{\partial h_2}{\partial x}\right] \quad (9)$$

$$k_{12} = \frac{1}{\sin\lambda}\left[-(k_1 + \frac{1}{h_1}\frac{\partial\lambda}{\partial x}) + \cos\lambda\left(k_2 + \frac{1}{h_2}\frac{\partial\lambda}{\partial y}\right)\right] \quad (10)$$

$$k_{21} = \frac{1}{\sin\lambda}\left[-(k_2 + \frac{1}{h_2}\frac{\partial\lambda}{\partial y}) + \cos\lambda\ (k_1 + \frac{1}{h_1}\frac{\partial\lambda}{\partial x})\right] \quad (11)$$

For an orthogonal system, $\lambda = 90^o$, $k_1 = -k_{12}$, $k_2 = -k_{21}$.

Rather than solve the energy equation (4), for adiabatic flow the density at each point in the boundary layer is frequently assumed to be approximated by the Crocco relation

$$\frac{\rho e}{\rho} = 1 + \frac{\gamma-1}{2}\, r\, M_e^2 \left(1 - \frac{U^2}{U_e^2}\right) \quad (12)$$

where r, the recovery factor is taken as 0.84 or 0.85 for laminar flows and 0.89 for turbulent flow.

At the boundary layer edge equations 2 and 3 reduce to

$$\rho e\ (\frac{u_e}{h_1}\frac{\partial u_e}{\partial x} + \frac{v_e}{h_2}\frac{\partial u_e}{\partial y} - \cot\lambda k_1 u_e^2 + \csc\lambda k_2 v_e^2 + k_{12} u_e v_e)$$

$$= -\frac{\csc^2\lambda}{h_1}\frac{\partial p}{\partial x} + \frac{\cot\lambda\csc\lambda}{h_2}\frac{\partial p}{\partial y} \quad (13)$$

and

$$\rho e\ (\frac{u_e}{h_1}\frac{\partial v_e}{\partial x} + \frac{v_e}{h_2}\frac{\partial v_e}{\partial y} - \cot\lambda k_2 v_e^2 + \csc^2\lambda k_1 u_e^2 + k_{21} u_e v_e)$$

$$= \frac{\cot\lambda\csc\lambda}{h_1}\frac{\partial p}{\partial x} - \frac{\csc^2\lambda}{h_2}\frac{\partial p}{\partial y} \quad (14)$$

The two Euler equations, 13 and 14, may be used to eliminate p from equations 2 and 3, or for a given pressure distribution p (x, y) may be solved for u_e and v_e subject to appropriate initial and boundary conditions.

3 INTEGRAL METHODS

The earliest and simplest attempts to find solutions to the general equations 1-3 involved the use of integral methods in which the equations are integrated term by term across the boundary layer from the wall to the outer edge. These integral equations were first derived by Myring[12]. To proceed further it is then necessary to assume some form for the velocity profiles in the boundary layer. All integral methods make use of streamwise and crosswise profiles. This is profiles in the directions of the external streamlines (s) and their orthogonal trajectories

(n) respectively. It was therefore necessary for Myring to derive the transformation relationships between the integral thicknesses in the x,y coordinate system and the corresponding thicknesses in the s,n coordinate system. The equations may however be derived rather more directly by starting from the integral equations in streamline coordinates[13] and transforming the independent variables s,n to x,y as shown below. This derivation will be useful when we come to discuss inverse methods.

The integral equations in streamline coordinates are:
s momentum

$$\frac{\partial\theta_{11}}{\partial s} + \frac{\partial\theta_{12}}{\partial n} + \theta_{11}\left(\frac{H+2-M_e^2}{U_e}\frac{\partial U_e}{\partial s} - k_s\right) + \theta_{12}\left(\frac{2-M_e^2}{U_e}\frac{\partial U_e}{\partial n} - 2\,k_n\right)$$

$$+ \delta_2\left(\frac{1}{U_e}\frac{\partial U_e}{\partial n} - k_n\right) + k_s\theta_{22} = \frac{C_{fs}}{2} \tag{15}$$

n momentum

$$\frac{\partial\theta_{21}}{\partial s} + \frac{\partial\theta_{22}}{\partial n} + \theta_{21}\left(\frac{2-M_e^2}{U_e}\frac{\partial U_e}{\partial s} - 2k_s\right) + k_n\theta_{11}(H+1) +$$

$$+ \theta_{22}\left(\frac{2-M_e^2}{U_e}\frac{\partial U_e}{\partial n} - k_n\right) = \frac{C_{fn}}{2} \tag{16}$$

Continuity

$$\frac{\partial(\delta-\delta_1)}{\partial s} - \frac{\partial\delta_2}{\partial n} + (\partial-\delta_1)\left[\frac{1-M_e^2}{U_e}\frac{\partial U_e}{\partial s} - k_s\right] -$$

$$- \delta_2\left(\frac{1-M_e^2}{U_e}\frac{\partial U_e}{\partial n} - k_n\right) = C_E \tag{17}$$

These equations are transformed to the general x,y non-orthogonal coordinate system by using the operators

$$\frac{\partial}{\partial s} = \frac{\sin(\lambda-\alpha)}{\sin\lambda}\frac{1}{h_1}\frac{\partial}{\partial x} + \frac{\sin\alpha}{\sin\lambda}\frac{1}{h_2}\frac{\partial}{\partial y} \tag{18}$$

$$\frac{\partial}{\partial n} = -\frac{\cos(\lambda-\alpha)}{\sin\lambda}\frac{1}{h_1}\frac{\partial}{\partial x} + \frac{\cos\alpha}{\sin\lambda}\frac{1}{h_2}\frac{\partial}{\partial y} \tag{19}$$

where λ is the angle between the x and y directions and α the angle between the x and s directions so that

$$\sin\alpha = \frac{v_e}{U_e}\sin\lambda \tag{20}$$

The curvature terms k_s and k_n are derived as follows. For k_s we equate the continuity equation (1) for the external flow in the x,y coordinate system to its counterpart in the s,n coordinate system. The result is

$$k_s = - \frac{1}{h_1 h_2 \sin\lambda} \left[\frac{\partial}{\partial x} [h_2 \sin(\lambda - \alpha)] + \frac{\partial}{\partial y} (h_1 \sin\alpha) \right] \tag{21}$$

For k_n we equate the equation for the component of vorticity of the external flow in the direction normal to the surface[14].

$$\xi = \frac{1}{h_1 h_2 \sin\lambda} \left[\frac{\partial}{\partial x} (h_2 U_e \cos(\lambda - \alpha)) - \frac{\partial}{\partial y} (h_1 U_e \cos\alpha) \right] \tag{22}$$

to its counterpart in the s,n coordinate system

$$\xi = k_n U_e - \frac{\partial U_e}{\partial n} \tag{23}$$

with the result that

$$k_n = \frac{1}{h_1 h_2 \sin\lambda} \left[\frac{\partial}{\partial x} (h_2 \cos(\lambda - \alpha)) - \frac{\partial}{\partial y} (h_1 \cos\alpha) \right] \tag{24}$$

We note from equation 23 that if the external flow is irrotational then $k_n = \frac{1}{U_e} \frac{\partial U_e}{\partial n}$ and some simplification of equations 15-17 results.

Up to this point no approximations have been made but to proceed further it is necessary to assume forms for the velocity profiles in order to derive relationships between the unknowns so that they may be reduced in number to three. The various assumptions which have been made are given below.

Smith[2] assumed power law profiles for the streamwise flow and Mager[15] profiles for the crossflow. Streamwise skin friction and the entrainment coefficient were derived from auxiliary relationships. Later the method was modified so that the entrainment coefficient was calculated by an extension of Green's[16] lag entrainment method.

Stock[17] modified Smith's method so that the streamwise profile was represented by Coles law of the wall plus wake. This removed the need for an empirical skin friction relationship.

In addition the entrainment was calculated from Horton's lag entrainment relationship.

Cross[18] uses a three-dimensional form of the law of the wall and wake together with Thompson's entrainment relationship.

Cousteix[19,20] and Aupoix use velocity profiles derived from an analysis of similarity solutions.

The methods are listed above in order of increasing complexity. They all use very similar numerical methods to solve the equations and it is interesting to note that this increase in complexity is reflected in increased computing times as listed by Lindhout[7] for the Amsterdam Workshop Test case. The times are shown in Table 1 in terms of effort measured in Mega Floating Point Operations per surface point.

TABLE 1

Method	Effort per point (MFLOP)
Smith	0.007
Stock	0.013
Cross	0.018
Cousteix and Aupoix	0.033

These figures must be interpreted with caution. The methods have usually been written with clarity and ease of alteration rather than speed as the prime objective and in addition the numerical method of Cousteix and Aupoix is more complicated than those used by the others. The important point to note is that all the methods are so fast that realistic configurations can be calculated in very acceptable run times and even the slowest integral method remains an order of magnitude faster than a finite difference method.

The assumption of velocity profiles and where necessary additional empirical relationships reduces the number of unknowns to three (A,B,C say) and, typically, three equations of the form

$$\frac{\partial A}{\partial x} = g1(A,\ B,\ C,\ \frac{\partial A}{\partial y},\ \frac{\partial B}{\partial y},\ \frac{\partial C}{\partial y},\ U_e,\ \alpha,\ \frac{\partial U_e}{\partial x},\ \frac{\partial U}{\partial y},\ \frac{\partial \alpha}{\partial x},\ \frac{\partial \alpha}{\partial y},$$

$$h_1,\ h_2,\ \lambda,\ \frac{\partial h_1}{\partial x},\ \frac{\partial h_1}{\partial y},\ \frac{\partial h_2}{\partial x},\ \frac{\partial h_2}{\partial y},\ \frac{\partial \lambda}{\partial x},\ \frac{\partial \lambda}{\partial y})$$

$$\frac{\partial B}{\partial x} = g2\ (A\ etc)$$

$$\frac{\partial C}{\partial x} = g3\ (A\ etc) \tag{25}$$

where all the boundary layer parameters are now known functions of A,B,C and U_e.

The equations are hyperbolic[12] with three real characteristics and it is necessary to take account of these characteristic directions when evaluating the y derivatives as shown in Fig 1. To a good approximation the upper and lower bounds of the characteristic directions with respect to the x axis are given by α and $\alpha+\beta$ respectively where β is the direction of the wall streamline with respect to s. With the y derivatives evaluated in this manner the solution is advanced in the x direction by a simple explicit method, Smith[2] for example uses a two step Euler method whilst Cousteix and Aupoix[19,20] use a fourth order Runge Kutta method. The explicit nature of the method restricts the forward step size to that allowed by the Courant Friedrich's Levy condition but as noted above run times are short and no attempt has been made to remove this restriction by developing an implicit method.

4 DISPLACEMENT THICKNESS AND EQUIVALENT SOURCE DISTRIBUTION

For calculations involving viscous-inviscid interactions the effect of the boundary layer upon the external flow may be represented as a change in boundary condition for the inviscid calculation from one of zero velocity normal to the surface to one in which the velocity component normal to the surface, m, is given by

$$m = \frac{1}{\rho_e h_1 h_2 \sin\lambda}\left[\frac{\partial}{\partial x}(\rho_e U_e h_2(\delta_1 \sin(\lambda-\alpha)-\delta_2\cos(\lambda-\alpha)) + \frac{\partial}{\partial y}[\rho_e U_e h_1(\delta_1 \sin\alpha + \delta_2 \cos\alpha)]\right] \tag{26}$$

m is often called the transpiration velocity or equivalent source strength.

Alternatively the effect of the boundary layer may be represented as a displacement of the surface through a distance δ* normal to the surface where δ* is given by

$$\frac{\partial}{\partial x}(\rho_e U_e h_2 \sin\lambda\delta^*) + \frac{\partial}{\partial y}(\rho_e v_e h_1 \sin\lambda\delta^*) = \rho_e h_1 h_2 \sin\lambda m \tag{27}$$

an additional partial differential equation which must be solved if this approach is adopted.

5 INVERSE INTEGRAL METHODS

For the calculation of separated flows it is necessary to invert the method so that the boundary layer development is specified and the velocity distribution is calculated. We can readily see that if two of the boundary layer thicknesses (B,C say) are specified as functions of x and y in equations 25 then the equations may be rearranged to the form

$$\frac{\partial A}{\partial x} = f1(A, U_e, \alpha, \frac{\partial A}{\partial y}, \frac{\partial U_e}{\partial y}, \frac{\partial \alpha}{\partial y}, B, C, \frac{\partial B}{\partial x}, \frac{\partial B}{\partial y}, \frac{\partial C}{\partial x}, \frac{\partial C}{\partial y}, h_1, h_2 \text{ etc})$$

$$\frac{\partial U_e}{\partial x} = f2(A \text{ etc})$$

$$\frac{\partial \alpha}{\partial x} = f3(A \text{ etc}) \tag{28}$$

Solutions of this type for infinite yawed wings, in which $\frac{\partial}{\partial y} = 0$, have been published by Cousteix[21] and Stock[22]. In general, however, the inverse method will be part of an interactive calculation and only the source strength m or the displacement thickness δ* will be available as input to the inverse method. An additional equation is then required and this is given by equation 22. The inverse method is then a simultaneous solution of equations 25, 26 or 27 and 22. Whilst this is obviously possible in principle the characteristic directions may cause difficulties for any marching method and no working method of this type has yet been published.

6 FINITE-DIFFERENCE METHODS

The alternative to the integral approach is to approximate the terms of equations 1 to 4 by finite differences. If this is done in a consistent manner the solution of the difference equations will hopefully converge to the solution of the differential equations as the step sizes are reduced to zero. The earliest methods strongly resembled the numerical techniques adopted for integral methods in that they were explicit. That is to obtain the solution at the point x_i, y_j, z_k the solution is written in terms of the unknowns at x_i, y_j, z_k and known values in, the y, z plane at x_{i-1}. Upwind differencing is usually used in the y direction as shown in Fig 1. More recently the trend has been to implicit methods in which the solution is sought simultaneously along a complete z column at x_i, y_j. Once again upwind differencing is used in the y direction and central differences are used to evaluate the second derivative which occurs in the z direction. Typical of this type of method is that of Chang and Patel[25]. It will be seen from Fig 1 that no stability restrictions arise from the CFL condition for this approach but the method is restricted to flows in which v, the velocity in the y direction, is everywhere positive. If v is positive the methods of Lindhout[26] and Nash and Scruggs[27] use differencing techniques identical to those of Chang and Patel. Those used by Mclean[28] are similar but in addition Mclean uses three point backward differences in an attempt to preserve second order accuracy. Once v becomes negative it is necessary to evaluate the y derivative at the $_{i-1}$ station and then these methods are subject to the stability restriction that the characteristic (streamline) through x_i, y_j, z_k must intersect x_{i-1}, $j<y<j+1$. The zig zag technique used by Kordulla[29], and originally formulated by Krause[31], is similar but here rather than switch operators as v changes sign the same

second order operator is used throughout as shown in Fig 2. In addition second order accuracy in the x direction is achieved by centering the difference equations on M, $(x_{i-\frac{1}{2}}, y_j, z_k)$ as shown in Fig 2. Hence derivatives for constant Δ_y may be written

$$\left(\frac{\partial(\ )}{\partial x}\right)_M = \frac{1}{\Delta x}\Big[(\ )_{i,j} - (\ )_{i-1,j}\Big]_K$$

$$\left(\frac{\partial(\ )}{\partial y}\right)_M = \frac{1}{2\Delta_y}\Big[(\ )_{i,j} - (\ )_{i,j-1} + (\ )_{i-1,j+1} - (\ )_{i-1,j}\Big]_K \qquad (29)$$

If varying step sizes Δy_1, Δy_2 say are used then second order accuracy is only preserved if $\Delta y_2 - \Delta y_1$ is kept of order $\Delta y_1 . \Delta y_2$. Similar restrictions apply to varying step sizes in the z direction. In order to overcome these restrictions and permit arbitrarily varying step sizes whilst preserving second order accuracy Cebeci[31] has used the box scheme. Here as shown in Fig 3 the equations are centred on $M(x_{i-\frac{1}{2}}, y_{j-\frac{1}{2}}, z_{k-\frac{1}{2}})$ and derivatives are written as

$$\left[\frac{\partial(\ )}{\partial x}\right]_M = \frac{1}{2\Delta x}\Big[(\ )_{i,j} + (\ )_{i,j-1} - (\ )_{i-1,j} - (\ )_{i-1,j-1}\Big]_{k-\frac{1}{2}}$$

$$\left[\frac{\partial(\ )}{\partial y}\right]_M = \frac{1}{2\Delta y}\Big[(\ )_{i,j} + (i)_{i-1,j} - (\ )_{i,j-1} - (\ )_{i-1,j-1}\Big]_{k-\frac{1}{2}}$$

$$\left[\frac{\partial(\ )}{\partial z}\right]_M = \frac{1}{4\Delta z}\Big[(\ )_{i,\ ,k} + (\ )_{i-1,j,k} + (\ )_{i,j-1,k} + (\ )_{i-1,j-1,k} - (\ )_{i,j,k-1} - (\ )_{i-1,j,k-1} - (\ )_{i,j-1,k-1} - (\ )_{i-1,j-1,k-1}\Big] \qquad (30)$$

where

$$\Delta x = x_i - x_{i-1}$$

$$\Delta y = y_j - y_{j-1}$$

$$\Delta z = z_k - z_{k-1}$$

and $(\ )_{i-\frac{1}{2},j-\frac{1}{2},k-\frac{1}{2}}$ is taken as the average of the eight corner values. The second derivative in the z direction is obtained by defining new unknowns f,g say equal to $\partial u/dz$ and $\partial v/dz$. Then $\partial^2 u/\partial z^2 = \partial f/\partial z$ and $\partial^2 v/\partial 2^2 = \partial g/\partial z$.

The box technique can only be used if v is positive and originally Cebeci[31] proposed a solution technique in which the solution marched in either the positive or negative y directions according to the sign of v. This technique poses programming problems (which appear to have been overcome in the method of Lindhout[26] to which we will shortly return) and so Cebeci et al[32] proposed two alternative techniques. The first of these, the zig-zag box scheme, combines the zig-zag and box methods when v is negative so that M becomes $x_{i-\frac{1}{2}}$, y_i, $z_{k-\frac{1}{2}}$ and now derivatives are written as

$$\left[\frac{\partial(\ )}{\partial x}\right]_M = \frac{1}{2\Delta x_{i-1}}\Big[(\ )_{i,j,k} + (\ )_{i,j,k-1} - (\ )_{i-1,j,k} - (\ )_{i-1,j,k-1}\Big]$$

$$\left[\frac{\partial(\ )}{\partial y}\right]_M = \frac{1}{4}\Big[\{(\ )_{i,j,k} + (\ )_{i,j,k-1} - (\ )_{i,j-1,k} - (\ )_{i,j-1,k-1}\}/\Delta y_{j-1} + \{(\ )_{i-1,j+1,k} + (\ )_{i-1,j+1,k-1} - (\ )_{i-1,j,k} - (\ )_{i-1,j,k-1}\}/\Delta y_{ji}\Big]$$

$$\left[\frac{\partial(\ )}{\partial z}\right]_M = \frac{1}{2\Delta z_{k-1}}\Big[(\ )_{i,j,k} + (\ )_{i-1,j,k} - (\ )_{i,j,k-1} - (\ )_{i-1,j,k-1}\Big] \tag{31}$$

and

$$(\ )_{i-\frac{1}{2},j,k-\frac{1}{2}} = \frac{1}{4}\Big[(\ )_{i,j,k} + (\ )_{i,j,k-1} + (\ )_{i-1,j,k} + (\ )_{i-1,j,k-1}\Big]$$

The use of these approximations does however re-introduce the stability restriction mentioned above together with the restriction on the variation of the y step sizes if second order accuracy is to be preserved. Cebeci et al's other method is the characteristic difference scheme, Fig 4, in which the inertia terms in the momentum equations are written as

$$\frac{u}{h_1}\frac{\partial}{\partial x} + \frac{v}{h_2}\frac{\partial}{\partial y} = \frac{u}{h_1}\frac{\partial}{\partial \Gamma}\frac{\partial \Gamma}{\partial x} \tag{32}$$

where Γ is the direction of the streamline or characteristic through x,y,z. If $\Gamma_{i,j,k-\frac{1}{2}}$ cuts x_{i-1} constant $z_{k-\frac{1}{2}}$ constant at Q the equations centred now on PQ/2 where $p = x_i, y_j, z_{k-\frac{1}{2}}$ can be differenced as for example

$$\left[\frac{u}{h_1}\frac{\partial u}{\partial x} + \frac{v}{h_2}\frac{\partial u}{\partial y}\right] = \left[\frac{u}{h_1}\frac{\partial u}{\partial \Gamma}\frac{\partial \Gamma}{\partial x}\right] = \frac{1}{2}\left(\frac{u_p}{h_{1p}} + \frac{u_q}{h_{1q}}\right)\left(\frac{u_p - u_q}{\Delta x}\right)$$

where

$$u_p = \frac{1}{2}\left[u_{i,j,k} + u_{i,j,k-1}\right]$$

$$u_q = \frac{1}{2}\left[\alpha(u_{i-1,j-1,k} + u_{i-1,j-1,k-1}) + (1-\alpha)(u_{i-1,j,k} + u_{i-1,j,k-1})\right] \quad y_{j-1} < y_q < yj$$

$$u_q = \frac{1}{2}\left[-\alpha(u_{i-1,j-1,k} + u_{i-1,j-1,k-1}) + (1-\alpha)(u_{i-1,k,k} + u_{i-1,j,k-1})\right] \quad y_{j-1} < y_q < y_{j+1}$$

$$\alpha = (\Delta x v h_1)/(\Delta y u h_2)$$

This scheme cannot however be simply applied to the continuity equation which is usually dealt with by a zig-zag scheme. For the characteristic scheme to remain second order accurate α must be evaluated at PQ/2 and u_q must be accurate to within O (Δ^3). This last condition can be satisfied by using quadratic interpolation through three points say y_{j-1}, y_j and y_{j+1}. Boyd[8] has compared these three Cebeci difference schemes for the Trondheim Trials[5] and NLR test cases and finds the characteristic scheme to be less robust and to require roughly three times the computation time. Boyd's preferred strategy is to use the box scheme unless any point on a z column requires the zig-zag in which case the zig-zag scheme is used for all that column. Cebeci et al found that the characteristic scheme allowed them to find solutions closer to separation than the other schemes but Boyd found that for the NLR test case these differences were very slight.

It was mentioned above that Cebeci originally adopted a marching scheme which depended upon the sign of v. Lindhout[26] uses just such a scheme and has devised an algorithm to define the optimal marching scheme. The advantage of this scheme is that the stability restriction on the x step is minimised, since the explicit operators tend only to be used as v changes sign and is therefore close to zero.

The end result of any of the implicit methods is a set of non-linear algebraic equations which are linearised by the use of iteration and then solved by standard matrix techniques. The need to use iteration makes comparisons of timings between the methods particularly difficult. Some comparative timings are given in Ref 7 but they are not reproduced here as Boyd[8] has found that the number of x steps in one example could be doubled with little or no increase in run times. This apparent anomaly occurs because the smaller step size produces a better initial guess at the next x station and hence fewer iterations are required. A similar mechanism may explain why Lemmerman[33] found that even the old explicit methods remained competitive.

The stability restriction may be removed altogether if the method is made implicit in both the y and z directions as is done by Nash and Scruggs[27], Patankar and Spalding[34] and Baker[35]. A simultaneous solution technique is then required for the whole y,z plane. Nash and Scruggs use an alternative direction implicit method, Patankar and Spalding their simple algorithm and Baker a finite element approach. The use of such a fully implicit technique allows the inclusion of the diffusion in the y direction and this has been done by both Baker and Spalding. In addition if one includes the z direction momentum equation the pressure may be allowed to vary with z although it becomes necessary to uncouple the pressure in the x momentum equation from that in the y and z momentum equations. For further details see Ref 34.

Comparisons between the results of the various difference methods may be found in Refs 5,6,7,8 and 33 but as mentioned above all timings contained therein must be treated with extreme caution and certainly no clear "best buy" has emerged to date.

No inverse finite-difference method has been published but Boyd[8] has suggested how a method using equations 1,2,3,4 or 12,22 and 26 could be devised.

CONCLUSIONS

Calculation methods for three-dimensional turbulent boundary layers on practical shapes are now fairly widely available. Of the two types, integral and finite-difference, the integral methods retain their advantages of speed and robustness which make them particularly suitable for viscous-inviscid interaction calculations. The finite-difference methods are however not restricted by any velocity profile assumptions and allow more advanced turbulence models to be more readily incorporated. In addition they hold the promise that they rather than experiment may suggest new profile families for integral methods. The emergence of inverse methods both integral and finite-difference would appear to be imminent.

NOTE ADDED IN PROOF

No sooner were the above words committed to print than an inverse finite-difference method appeared. This method[36] by Formery and Delery like the integral methods described in Section 5 requires two boundary layer thicknesses to be specified. This may make its application to viscous inviscid interactive calculations difficult. It is not however restricted to infinite yawed wing flows and good agreement is shown with both infinite yawed wing and fully three-dimensional experiments. Difficulties produced by the characteristics coming from downstream are avoided in this method by using the FLARE approximation in which for example the term $\partial u/\partial x$ is set to zero if u is negative.

REFERENCES

1 Fernholz, H. Three-dimensional turbulent boundary layers: a report on EUROMECH 33. J. Fluid Mech., Vol 58, Part 1 pp 177-186 (1973).

2 Smith, P.D. An integral prediction method for three-dimensional compressible turbulent boundary layers, ARC R&M 3739 (1972).

3 Wesseling, P. Lindhout, J.P.F. A calculation method for three-dimensional incompressible turbulent boundary layers. AGARD CP-73 (1971).

4 Bradshaw, P. Calculation of three-dimensional turbulent boundary layers. J. Fluid Mech., Vol 46, pp 417-445 (1971).

5 East, L.F. Computation of three-dimensional turbulent boundary layers. Euromech 60 Trondheim 1975. F.F.A. TN AE-1211 (1975).

6 Humphreys, D.A. Comparison of boundary layer calculations for a wing: The May 1978 Stockholm Workshop test case. F.F.A. TN AE-1522 (1979).

7 Lindhout, J.P.F. van den Berg, B. Elsenaar, A. Comparison of boundary layer calculations for the root section of a wing. The September 1979 Amsterdam Workshop test case. NLR MP 80028U (1980).

8 Boyd, C. A finite-difference calculation method for the three-dimensional boundary layer. RAE Report to be published.

9 Squire, L.C. The three-dimensional boundary layer equations and some power series solutions. ARC R&M 3001 (1957).

10 Hansen, A.G. Compressible, three-dimensional laminar boundary layers - a survey of current methods of analysis. Douglas Paper No 3105 (1964).

11 Cebeci, T. Kaups, K. Ramsey, J.A. A general method for calculating three-dimensional compressible laminar and turbulent boundary layers on arbitrary wings. NASA CR 2777 (1977).

12 Myring, D.F. An integral prediction method for three-dimensional turbulent boundary layers in incompressible flow. RAE Technical Report 70147 (1970).

13 Cooke, J.C. Hall, M.G. Boundary layers in three dimensions. Progress in Aeronautical Sciences Vol 2 (1962). Pergamon Press.

14 Hirschel, E.H. Kordulla, W. Shear-flow in surface oriented coordinates. Notes on Numerical Fluid Mechanics Vol 4 Vieweg (1981).

15 Mager, A. Generalisation of boundary layer momentum integral equations to three-dimensional flows including those of rotating systems. NACA Report 1067 (1952).

16 Green, J.E. Weeks, D.J. Brooman, J.W.F. Prediction of turbulent boundary layers and wakes in compressible flow by a lag entrainment method. ARC R&M No 3791 (1973).

17 Stock, H.W. Calculation of three-dimensional boundary layers on wings and bodies of revolution. Proceedings DEA Meeting "Viscous and interacting flow field effects'. Meersburg, April 1979 (1979).

18 Cross, A.G.T. Calculation of compressible three-dimensional turbulent boundary layers with particular reference to wings and bodies. British Aerospace Brough YAD 3379 (1979).

19 Cousteix, J. Analyse theorique et moyens de prevision de la couche limite turbulente tridimensionelle. ONERA Publ No 157 (1974). English translation ESA TT-238 (1976).

20 Cousteix, J. Progres dans les methodes de calcul des couche limites turbulentes bi et tridimensionelles. 13 eme Colloque d'Aerodynamique Appliquee, Lyon 8-10 Nov 1976 (1976).

21 Cousteix, J. Turbulence modelling and boundary layer calculation methods. ONERA Rapport Technique OA 43/2259 AYD (DERAT 27/5004 DY) (1981).

22 Stock, H.W. Computation of the boundary layer and separation lines on inclined ellipsoids and of separated flows on infinite swept wings. AIAA-80-1442 (1980).

23 Nash, J.F. An explicit scheme for the calculation of three dimensional turbulent boundary layers. Trans ASME J. Basic Eng 94D No 1 1 131 (1972).

24 Elsenaar, A. Addendum to report TR 74159U. NLR AJ-76-023 in Dutch. (1976).

25 Chang, K.C. and Patel, V.C. Calculation of three-dimensional boundary layers on ship forms. Iowa Institute of Hydraulic research. The University of Iowa Report No 178 (1975).

26 Lindhout, J.P.F. Moek, G. de Boer, E, and van den Berg,B. A method for the calculation of 3D boundary layers on practical wing configurations. Trans ASME Jnl Fluids Eng, Vol 103 pp 104-111 (1981).

27 Nash, J.F. and Scruggs, R.M. An implicit method for the calculation of three dimensional boundary layers on fuselage configurations. Rept LG76ER0199 Sybucon Inc (1976).

28 McLean, J.D. Three dimensional turbulent boundary layer calculations for swept wings. AIAA 77-3 (1977).

29 Kordulla, W. Investigations related to the inviscid-viscous interaction in transonic flows about finite 3D wings. AIAA 77-209 (1977).

30 Krause, E. Hirschel, E.H. Bothmann, Th. Die numerische integration der bewegungsgleichungen dreidemensionaler laminarer kompressibler grenschichten. Fachtagung aerodynamik. Berlin 1968 DGLR-Fachbuchr iche Band 3 Braunschweig (1969).

31 Cebeci, T. Kaups, K. Ramsey, J.A. A general method for calculating three dimensional compressible laminar and turbulent boundary layers on arbitrary wings, NASA CR 2777 (1977).

32 Cebeci, R. Khattab, A.A. Stewartson, K. Prediction of three dimensional laminar and turbulent boundary layers on bodies of revolution at high angles of attack. 2nd Symposium on Turbulent Shear Flows, London, July 3-4 1979.

33 Lemmerman, L.A. Atta, E.H. A comparison of existing three dimensional boundary layer calculation methods. AIAA 80-0133 (1980).

34 Patankar, S.V. Spalding, D.B. A calculation procedure for heat mass and momentum transfer in three-dimensional parabolic flows. Int Jnl Heat and Mass Transfer, Vol 15, No 10, pp 1787-1806 (1972).

35 Baker, A.J. Finite element solution theory for three-dimensional boundary flows. Computer methods in applied mechanics and engineering, Vol 4, pp 367-386 (1974).

36 Formery, M. Delery, J. Finite difference method for inverse mode computation of a three-dimensional turbulent boundary layer. La Recherche Aérospatiale No 1981-5 English edition pp 11-21 (1981).

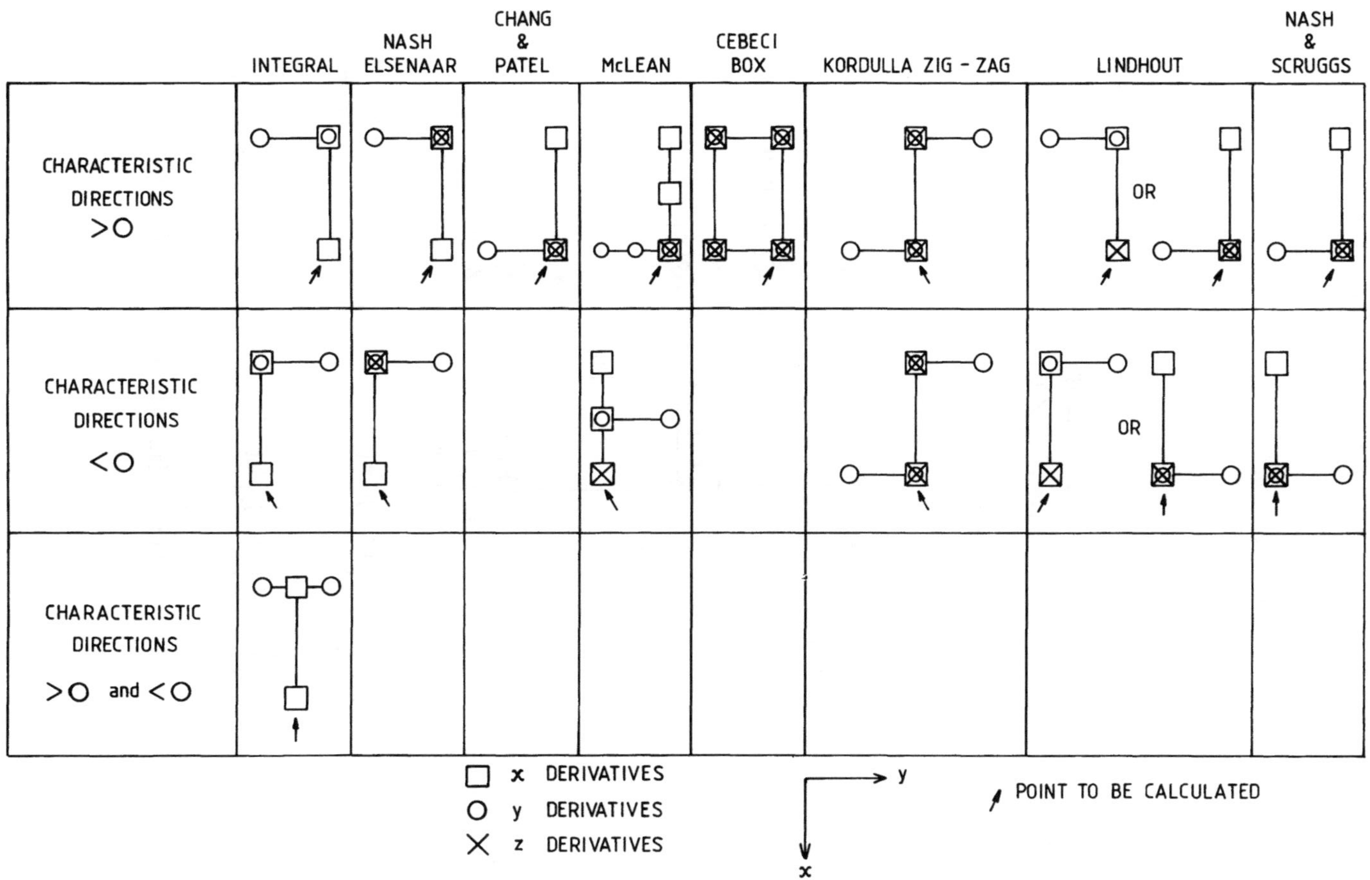

FIG.1 FINITE DIFFERENCE MOLECULES

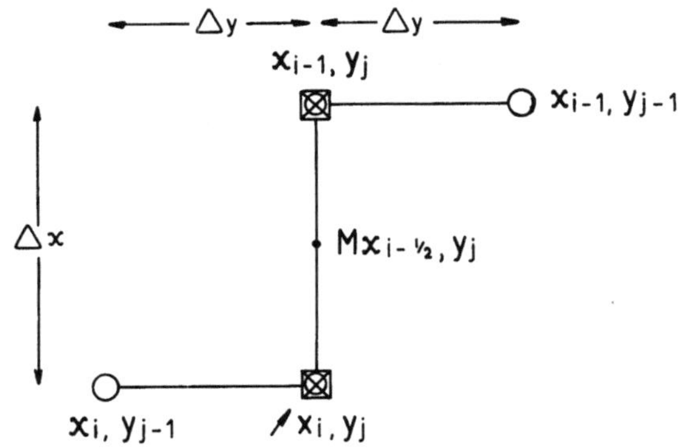

DIFFERENCE EQUATIONS CENTRED ON M

$$\left[\frac{\partial(\ )}{\partial x}\right]_M = \frac{1}{\Delta x}\left[(\ )_{i,j} - (\ )_{i-1,j}\right]$$

$$\left[\frac{\partial(\ )}{\partial y}\right]_M = \frac{1}{2\Delta y}\left[(\ )_{i,j} - (\ )_{i,j-1} + (\ )_{i-1,j+1} - (\ )_{i-1,j}\right]$$

FIG. 2 ZIG-ZAG DIFFERENCE SCHEME

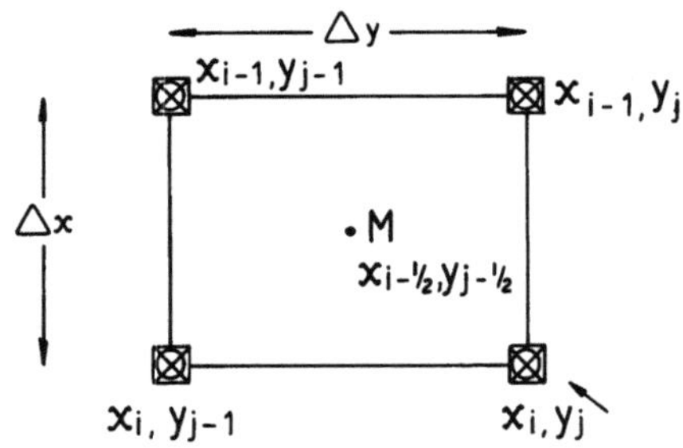

DIFFERENCE EQUATIONS CENTRED ON M

$$\left[\frac{\partial(\ )}{\partial x}\right]_M = \frac{1}{2\Delta x}\left[(\ )_{i,j} + (\ )_{i,j-1} - (\ )_{i-1,j} - (\ )_{i-1,j-1}\right]_{k-1/2}$$

$$\left[\frac{\partial(\ )}{\partial y}\right]_M = \frac{1}{2\Delta y}\left[(\ )_{i,j} + (\ )_{i-1,j} - (\ )_{i,j-1} - (\ )_{i-1,j-1}\right]_{k-1/2}$$

FIG.3 BOX DIFFERENCE SCHEME

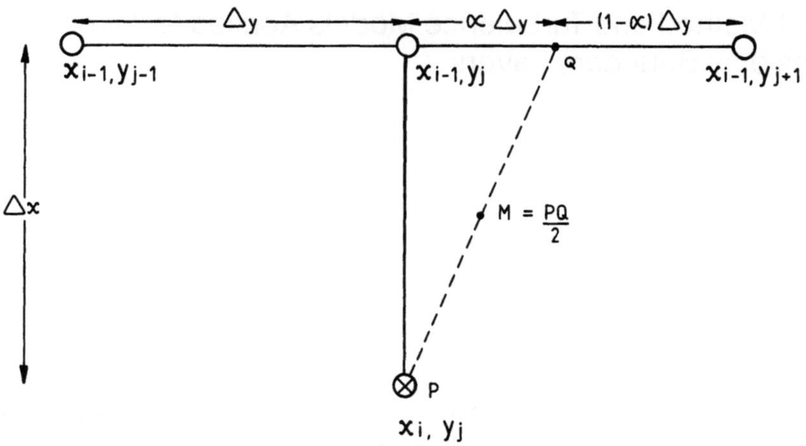

DIFFERENCE EQUATIONS CENTRED ON M

$$\left[\frac{u}{h_1}\frac{\partial u}{\partial x}+\frac{v}{h_2}\frac{\partial u}{\partial y}\right]_M=\left[\frac{u}{h_1}\frac{\partial u}{\partial \Gamma}\cdot\frac{d\Gamma}{dx}\right]_M=\frac{1}{2}\left(\frac{u_P}{h_{1P}}+\frac{u_Q}{h_{1Q}}\right)\cdot\left(\frac{u_P-u_Q}{\Delta x}\right)$$

$$u_Q=\frac{1}{2}\left[-\alpha\left(u_{i-1,j+1,k}+u_{i-1,j+1,k-1}\right)+(1-\alpha)\left(u_{i-1,j,k}+u_{i-1,j,k-1}\right)\right]$$

$$\alpha=(-\Delta x v h_1)/(\Delta y u h_2)$$

FIG.4 CHARACTERISTIC DIFFERENCE SCHEME

Integral Method and Turbulence Models Applied to Three-Dimensional Boundary Layers

J. COUSTEIX

ONERA/CERT
2 avenue Edouard Belin
31055 TOULOUSE Cedex (FRANCE)

SUMMARY

The paper summarizes work done at ONERA/CERT on three-dimensional turbulent boundary-layer calculations. Applications of transport equation models and of an integral method are presented. The use of isotropic eddy viscosity is discussed in connection with extra rates of strain caused by the deflection of the boundary-layer streamlines.

Finally, the results of a study of possible singularities in boundary-layer calculations are given. When separation is involved, this leads to the use of an inverse procedure which is described and applied to an experimental case.

NOTATIONS

U, W : streamwise and crosswise velocity components
C_f : skin friction coefficient
β_o : angle between the wall streamline and the external streamline
K_2, K_1 : geodesic curvatures of external streamlines and of their orthogonal
R_L : Reynolds number $R_L = \rho_e U_e L / \mu_e$

$$\delta_1 = \int_o^\delta \left(1-\frac{U}{U_e}\right)dy, \quad \theta_{11} = \int_o^\delta \frac{U}{U_e}\left(1-\frac{U}{U_e}\right)dy, \quad \delta_2 = \int_o^\delta -\frac{W}{U_e}dy, \quad \theta_{21} = \int_o^\delta -\frac{WU}{U_e U_e}dy, \quad \theta_{12} = \int_o^\delta \frac{W}{U_e}\left(1-\frac{U}{U_e}\right)dy, \quad \theta_{22} = \int_o^\delta -\frac{W^2}{U_e^2}dy$$

' : denotes a fluctuating quantity
e : refers to external flow

INTRODUCTION

In addition to the problems posed by the prediction of two-dimensional turbulent boundary-layers, the three-dimensional case brings a new problem which is the calculation of the direction of the shear stress. In two-dimensional flow, the concept of an eddy viscosity coefficient is often used and the most straightforward extension to the three-dimensional case consists of assuming that this coefficient is the same in any direction parallel to the wall : then, the eddy viscosity is said to be isotropic.

Unfortunately, such a hypothesis is at variance with experimental results. A solution has been proposed by ROTTA /1/ to take into account the effects of non-isotropic eddy viscosity. This will be examined by means of comparisons between calculations and experiments.

Another possible explanation of these effects lies in extra rates of strain caused by the deflection of the boundary-layer streamlines. An analysis is given in the case where the deflection is due to wall rotation.

Undoubtedly, the effects of non-isotropic eddy viscosity deserve to be considered but it is believed that in certain circumstances, a correct interpretation of the results can be obscured by singularities occuring in calculations. The results of an analysis of these difficulties will be recalled.

Finally, a solution for avoiding these problems will be presented : it consists of working in an inverse mode.

I. TRANSPORT EQUATION MODELS

From the concept of isotropic eddy viscosity, the shear stress components are calculated by :

(1) $$-\overline{u'v'} = \nu_t \frac{\partial U}{\partial y} \quad , \; -\overline{w'v'} = \nu_t \frac{\partial W}{\partial y}$$

The coefficient ν_t can be expressed from a mixing length hypothesis :

(2a) $$\nu_t = l^2 \left[\left(\frac{\partial U}{\partial y}\right)^2 + \left(\frac{\partial W}{\partial y}\right)^2 \right]^{1/2}$$

where l is a function of the distance from the wall :

(2b) $$\frac{l}{\delta} = \tanh\left(\frac{0.41}{0.085} \cdot \frac{y}{\delta}\right)$$

If a $k-\varepsilon$ two-equation model is used, ν_t is expressed by :

(3) $$\nu_t = C_\mu k^2/\varepsilon \qquad C_\mu = 0.09$$

An alternative to the eddy viscosity hypothesis consists of calculating the shear stress components from transport equations. If the boundary-layer hypotheses are applied to the Reynolds stress equations modelled by LAUNDER, REECE, RODI /2/, the following equations are obtained :

(4) $$\begin{aligned} \frac{D}{Dt}\overline{u'v'} &= C_{\tau_1} k \frac{\partial U}{\partial y} - C_1 \frac{\varepsilon}{k}\overline{u'v'} + \frac{\partial}{\partial y}\left(\frac{\nu_t}{\sigma_\tau}\frac{\partial}{\partial y}\overline{u'v'}\right) \\ \frac{D}{Dt}\overline{w'v'} &= C_{\tau_1} k \frac{\partial W}{\partial y} - C_1 \frac{\varepsilon}{k}\overline{w'v'} + \frac{\partial}{\partial y}\left(\frac{\nu_t}{\sigma_\tau}\frac{\partial}{\partial y}\overline{w'v'}\right) \end{aligned} \qquad C_{\tau_1} = \frac{2}{5} - \frac{\overline{v'^2}}{k}$$

In addition, if the transport equations for $\overline{v'^2}$ and k are simplified by neglecting the advection and diffusion terms, it can be shown that $\overline{v'^2}/k$ is a constant and the coefficient C_{τ_1} is a constant too.

System (4) is not a priori equivalent to eq. (1) but calculations we have performed /3/ have shown that the advection and diffusion terms of system (4) are often negligible. Then, the first and the second terms in the right hand side of system (4) balance. This means that the shear stress components are described by equations nearly equivalent to eqs (1). Even if the initial conditions on $\overline{u'v'}$ and $\overline{w'v'}$ are such that eqs (1) are not satisfied, a rapid decrease in the advection and diffusion terms is observed : thus for distances greater than a few boundary-layer thicknesses from the starting station, the shear stress components obey eqs (1) approximately.

Unfortunately, a number of experimental results do not confirm the hypothesis of isotropic eddy viscosity and therefore, system (4) is not sufficient to calculate the direction of the shear stress accurately.

The governing parameters of these problems are not yet very well known. An attempt at taking into account the non-isotropy of the eddy viscosity has been proposed by ROTTA /1/ and we have, ourselves, reanalysed the problem/3/. This is introduced in the modelling of the pressure-strain terms of the Reynolds stress equations through an extra parameter T . If the resulting equations are simplified by neglecting the advection and diffusion terms, algebraic expressions of the shear stress components are deduced .

(5) $$-\overline{u'v'} = \nu_t\left(\frac{a_{pxx}}{C_{\tau_1}}\frac{\partial U}{\partial y} + \frac{a_{pxz}}{C_{\tau_1}}\frac{\partial W}{\partial y}\right) \; , \; -\overline{w'v'} = \nu_t\left(\frac{a_{pzx}}{C_{\tau_1}}\frac{\partial U}{\partial y} + \frac{a_{pzz}}{C_{\tau_1}}\frac{\partial W}{\partial y}\right)$$

with $$a_{pxx} = C_{\tau_1}\frac{U^2 + TW^2}{U^2 + W^2} \; ; \; a_{pzz} = C_{\tau_1}\frac{TU^2 + W^2}{U^2 + W^2} \; ; a_{pxz} = a_{pzx} = C_{\tau_1}(1-T)\frac{UW}{U^2 + W^2}$$

Let us notice that if $T = 1$, the classic model of isotropic eddy viscosity is obtained.

The coefficient ν_t can be calculated from a $k-\varepsilon$ model by :

(6) $$\nu_t = C_\mu k^2/\varepsilon$$

or by using a mixing length model. In this case, the following expression is derived /1/ :

(7) $$\nu_t = l^2\left[\left(\frac{\partial U}{\partial y}\right)^2 + \left(\frac{\partial W}{\partial y}\right)^2 + (T-1)\left(W\frac{\partial U}{\partial y} - U\frac{\partial W}{\partial y}\right)^2 / (U^2 + W^2)\right]^{1/2}$$

and l is given by eq. (2b).

If the shear stress components are calculated from the algebraic relationships (5), the parameter T can be identified with the parameter N introduced by EAST /4/ :

(8) $$N = \tan(\gamma_\tau - \gamma)/\tan(\gamma_g - \gamma)$$

where γ , γ_g , γ_τ are the directions of $\vec{V}$, $\frac{\partial \vec{V}}{\partial y}$ and of the shear stress with respect to an arbitrary axis.(let us notice that the identification between T and N is not possible if the shear stress is calculated from the complete transport equations).

As ROTTA /1/ has himself noted, T is dependent on the frame of reference (except if $T = 1$). In two axis-systems moving in uniform translation one with respect to the other, the direction of the velocity $\vec{V}$ is generally not the same, whereas γ_g and γ_τ have the same values. Therefore, it is important to specify that N and T are defined in the axis system tied to the wall along which the boundary-layer develops. In addition, existing experimental data do not show any trend for T to be a universal constant.

Applications of transport equation models will be presented in section IV.

II - INTEGRAL METHOD

The basic method, described in Ref. /5/, is based on the solution of the integral equations of continuity and of momentum. In a streamline coordinate system for example, the equations read :

$$\frac{C_{fs}}{2} = \frac{\partial \theta_{11}}{\partial s} + \theta_{11}\left(\frac{H+2}{U_e}\frac{\partial U_e}{\partial s} - K_1\right) + \frac{\partial \theta_{12}}{\partial n} + K_1 \theta_{22} \tag{9}$$

$$\frac{C_{fs}}{2}\tan\beta_o = \frac{\partial \theta_{21}}{\partial s} + 2\theta_{21}\left(\frac{1}{U_e}\frac{\partial U_e}{\partial s} - K_1\right) + \frac{\partial \theta_{22}}{\partial n} + K_2\theta_{22} + K_2\theta_{11}(H+1)$$

$$\frac{\partial \delta}{\partial s} - \frac{V_e}{U_e} = \frac{1}{\rho_e U_e}\frac{\partial}{\partial s}[\rho_e U_e(\delta - \delta_1)] - K_1(\delta - \delta_1) - \frac{\partial \delta_2}{\partial n} \tag{10}$$

These three equations are a system for ten unknowns : C_{fs}, θ_{11}, H, β_o, θ_{12}, θ_{21}, θ_{22}, δ_2, δ, $\frac{\partial \delta}{\partial s} - \frac{V_e}{U_e}$. The necessary closure relationships have been obtained from self-similar solutions. The principle of these solutions is identical to the FALKNER-SKAN solutions in laminar flow. It is assumed that the streamwise velocity profile U/U_e, which is a priori a function of y, s, n is a function of a single variable $\eta = y/\delta$; a comparable hypothesis is used for the crosswise velocity profile W/U_e. Then, the partial differential boundary-layer equations become ordinary differential equations where the independent variable is η . To solve these similarity equations, a turbulence model is needed : a mixing length scheme has been used.

In addition, these solutions have been calculated with a small crossflow hypothesis. Then, the streamwise flow behaves like a two-dimensional flow ; from available experimental data, this approximation seems to be reasonable. Therefore, the streamwise velocity profiles depend on two parameters which can be chosen as the shape parameter H and the Reynolds number $R_{\theta_{11}}$. The crossflow depends on a third parameter : $\delta_2 / K_2 \delta^2$.

The closure relationships, which are, in fact, relationships between the unknowns, are deduced from the properties of the families of self-similar velocity profiles. Therefore, the closure relationships for the streamwise quantities depend on two parameters, whereas the relationships for the crosswise quantities depend on three parameters. In particular, the law for β_o (angle between the external streamline and the wall streamline) includes the effect of the third parameter by contrast with other laws often presented in the literature which do not contain the effect of such a parameter.

Let us notice that the numerical solution of global equations is available for any non-orthogonal coordinate system tied to the surface and is not restricted to a streamline coordinate system. The method has also been extended to compressible flows on adiabatic walls for Mach numbers less than about 4.

Moreover, the self-similar solutions have been revised so as to take into account the effect of the parameter T . For this, the modified mixing length formula (7) has been used in the self similar solutions. In this way, the closure relationships depend on the value of T /3/.

III. APPLICATIONS OF CALCULATION METHODS

The presented examples of application have been chosen to discuss the problem of non-isotropic eddy viscosity.

The first example is based on the experiments of VAN DEN BERG-ELSENAAR /6/ who have studied a boundary-layer developing on an "infinite" swept wing. This case involves the separation of the boundary-layer in the rear part of the studied zone. The second example is an experiment on a three-dimensional wake developing behind a swept wing /3/.

The calculations have been performed with a four-equation model $k-\varepsilon-\overline{u'v'}-\overline{w'v'}$ and the influence of the parameter T has been studied.

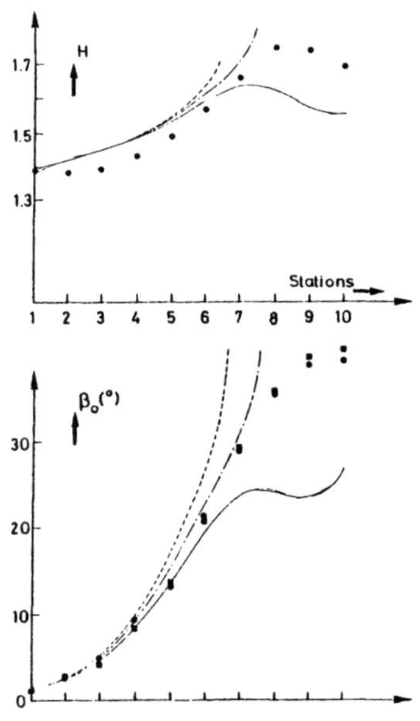

Fig. 1-Calculations of VAN DEN BERG-ELSENAAR's experiments - Exp. : ● ■ Calc. : $k-\varepsilon-\overline{u'v'}-\overline{w'v'}$ *model* ——T=1 —·—T=0.7 - - - - T=0.5

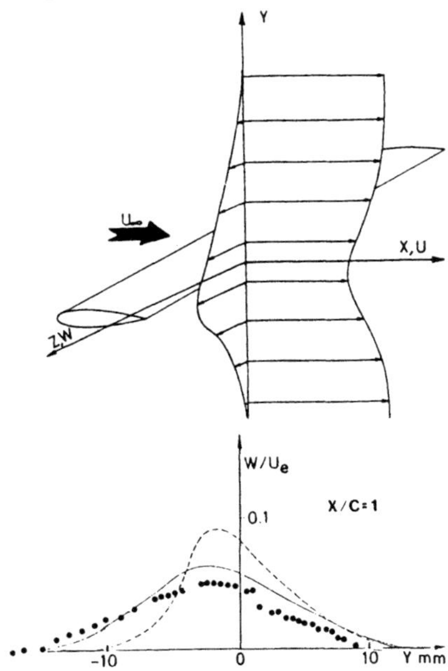

Fig. 2-Crossflow velocity profile in a wake behind a swept wing /3/ Exp. : ● ; Calc. : $k-\varepsilon-\overline{u'v'}-\overline{w'v'}$ *model* ——T=1 - - - - T= 0.5

Fig. 1 shows comparisons with VAN DEN BERG-ELSENAAR's data. The evolutions of the streamwise shape factor and of the angle β_o between the wall streamline and the external streamline are plotted along the line of measurement stations. Fig. 2 shows comparisons with the wake data. The crosswise velocity is plotted as a function of a distance y measured along a normal to the mean surface of the wake. The data are obtained at a station located one chord length downstream of the trailing edge.

In the case of the VAN DEN BERG-ELSENAAR's experiments, the effect of T is important on the crosswise flow as well as on the streamwise flow. The value $T = 1$ does not indicate separation of the boundary-layer. Smaller values of T are necessary to predict a separation but it is impossible to obtain calculated results which fit the experimental evolutions : at separation the slope of β_o becomes very large, which is not true in experiments. Let us notice that similar calculations have been performed with the integral method and the results are roughly the same as those obtained with the four-equation model.

In the case of the wake experiments, the effects of T are significant mainly on the crossflow evolution. In this case, the secondary velocities are rather small and the streamwise flow is not very much affected. A value of $T = 1$ gives approximately the right decay of the crossflow. By contrast, smaller values of T give a too slow decay and a value of T slightly greater than 1 would give even better results.

From these comparisons, it results that the influence of T is important on the evolution of the crossflow and can be significant for the streamwise flow if the secondary velocities are large enough. It is also concluded that a universal value of T is not a good solution for taking into account correctly the effects of non-isotropic eddy viscosity. In fact, it seems more logical to keep $T = 1$ and to seek another approach for calculating the direction of the Reynolds stress.

IV. EXTRA RATES OF STRAIN : DEFLECTION OF BOUNDARY-LAYER STREAMLINES

A possible solution could be to retain in the Reynolds stress equations not only the terms in $\frac{\partial U}{\partial y}$ and $\frac{\partial W}{\partial y}$ but also terms with other velocity derivatives reflecting the deflection of the boundary-layer streamlines. Such an idea is not at variance with experimental data because it seems that the more curved are the boundary-layer streamlines, the more different are the directions between the shear stress $\vec{\tau}\ (-\rho\overline{u'v'}, -\rho\overline{w'v'})$ and the velocity derivative $\frac{\partial \vec{V}}{\partial y}\left(\frac{\partial U}{\partial y}, \frac{\partial W}{\partial y}\right)$.

The curvature of the boundary-layer streamlines is calculated from the derivatives $\frac{\partial W}{\partial x}, \frac{\partial W}{\partial z}, \frac{\partial U}{\partial x}, \frac{\partial U}{\partial z}$. Therefore, all these terms should be retained in the Reynolds stress equations. In addition from the continuity equations, $\frac{\partial V}{\partial y}$ is of the same order of magnitude. If the Reynolds stress equations are written with all these terms, it is not easy to obtain an explicit algebraic expression of the Reynolds stresses and a conclusion does not emerge simply.

For these reasons, and to support our assumption that the deflection of boundary-layer streamlines can be an important parameter, another problem has been studied in which the three-dimensionality is generated by a wall rotation.

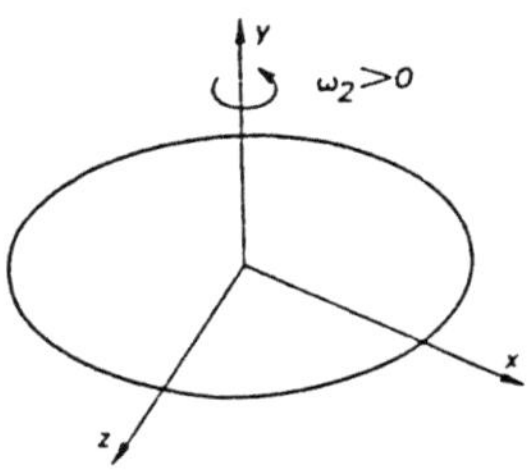

Fig. 3 - Boundary-layer on a revolving disk

The Reynolds stress equations modelled by LAUNDER, REECE, RODI /2/ have been written in a rotating system and the particular flow on a disk revolving around an axis perpendicular to the wall (Fig. 3) has been studied /7/. In this case, only the component ω_2 of the rotation vector is non zero.

This particular flow is studied by keeping in the Reynolds stress equations the terms in $\frac{\partial U}{\partial y}$ and $\frac{\partial W}{\partial y}$, the other velocity derivatives being assumed to be small. The terms containing ω_2 are also kept and they play a role of extra rates of strain. In addition, if the advection and diffusion terms of equations (10) are neglected, an algebraic system is obtained for the six components of the Reynolds stress tensor.

In the solution of this system, a dimensionless rotation parameter R_2 appears. R_2 is defined as :

$$R_2 = -\omega_2 / [(\frac{\partial U}{\partial y})^2 + (\frac{\partial W}{\partial y})^2]^{1/2} \tag{11}$$

If R_2 is assumed to be small compared with 1 and if the algebra is carried out up to the first order in R_2, the Reynolds stresses are expressed with simple formulae :

$$-\overline{u'v'} = C_\mu \frac{k^2}{\varepsilon}(\frac{\partial U}{\partial y} + \lambda R_2 \frac{\partial W}{\partial y}); -\overline{w'v'} = C_\mu \frac{k^2}{\varepsilon}(\frac{\partial W}{\partial y} - \lambda R_2 \frac{\partial U}{\partial y}); \frac{\tau}{k} = [(\frac{\overline{u'v'}}{k})^2 + (\frac{\overline{w'v'}}{k})^2]^{1/2} = C_\mu^{1/2} \tag{12}$$

where all the quantities are expressed in the rotating axis system. With the classic values of the model constants, λ equals 1.892.

Expressions (12) of the Reynolds stresses are to be compared with the hypothesis in which the influence of rotation is ignored. This hypothesis leads to :

(13) $$-\overline{u'v'} = C_\mu \frac{k^2}{\varepsilon}\frac{\partial U}{\partial y} \quad ; \quad -\overline{w'v'} = C_\mu \frac{k^2}{\varepsilon}\frac{\partial W}{\partial y}$$

Thus, it will be noticed that the modulus of the turbulent shear stress τ is not affected by rotation effects (up to the first order in R_2). In the same way, the ratio τ / k is not modified either. By contrast, the rotation introduces a difference of orientation between $\vec{\tau}\,(-\rho\overline{u'v'}, -\rho\overline{w'v'})$ and $\frac{\partial \vec{V}}{\partial y}\,(\frac{\partial U}{\partial y}, \frac{\partial W}{\partial y})$. Let θ be the angle between $\vec{\tau}$ and $\frac{\partial \vec{V}}{\partial y}$. From eqs(12), θ is given by :

(14) $$\tan\theta = -\lambda R_2$$

Therefore, it appears that vectors $\vec{\tau}$ and $\frac{\partial \vec{V}}{\partial y}$ have increasingly different orientations as the rotation parameter R_2 becomes larger.

If now the case of a three-dimensional flow on a fixed wall is considered, an analogy between the effects of deflection of boundary-layer streamlines and the effects of rotation can be imagined. This would lead us to express the difference of orientation between $\vec{\tau}$ and $\frac{\partial \vec{V}}{\partial y}$ as a function of a curvature parameter similar to the rotation parameter R_2. Therefore, this curvature parameter would be the relevant parameter governing the isotropy or non-isotropy of eddy viscosity.

Finally, it should be noticed that the above analysis of rotation effects remains only partial since, to these effects, must generally be added the effects of curvature of the boundary-layer streamlines (these streamlines being defined in the moving axis system).

V. SINGULARITIES IN BOUNDARY-LAYER CALCULATIONS - INVERSE MODE

In addition to the problems related to the accurate calculation of the direction of the shear stress, other difficulties can occur when a boundary-layer is calculated with a prescribed pressure distribution. In the case of a two-dimensional laminar flow, this is now a classic problem : the boundary-layer equations become singular when the skin-friction vanishes. Moreover, solutions for avoiding the singularity are known : one of them consists of working in the so-called inverse mode in which a boundary-layer quantity becomes a prescribed datum of the equations and the pressure becomes an unknown.

In the three-dimensional case, an approach to these problems has been proposed by analysing the properties of a system of integral equations /8/. The analysis will not be repeated here and the main results only will be recalled.

The system of integral equations consists of the global continuity and momentum equations closed as described in section II. This system is a first-order partial-differential system for three main unknowns which can be chosen as δ_1, θ_{11} and θ_{21}. With a few minor assumptions on the closure relationships, it has been shown that the characteristic equation of the system has three roots. If the angle between one of the characteristic lines and the external streamline is defined by $\tan^{-1}(1/\lambda)$, the three roots are :

$$(15) \qquad -\lambda_1\frac{\theta_{21}}{\theta_{11}}=\frac{1-\alpha_1 H}{H-1}\;;\; -\lambda_2\frac{\theta_{21}}{\theta_{11}}=\frac{1}{H-1}\;;\; -\lambda_3\frac{\theta_{21}}{\theta_{11}}=\frac{(2\alpha+\alpha_1)H+1}{H-1}$$

with

$$\alpha_1=-\alpha+(\alpha^2+\alpha)^{1/2} \qquad \alpha=0.631 \qquad \alpha_1=0.383$$

Therefore, it appears that the three roots are always real and distinct. The system is hyperbolic.

In addition, by comparing the expression of λ_1 with the closure relationships for β_o (β_o = angle between the external streamline and the wall streamline), it has been shown that the λ_1 -characteristic line is approximately a wall streamline.

In the range $H>1$, λ_2 and λ_3 have always the same sign whereas λ_1 equals zero when $H=H_c=1/\alpha_1=2.6$. In two-dimensional flow, the point $H=H_c$ is singular : the derivative of the displacement thickness $\frac{d\delta_1}{dx}$ becomes infinite but this is not the case in three-dimensional flow. In this latter case, the passage through a point $H=H_c$ implies only the incipient occurence of an influence from downstream to upstream because the wall streamline becomes orthogonal to the external streamline.

Generally, in the three-dimensional case, the singularities are not local. They occur as a focusing of characteristic lines belonging to the same family. More precisely, they are the lines defined by λ_1 , i.e. the wall streamlines which are likely to focus. Moreover, these lines of focusing are lines of discontinuity for some boundary-layer quantities.

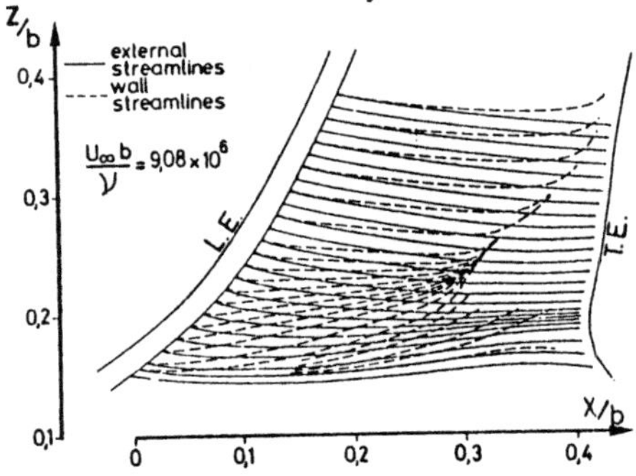

Fig. 4 - Example of calculation with singularities - Boundary-layer on the root section of a wing /9/ (L.E.: leading edge - T.E.: trailing edge)

An example of this phenomenon is given in Fig. 4. The basis of calculations is an experimental study by LINDHOUT et al. which has been used as a test case for a workshop on three-dimensional turbulent boundary-layers /9/.

Such lines of focusing must not be confused with separation lines, because separation lines must have properties of regularity /10/. In fact, the singularities originate from the use of boundary-layer equations with a prescribed pressure distribution. In addition, they are associated with numerical difficulties which can lead to misinterpretations of the results as is the case in two-dimensional flow. Finally, it should be noticed that the singularity occuring in the two-dimensional case appears as a degenerate form of the three-dimensional singularities.

These singularities can be avoided if the boundary-layer equations are solved in the inverse mode. A great variety of inverse methods can be invented according to the choice of the prescribed functions. The method which has been implemented is an integral method in which the closure relationships are those described in section II (with $T=1$) ; the distributions $\delta_1(x,z)$ and $\delta_2(x,z)$ are prescribed whereas the modulus and the direction of the external velocity become unknowns. This method has been applied to VAN DEN BERG-ELSENAAR's experiments and the distributions $\delta_1(x,z)$ and $\delta_2(x,z)$ have been obtained from experimental data.

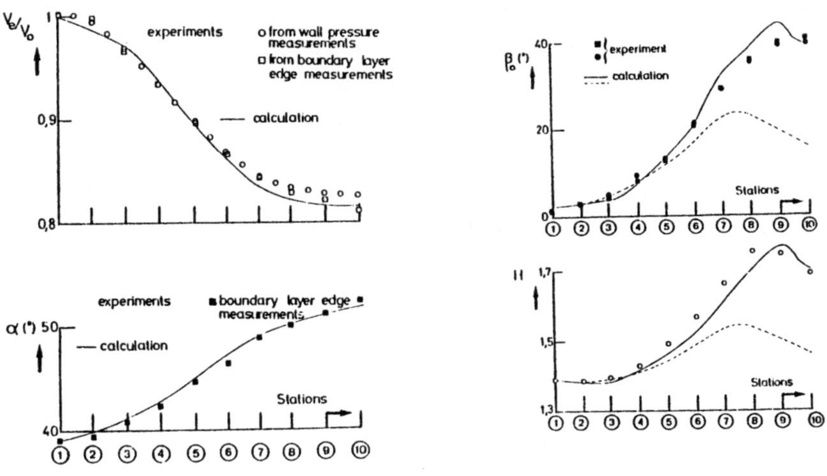

Fig. 5 - Calculation of VAN DEN BERG-ELSENAAR's experiments with an integral method —— inverse mode ---- direct mode in which the external velocity distribution is prescribed from the results of the inverse mode calculations.

Comparisons between experiments and calculations are shown in Fig. 5. A fairly good agreement is observed even near the separation line located between stations 8 and 9. In the case of an infinite swept wing, the separation line is observed when the wall streamlines become parallel to the leading edge. If α is the direction of the external streamline with respect to a line normal to the leading edge, this condition will be $\alpha + \beta_o = 90$ deg. It should be noted that a singularity would occur in a calculation in the direct mode if this condition were fulfilled. In the inverse mode, no singularity is observed.

Finally, it seems that the results are better than those obtained in the direct mode. However, it must be noted that the calculated external velocity distribution is not exactly the same as the experimental distribution. Therefore, a calculation in the direct mode has been carried out by prescribing the external velocity distribution calculated from the inverse mode. Surprisingly, the calculation in the direct mode does not indicate separation (Fig.5). It is believed that the reason is that the singularities of the direct mode imply numerical difficulties. Therefore, it is concluded that any attempt at predicting separation in the direct mode can be very misleading.

CONCLUSIONS

An accurate determination of the direction of the shear stresses is needed to calculate correctly the evolution of crossflow in three-dimensional turbulent boundary-layers. If the crossflows are strong enough, the influence on the streamwise flows can be significant.

The evaluation of effects of non-isotropic eddy viscosity through the parameter T is not a good solution as T is not a universal constant. Another approach could consist of keeping in the Reynolds stress equations extra rates of strain reflecting the deflection of the boundary-layer streamlines.

The analysis of the properties of boundary-layer equations has shown that singularities can occur as focusing of wall streamlines. Numerical difficulties are certainly associated with these singularities and can dim the discussion on turbulence modelling. To avoid these problems, a solution consists of working in the inverse mode.

REFERENCES

/1/ ROTTA J.C. : "A family of turbulence models for three-dimensional thin shear layers" - First Symp. on Turbulent Shear Flows,1977

/2/ LAUNDER B.E., REECE G.J., RODI W. : "Progress in the development of a Reynolds-stress turbulence closure" - J. of Fl. Mech., Vol. 68, Part 3, pp. 537-566

/3/ COUSTEIX J., AUPOIX B., PAILHAS G.: "Synthèse de résultats théoriques et expérimentaux sur les couches limites et sillages turbulents tridimensionnels" - ONERA NT 1980-4

/4/ EAST L.F. : "Measurements of the three-dimensional incompressible turbulent boundary-layer induced on the surface of a slender delta-wing by the leading edge vortex" - ARC RM 3768, 1973

/5/ COUSTEIX J. : "Analyse théorique et moyens de prévision de la couche limite turbulente tridimensionnelle" - ONERA Publication n° 157 (English translation ESA TT 238)

/6/ VAN DEN BERG B., ELSENAAR A. : "Measurements in a three-dimensional incompressible turbulent boundary-layer in an adverse pressure gradient under infinite swept wing conditions" - NLR TR 720920

/7/ COUSTEIX J., AUPOIX B. : "Reynolds stress equations modelling in a rotating reference system" - La Recherche Aérospatiale - English version n° 1981-4

/8/ COUSTEIX J., HOUDEVILLE R. : "Singularities in three-dimensional turbulent boundary-layer calculations and separation phenomena" - AIAA Journal Vol. 19 n° 8 Aug. 1981

/9/ LINDHOUT J.P.F., VAN DEN BERG B., ELSENAAR A. : "Comparison of boundary-layer calculations for the root section of a wing" - The September 1979 AMSTERDAM Workshop test case - NLRMP 800280

/10/LEGENDRE R. : "Lignes de courant d'un écoulement permanent - décollement et séparation" - La Recherche Aérospatiale n° 1977-6 - English translation ESA TT 476

A Method for Three-Dimensional Boundary Layer Calculations on Arbitrary Bodies: Some Results on Aircraft Wings and Engine Cowls [1)]

J.P.F. LINDHOUT, A.C. DE BRUIN, J.H.M. GOODEN

Summary

The computation is described of three-dimensional boundary layers in non-orthogonal coordinate systems. The turbulence model is a simple eddy-viscosity formulation. The boundary layer is computed in the region of determinacy of the initial data; no extra information is required along lateral boundaries. An hybrid difference scheme is used. Applications are shown of the computation of the boundary layers over engine cowls and over airplane wings.

Introduction

Since the sixties there is a contineously growing interest in the development of computational methods for 3D boundary layers. At this moment a number of computational methods have reached such a matureness they can be used for routine calculations of attached flow on wings, cowls and fuselages. The NLR has contributed in this process by developing two field method [1, 2, 3]. With the earlier method it is possible to compute practical cases, though with considerable human and computer effort. In the latest method, which will be the subject of this paper, lessons are implemented learned by the application of the first method.

Comparing field methods with integral methods one often attributed to field methods lack of robustness and long run times. In the present method the first problem is overcome by applying carefully design procedures based on informatics principle [4]. The Euro Visc Amsterdam-Workshop [5] demonstrated that the run times required by this method are of the same order of magnitude as of the more advanced integral methods.

1) This investigation has been performed under contract with the Netherlands Agency for Aerospace Programs (NIVR).

The main features of the present method are given below:

- An eddy viscosity turbulence model is implemented. Although there exist more advanced turbulence models it is difficult to do better than simple eddy viscosity models [6].
- Flexibility is attained by describing the governing equations in a general coordinate system. The geometry coefficients are evaluated in separate subroutines, attached to the program for each new configuration.
- The computations are performed in a region in which the results are completely determined by the initial data.
  No redundant information has to be given on lateral boundaries with outflow conditions, neither data are found by extrapolations on inflow boundaries.
- A linear combination of first and second order difference schemes is used to approximate tangential derivatives for internal as well as for boundary points.
- The program rejects points which are near a breakdown, e.g. close to separation. The computation is then continued with the remaining 'healthy' points.

Applications for boundary layers over engine cowls will be shown. The influence of the initial data on the computed flow pattern will be discussed.

Also examples of boundary layers over aircraft wings will be dealt with. It will be reasoned that it is possible to compute the flow, from the given pressure distribution, with a minimum of initial data, prescribed at the wing-body junction near the leading edge.

Basic equations

The boundary layer equations are solved in a general coordinate system (x_1, x_2, Z); the basic equations are:

Continuity:

$$\partial(m\rho u_1/h_1)/\partial x_1+\partial(m\rho u_2/h_2)/\partial x_2+m\partial(\rho w)/\partial Z=0 \tag{1}$$

Momentum in x_α-direction (α=1,2):

$$u_1/h_1.\partial u_\alpha/\partial x_1+u_2/h_2.\partial u_\alpha/\partial x_2+w.\partial u_\alpha/\partial z+R^\alpha_{11}u_1^2+R^\alpha_{12}u_1u_2+R^\alpha_{22}u_2^2=$$

$$\tfrac{1}{2}\rho.h_1h_2/m^2.\{-h_\beta\partial C_p/\partial x_\alpha+a_{12}/h_\beta\partial C_p/\partial x_\beta\},\quad(\beta=3-\alpha) \tag{2}$$

The coordinate z is normal to the surface of the body. (x_1,x_2) is a non-orthogonal coordinate system generating parameter lines over the surface. (u_1,u_2,w) is the velocity vector, ρ is the density and C_p is the given pressure. If $\underline{X}$ is the position vector in a Cartesian system then are $\underline{B}_\alpha=\partial\underline{X}/\partial x_\alpha$ the tangent vectors to the parameter lines. The metric coefficients are defined as their inner products:

$$a_{\alpha\beta}=\underline{B}_\alpha\cdot\underline{B}_\beta\quad(\alpha,\beta=1,2) \tag{3}$$

The geometrical quantities $h_\alpha, R^\alpha_{\beta\gamma}, m$ are functions of the metric coefficients and their derivatives. To apply the method in a specific coordinate system the user has to implement in a subroutine the relations:

$$\underline{B}_\alpha=\underline{f}_\alpha(x_1,x_2);\quad \partial_\beta\underline{B}_\alpha=\underline{g}_{\alpha\beta}(x_1,x_2) \tag{4}$$

It is assumed that the density variation through the boundary layer can be described by the Crocco-relation. For turbulent boundary layers an isotropic eddy viscosity model is assumed:

$$\tau_\alpha=[\mu+\rho F^2 l^2|\partial u/\partial z|]\quad(\partial u_\alpha/\partial z), \tag{5}$$

F is a van Driest damping function and l is the mixing length [3].

Similarity type transformations are used by scaling the flowquantities with appropriate values at the edge of the boundary layer. It can overcome the effect of drastic changes of the velocity profiles in streamwise direction due to pressure gradients.

The normal coordinate z is scaled by δ, the value of z at which the edge of the boundary layer is placed. δ is chosen to be proportional with an integral thickness of the boundary layer [3]. Defining δ in terms of the velocity defect in the outer region would make the scaling too sensible for details of the solution. To stretch the wall region for turbulent flow the following transformation is used:

$$\zeta=z.(a+b.z/\delta)/(1+c.z/\delta) \tag{6}$$

A smooth transition from laminar to turbulent flow in the computational domain is however not possible with this transformation. Work is in progress on this subject.

The boundary layer equations (1) and (2) are solved in the region of determinacy of the initial data from the given pressure distribution. The external flow is determined simultaneously. This facilitates the computation of boundary layers from experiments of which not the outer flow angle is measured but only the pressure distribution. In principle a Cauchy problem is solved for the outer flow angle. Because the equations to be solved are non-linear the characteristics, i.e. the streamlines, may intersect each other. In this case the solution is, of course, unphysical, indicating that the initial conditions for the outer flow angle do not fit to the given pressure distribution. A similar problem is envisaged by Slooff in reference 7 for the solution of the inverse flow problem.

Numerical solution

The method computes the flow successively in surfaces normal to the body along coordinate lines x_1=constant. A mesh is chosen of which $\Delta x_1, \Delta x_2, \Delta\zeta$ are stepsizes along the (x_1, x_2, ζ) coordinate lines.

For the approximation of the normal derivatives second order standard central difference formulations are employed [3]. For the approximation of the tangential derivatives an hybrid scheme is used which is a combination of a first order and a second order scheme. For example the first order scheme for the approximation of $\partial u/\partial x$ reads as:

$$\left(\frac{\partial u}{\partial x}\right)_i \sim \frac{u_i - u_{i-1}}{\Delta x} + a.\Delta x + b(\Delta x)^2 + \ldots.. \qquad (7)$$

The second and third term are the leading two truncation terms. This scheme is strongly stable but suffers from a rather large truncation error. A smaller error has the following second order scheme:

$$\left(\frac{\partial u}{\partial x}\right)_i \sim -\left(\frac{\partial u}{\partial x}\right)_{i-1} + 2\,\frac{u_i - u_{i-1}}{\Delta x} - b(\Delta x)^2 + \ldots.. \qquad (8)$$

This scheme however is weakly stable, because of the use of derivatives at 'old' levels.

Therefore we use a linear combination of schemes (7) and (8):

$$\left(\frac{\partial u}{\partial x}\right)_i \sim - D \left(\frac{\partial u}{\partial x}\right)_{i-1} + (1+D) \frac{u_i - u_{i-1}}{\Delta x} + (1-D)a \Delta x + (1-2D)b(\Delta x)^2 + .. (9)$$

For D<1 this scheme is strongly stable. Formally it has a first order truncation error. However for a practical computation the actual error is important. This error can be made small by chosing D sufficient close to 1, e.g. D=0.9. As an example the infinite swept wing case of reference 9 is computed with two different stepsizes: $\Delta x_1=0.005$ and $\Delta x_1=0.040$. Despite the large difference in stepsizes the results differed less than 1%.

The hybrid approximations of the derivatives are used to form explicit and implicit difference molucules [3]. For the explicit moleculesthe Δx_1 step is limited by the CFL criterion. These stability restrictions can be relieved by applying as much as possible implicit molecules. A computational sequence favoring these schemes is implemented.

Engine cowls

The present example concerns the boundary layer over the outer surface of engine cowls under real flight conditions. The aim is to determine the viscous resistance and to establish the existence of separations. A sketch of the coordinate system is given in figure 1. The position vector of a point on the surface is given by:

$$\underline{X}=[X^1(x_1),r(x_1,x_2)\sin(x_2),r(x_1,x_2)\cos x_2]^T \quad (10)$$

X^1 is a function of x_1, which is chosen as the pathlength along the top of the cowl, x_2 is the angle swept from X^3-axis. The pressure distribution is computed with the NLR panel method. In experiments the boundary layer is commonly tripped just downstream of the highlight. Turbulent boundary layer calculations were started there, using Coles' velocity profiles as initial data. Several runs were done varying the initial conditions. The results are rather indifferent for them, except for the flow angles. Figure 2 shows two sets of wall streamlines. One for an initial flow angle distribution taken from the potential flow, the other set zero. For the last case the boundary layer was found rather proximate to separation at the lower side. The momentum thickness at the trailing edge, a measure for the viscous resistance, is given in figure 3.

The cowl was set also under an angle of attack of 15°. Due to the high angle of attack the pressure gradient is mostly favorable at the lower side, while the pressure gradient at the upper side is adverse. The results (fig. 4, 5) show a strong convergence of outer streamlines to the upper side of the cowl, which effect is magnified in the wall streamlines due to the low skin friction.

Because of the favorable pressure gradients it is questionable whether the boundary layer will be completely turbulent from the highlight as was originally assumed. Computations were repeated assuming laminar flow. Initial conditions were generated from Falkner-Skan equations.
Figure 6 shows computed lines of constant Re_θ. At the upper side only a small region of laminar flow can exist: almost immediately the laminar flow separates and will become turbulent. The computation proceeds with the remaining stations still far from separation. For constant pressure laminar flow can exist for Re_θ up to 1000; in this case with favorable pressure gradients, laminar flow is likely to exist for Re_θ larger than 1000. Hence one can infer that the boundary layer may well be laminar for a substantial part on the lower side.

<u>Wings with leading edge region</u>

The boundary layer method is incorporated in the XFLO22 NLR program [8], which computes the transonic flow around wings taking into account the effect of the fuselage. The chord/span,=(x_1,x_2), coordinate system projected on the wing plan form is adopted in it:

$$\underline{X}=[l(x_2)+c(x_2).x_1,x_2,0]^T, \qquad (11)$$

l is the distance of the leading edge from the X^2-axis and c is the local chord. Initial conditions for the boundary layer quantities are estimated at 5% chord from very simple quadrature formulae. Standard meshes of $n_{span} x n_{chord} x n_{normal}$=20x40x21 points are employed. Computer effort for the boundary layer only required 176 MFLOP's (=88 seconds on a Cyber 175) for a complete wing, which is a fraction of the total computer effort.

It is desirable however to compute the 3D boundary layer flow in the leading edge region to provide the computations on lower and upper side with more accurate initial conditions. The inviscid and wall streamlines through the leading edge junction of body and wing cover almost the complete wing. Therefore the flow in the covered region can be computed from initial data given at that very position. In figure 7 a sketch is given of such a situation.

To describe the coordinate system (11) is used as a starting point: x_1 is replaced by a function $g(x_1)$; g is the chord-percentage and x_1 is a new variable stretching the nose region. It also discriminates between the upper and lower surface. The coordinate system is given as:

$$\underline{X}=[l^1+c.(g\cos\alpha+t\sin\alpha),x_2,l^3+c(t\cos\alpha-g\sin\alpha)]^T \qquad (12)$$

l^i, c, α are functions of the span, x_2, only; l^i are projections of the leading edge position on the X^i-axis, c is the local chord, α is the angle between the chord and X^1-axis; $t=t(x_1,x_2)$ gives the local profile shape.

The calculations are done in three stages: the computation marches from the l.e. wing-body junction in spanwise direction between two lines of constant chord. Then the program continues with the upper and lower surfaces employing the data at the chord lines as initial conditions.

This procedure is tested for an infinite swept wing geometry with lifting Karman-Trefftz wing sections. Initial conditions for turbulent flow were estimated with formulae of Cumpsty and Head [10] and Poll [11]. The flow angle was taken from the analytical solution of the outer flow.

In figure 8 results are given for θ_{ss} versus span for two initial values; within five steps the infinite swept wing solution is reached. Again the solution is found rather insensitive for initial values of c_f and θ_{ss}. It was also confirmed that the initial flow angles are much more critical. In figure 9 θ_{ss} and c_f are shown versus the pathlength along a wing section. The position is far enough from the starting line to be converged to the infinite swept wing solution. Results are computed for different widthes of the leading edge region. Only small differences can be observed. The changeover to the marching direction, parallel to the main flow, can be taken smoothly in both cases.

Conclusions

Elements of the NLR field method for boundary layer computations are described. Applications of the method on engine cowls and complete wings are shown. Work has still to be done to perform leading edge computations on a routine base.

References

1. P. Wesseling, J.P.F. Lindhout, A calculation method for three-dimensional incompressible turbulent boundary layers, AGARD Conference Proceedings, vol.93, 1971.

2. J.P.F. Lindhout, An ALGOL-program for the calculation of incompressible three-dimensional turbulent boundary layers, NLR TR 74159 U, 1974.

3. J.P.F. Lindhout, G. Moek, E. de Boer, B. van den Berg, A method for the calculation of 3D boundary layers on practical wing configurations, Joint ASME/CSME Applied Mechanics, Fluids Engineering and Bioengineering Conference, Niagara Falls, 1979.

4. J.P.F. Lindhout, B. van den Berg, Design of a calculation method for 3D turbulent boundary layers,third GAMM-Conference on Numerical Methods in Fluid Mechanics, Oct. 10-12, Cologne, 1979.

5. J.P.F. Lindhout, B. van den Berg, A. Elsenaar, Comparison of boundary layer calculations for the root section of a wing. The September 1979 Amsterdam workshop test case, NLR MP 80028 U, 1981.

6. Preprint of the 1980-81 AFOSR-HTTM-Stanford Conference on complex turbulent flows, 1981.

7. T.L. Holst, J.W. Slooff, H. Yoshihara, W.F. Ballhaus, Applied Computational Transonic Aerodynamics, AGARDograph to be published.

8. J.W. Slooff, J.T. van der Kolk, J. van der Vooren, XFLO22 NLR; A system for the numerical simulation of sub- and transonic, viscous attached flows about wing-body configurations, submitted to the AIAA/ASME Joint Fluids, Plasma, Thermophysics and Heat Transfer Conference, June 7-11, 1982, St. Louis, Missouri, USA.

9. B. van den Berg, A. Elsenaar, Measurements in a three-dimensional incompressible turbulent boundary layer in an adverse pressure gradient under infinite swept wing conditions, NLR TR 72092 U, 1972.

10. N.A. Cumpsty, M.R. Head, The calculation of three-dimensional turbulent boundary layers, The Aeronautical Quarterly, Sept. 1966.

11. D.I.A. Poll, Skin friction and heat transfer at an infinite swept attachment line, The Aeronautical Quarterly, Nov. 1981.

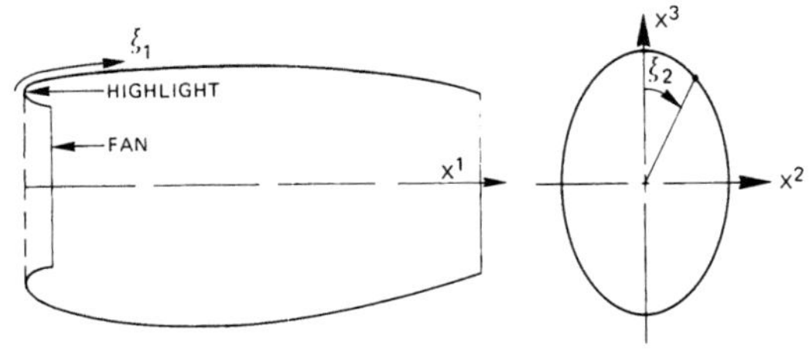

Fig.1. Engine cowl, coordinate system

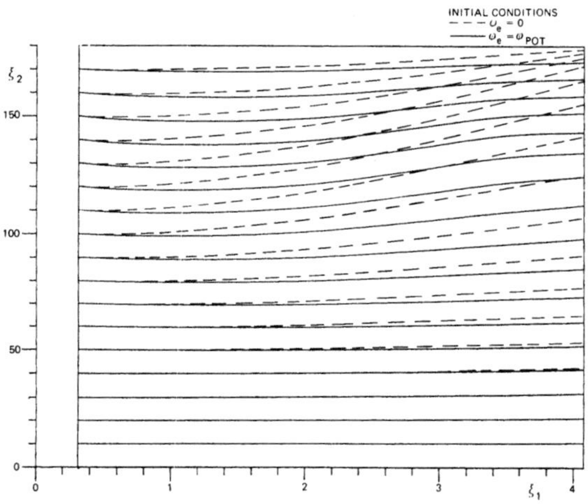

Fig.2. Engine cowl, wall streamlines for two distibutions of initial conditions for the flow angle.

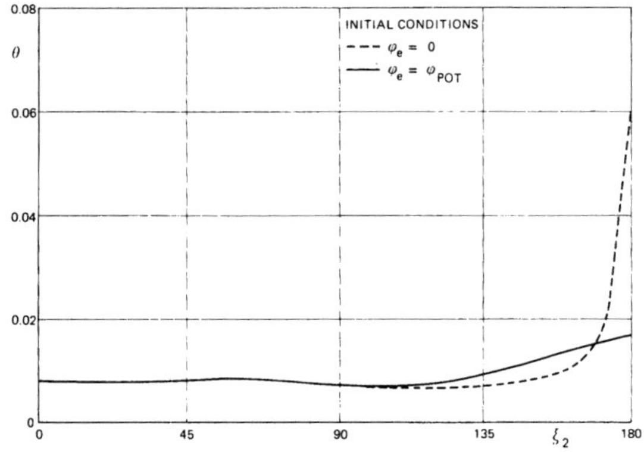

Fig.3. Engine cowl, momentum thickness at the trailing edge for two distributions of initial conditions for the flow angle.

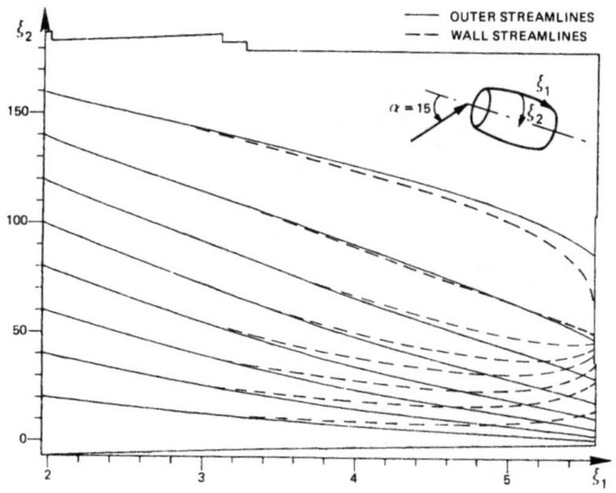

Fig.4. Engine cowl under angle of attack, outer streamlines and wall streamlines.

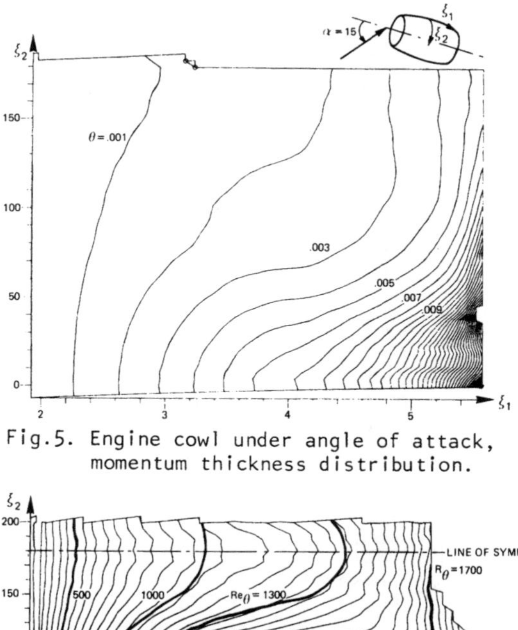

Fig.5. Engine cowl under angle of attack, momentum thickness distribution.

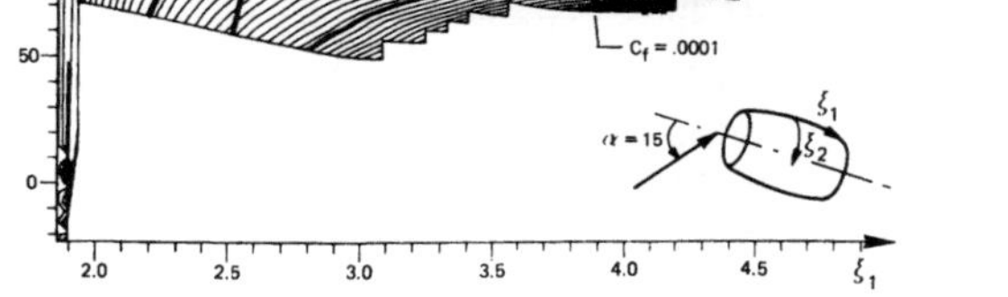

Fig.6. Engine cowl under angle of attack, laminar flow, distribution of Re_θ.

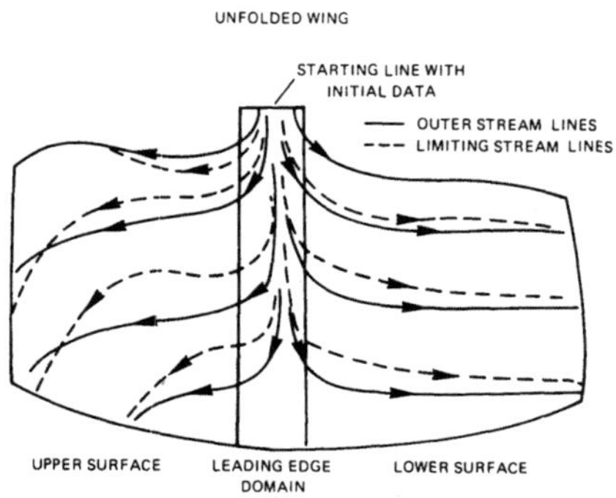

Fig.7. Wing with leading edge region, unfolded wing.

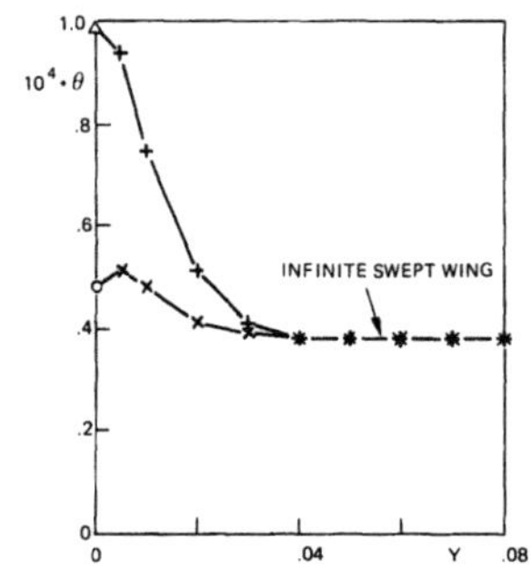

Fig.8. Wing with leading edge region, momentum thickness along leading edge.

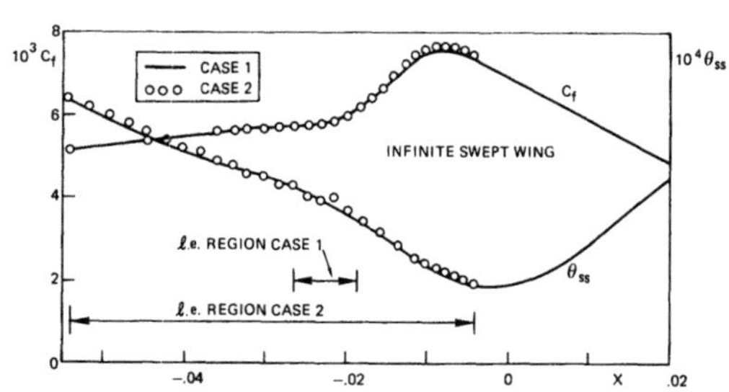

Fig.9. Wing with leading edge region, friction coefficient and momentum thickness along wing section.

Comparison of Three-Dimensional Turbulent Boundary-Layer Calculations with Experiment

U. R. MÜLLER*

Ames Research Center, NASA
Moffett Field, California

Summary

The incompressible, three-dimensional, turbulent boundary-layer (3DTBL) experiments of van den Berg and Elsenaar [1], Dechow [2], and Müller [3] were simulated numerically by integrating the boundary-layer equations together with an algebraic eddy-viscosity turbulence model. For the flow treated by van den Berg and Elsenaar, the downstream portion, where the crossflow was large, could not be predicted with the present computational method; we feel that this flow was significantly influenced by elliptic flow-field effects. Though Dechow's experiment also indicated departures from the boundary-layer concept, our calculations agreed reasonably well with the mean-flow development up to separation. In Müller's experiment the normal pressure gradients were found to be negligible in regions with large skewing and enabled us to test turbulence models using the boundary-layer equations. The simulation of this flow compared favorably with the experimental data throughout the flow field and suggested the applicability of algebraic eddy-viscosity models for 3DTBLs.

Introduction

Predictions of 3DTBLs are needed for many engineering purposes, for example, in designing swept, low-aspect-ratio wings and improving their performance. However, the development of computational methods, as well as our understanding of the turbulent momentum transfer in three-dimensional flows, has lagged substantially behind the state of art reached for two-dimensional flows. For most numerical computations of 3DTBLs, turbulence models developed for two-dimensional flows have been applied, together with an ad hoc closure assumption for the crossflow momentum equation.

The predictions of low-speed flows with moderate crossflow gen-

*National Research Council Associate.

erally compared well with experimental data, but flow fields with large crossflow often could not be simulated with reasonable accuracy [4], especially the flow field experiment performed by van den Berg and Elsenaar [1].

Recently, Marvin [5] suggested that eddy-viscosity turbulence models may not be adequate for computing flows with severe crossflow and that the models require considerable improvement. The present author feels that further comparisons of measured and predicted mean-flow and turbulence fields are needed before discarding simple, gradient-diffusion turbulence models. In fact, during recent years several documentations of 3DTBL developments, including the measurements of all components of the Reynolds stress tensor, were completed, for example, by Dechow [2] and Müller [3]. To the author's knowledge these experiments have not yet been used for comparing theory and experiment.

In the present study the three flow fields mentioned above were simulated numerically by solving the boundary-layer equations. We concentrated our efforts on investigating whether an algebraic turbulence model as developed for two-dimensional flows allows for reasonable predictions of 3DTBLs. To sort out the applicability of a particular turbulence model, we also had to study the sensitivity of the predictions to both modifications of the closure assumptions and to the accuracy of the pressure distribution prescribed.

Numerical Method and Turbulence Models

The equations for incompressible 3DTBLs were integrated in a Cartesian frame of reference:

$$\left.\begin{aligned} &\partial\bar{U}/\partial x + \partial\bar{V}/\partial y + \partial\bar{W}/\partial z = 0 \;; \quad \partial p/\partial y = 0 \\ &\bar{U}\partial\bar{U}/\partial x + \bar{V}\partial\bar{U}/\partial y + \bar{W}\partial\bar{U}/\partial z = -(1/\rho)\partial p/\partial x \\ &\qquad\qquad + \partial(\nu\partial\bar{U}/\partial y - \overline{uv})/\partial y \\ &\bar{U}\partial\bar{W}/\partial x + \bar{V}\partial\bar{W}/\partial y + \bar{W}\partial\bar{W}/\partial z = -(1/\rho)\partial p/\partial z \\ &\qquad\qquad + \partial(\nu\partial\bar{W}/\partial y - \overline{vw})/\partial y . \end{aligned}\right\} \quad (1)$$

Quasi-two-dimensional flows were computed setting $\partial/\partial z = 0$; for two-dimensional flows $\bar{W}$ was set to zero. The boundary conditions included no-slip conditions at the wall and prescribed outer-edge velocities, which were determined by smoothing the measured pressure distribution and solving the Euler equations. The baseline turbulence model used for closing Eqs. (1) was formed by the mixing-length formula of Michel et al. [6], together with the assumption of isotropic eddy viscosities:

$$\left.\begin{array}{l} -\overline{uv} = \nu_x \partial \bar{U}/\partial y \;;\; -\overline{vw} = \nu_z \partial \bar{W}/\partial y \;;\; \nu_t = \nu_x = \nu_z \\ \nu_t = \ell^2 F^2 \{(\partial \bar{U}/\partial y)^2 + (\partial \bar{W}/\partial y)^2\}^{\frac{1}{2}} \;;\; F = 1 - \exp\{-y u_\tau N/(26\nu)\} \\ \ell/\delta = \ell_e/\delta * \tanh\{0.41/(\ell_e/\delta) * (y/\delta)\} \;;\; u_\tau = \sqrt{|\vec{\tau}_w|/\rho} \;;\; \ell_e/\delta = 0.085 \end{array}\right\} \qquad (2)$$

We applied Cebeci's [7] pressure gradient correction to the van Driest damper F, with s being the coordinate along the outer-edge streamline:

$$N = (1 - 11.8\, p^+)^{\frac{1}{2}} \;;\quad p^+ = -\{1/(\rho u_\tau^3)\} \partial p/\partial s \;. \qquad (3)$$

Additionally, we used a closure assumption with anisotropic eddy viscosities. As suggested by Rotta [8] the Reynolds stresses were defined in a local coordinate system, the x-axis of which was aligned with the yaw direction of the resultant flow vector. Then the stresses were transformed into the frame of reference used for Eqs. (1).

$$\left.\begin{array}{l} -\overline{u_m v_m} = \tilde{\nu}_x \partial \bar{U}_m/\partial y \;;\; -\overline{v_m w_m} = T\tilde{\nu}_z \partial \bar{W}/\partial y \;;\; \tilde{\nu}_x = \tilde{\nu}_t \;; \\ \tilde{\nu}_z = T\tilde{\nu}_t \;;\; \tilde{\nu}_t = \ell^2 F^2 \{(\partial \bar{U}_m/\partial y)^2 + T(\partial \bar{W}_m/\partial y)^2\}^{\frac{1}{2}} \;. \end{array}\right\} \qquad (4)$$

The ratio T of the eddy viscosities defined the degree of anisotropy.

The initial conditions for the spanwise integration were obtained either from solving the plane-of-symmetry equations [7] or by integrating Eqs. (1) with Hall's difference scheme. The CFL-condition was applied in the wall-streamline direction. Then the integration was carried out with Crank-Nicholson differencing. For all computations we used step sizes $\Delta x = \Delta z = 8$ mm; the normal grid spacing Δy was smaller than $y^+ = 1$ at the wall and then was stretched toward the outer edge, yielding 70 to 150 node points across the boundary layer. As found from test calculations, grid effects were negligible for all results obtained.

The integration of the governing equations was based on the algorithm of Krause et al. [9]. Using the baseline turbulence model, the computational method was tested against case A4 of Ref. 4. All results obtained agreed well with those of other differential methods.

Results of Numerical Simulations

Experiment of van den Berg and Elsenaar [1] and Elsenaar and Boelsma [10]

This investigation of a 3DTBL on a swept flat plate was performed under "infinite swept wing conditions," that is, vanishing gradients of all flow variables in spanwise direction (Fig. 1). By means of a diffuser-type channel an adverse pressure gradient

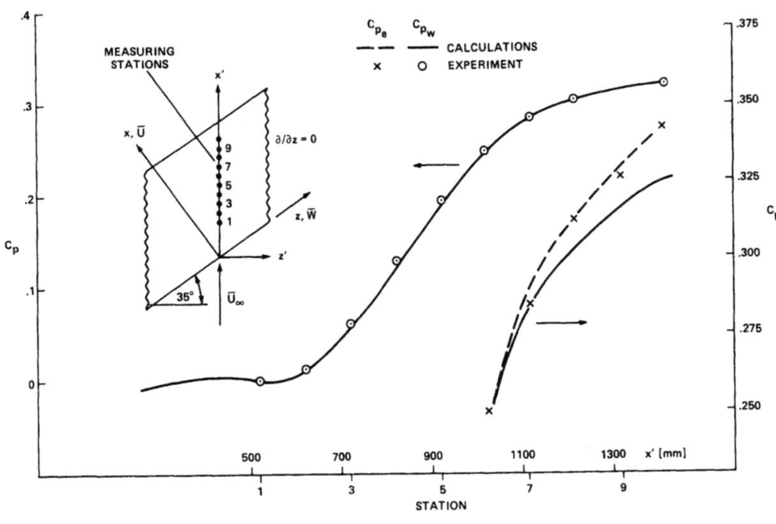

Fig.1. Experimental setup and pressure distribution (van den Berg and Elsenaar)

normal to the leading edge was generated, such that three-dimensional separation occurred in the downstream region.

We obtained the initial conditions for starting the numerical simulations by matching a two-dimensional, zero-pressure gradient, turbulent boundary-layer calculation with Reynolds number $Re = 2.42 \times 10^6$ ($\bar{U}_\infty$ defined at station 1; L = 1 m) to the measured momentum-thickness Reynolds number and skin-friction coefficient.

The computation of the downstream flow development was carried out under quasi-two-dimensional conditions with $\partial/\partial z = 0$. Using the pressure distribution measured at the wall, together with the baseline turbulence model, we obtained the result indicated by (run A1) in Figs. 2 and 3. Up to station 5 the measured and calculated mean-velocity profiles, as well as the skin-friction coefficients $c_f = 2\,\vec{\tau}_w/(\rho \bar{U}_e{}^2)$ and the wall turning-angles $(\alpha + \beta_w)$, agreed well. The angle α defined the direc-

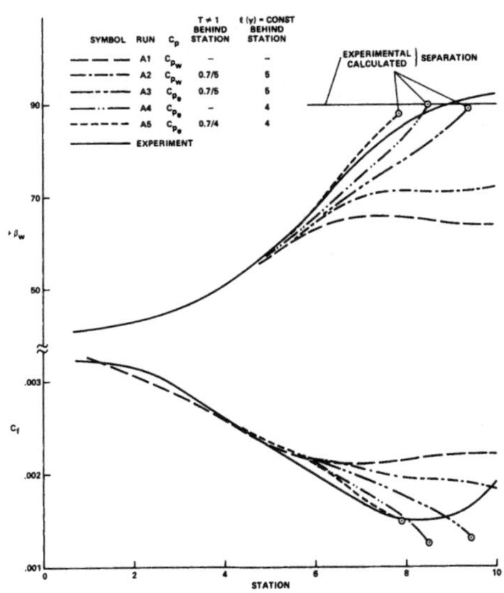

Fig.2. Comparison of measured and computed wall turning angles and skin-friction coefficients for different eddy viscosities and pressure distributions (van den Berg and Elsenaar)

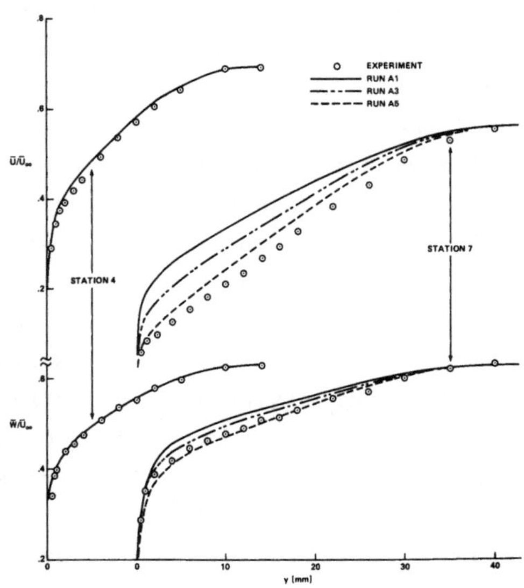

Fig.3. Comparison of measured and computed mean velocities (van den Berg and Elsenaar)

tion of the outer-edge flow with respect to the x-axis, and β_w was the turning angle between wall and outer-edge streamline direction. Further downstream the comparison indicated deviations that increased rapidly; at station 7 the computed wall turning-angle lagged 15° behind the measured value of 40°, and the mean-velocity component $\bar{U}$ was overpredicted by as much as 100% in the near-wall region. Beyond station 7 the computed values of c_f and β_w were approximately constant and did not lead to separation.

Such results - which had already been reported, for example, by Krause and Kordulla [11] - seemed to indicate an overestimation of the turbulent viscosities by the closure assumption used. That is why we reduced the magnitude of viscous forces in subsequent calculations. First we considered Rotta's definition of anisotropic eddy viscosities, Eqs. (4). In Ref. 10 the lowest values for the ratio of the local crosswise to streamwise viscosity ($T \approx 0.5$) were measured in the upstream, weak three-dimensional flow, but according to Müller's error analysis [12] the accuracy of these results is suspect. That is why for the present calculations the ratio T was set equal to one initially and then was gradually reduced beyond station 4 or 5, respectively, such that the prescribed experimental value was obtained at the next downstream measuring station. In our calculation with $T = 0.7$, introduced at station 5, we found negligible differences compared to run A1.

Second, we investigated the sensitivity of the results to the outer-layer mixing-length distribution, which in Ref. 10 was measured to be approximately constant downstream of station 4 or 5. When using Eq. (2) up to station 5 and then "freezing" $\ell(y)$, the results were close to those of run A2, Fig. 2. For run A2, $T = 0.7$ was used in addition to the frozen mixing length.

In further computations we investigated the sensitivity of the predictions to the prescribed static pressure distribution. In Ref. 1 the pressure at the outer edge of the boundary layer was derived from the mean-flow measurement and was found to be larger than that at the wall downstream of station 6 (Fig. 1). Our approximation of the outer-edge data, as indicated by the broken line in the figure, will be used throughout the calculations discussed hereafter. Using the baseline turbulence model, the results indicated an increase in β_W by 4°, while c_f stayed about the same. However, when we applied the larger pressure distribution together with the reduced eddy viscosities of run A2, we obtained a monotonic decrease of c_f (run A3 in Fig. 2), and separation occurred between stations 9 and 10. The velocity component $\bar{U}$ was still overpredicted (Fig. 3). Introducing $T = 0.7$ at station 4 had little effect compared with run A3 where $T = 0.7$ was introduced at station 5.

For run A4 we used isotropic eddy viscosities and kept the mixing-length distribution constant downstream of station 4; then we also introduced $T = 0.7$ at the same station, run A5. Compared with previous computations, c_f decreased further, and β_W increased substantially. For run A5, β_W even indicated an overshoot compared with the measurements, and both mean velocity components were close to the experimental data.

<u>Experiment of Dechow [2]</u>

Dechow investigated a 3DTBL approaching a tear-drop-shaped body extending between the lower wall (test surface) and upper wall of a channel with a height of 300 mm (Fig. 4). Most measurements of mean velocities and Reynolds stresses were taken at stations oriented along an external streamline. Since the present computational method was limited to attached flows, we could not simulate any measurements downstream of station 6.

The two-dimensional initial conditions at $x = 0$ were obtained from smoothing the measured velocity profile $\bar{U}(y)$ of station 1. We matched the momentum-thickness Reynolds number and the skin-friction coefficient to the experimental data. The free-stream Reynolds number was prescribed as 1.45×10^6 with $\bar{U}_\infty$ defined at station 1 and $L = 1$ m. The computations discussed below were based on the measured wall pressure distribution plotted in Fig. 5. From our set of curves, which we fitted by means of cubic splines, we encountered problems in obtaining smooth distributions of the pressure gradients, especially for the downstream region of the flow. Therefore, using the plane-of-symmetry solution as starting conditions for the crosswise integration was not sufficiently accurate. Instead, we solved the boundary-layer equations with Hall's scheme in the vicinity of the plane of symmetry. The downstream side boundary of the computational domain was prescribed with $z = 128$ mm.

The results obtained from calculations with the baseline turbulence model are indicated by solid lines in Figs. 6 and 7. Up to station 4, the inner-layer profile of the mean velocity $\bar{U}$ (Fig. 6) agreed well with the experimental data; the outer-layer profile was underpredicted by about 10% at station 4. Farther downstream the near-wall flow was predicted to be larger than the measurements suggesting that for this part of the flow the turbulent viscosities were overestimated by Eq. (2). Indeed, at stations 5 and 6 the measured mixing-length distributions could no longer be represented by $\ell = 0.41\ y$ but were much less. Flow separation occurred at $x = 508$ mm and $z = 44$ mm. In Fig. 7 the calculated Reynolds shear stresses $\overline{u_m v_m}$ and $\overline{v_m w_m}$

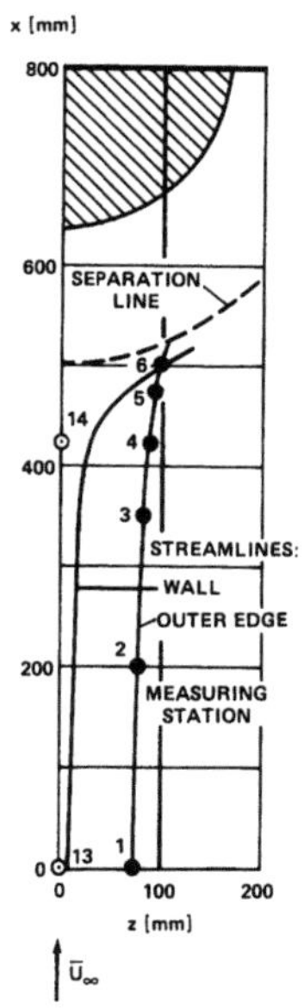

Fig.4. Experimental test setup (Dechow)

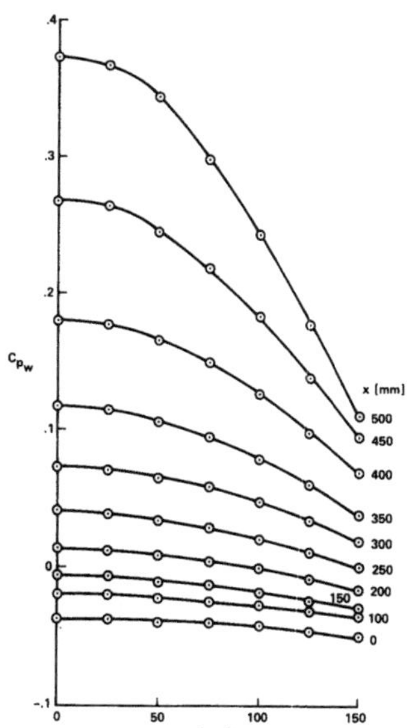

Fig.5. Measured wall pressure distribution (Dechow)

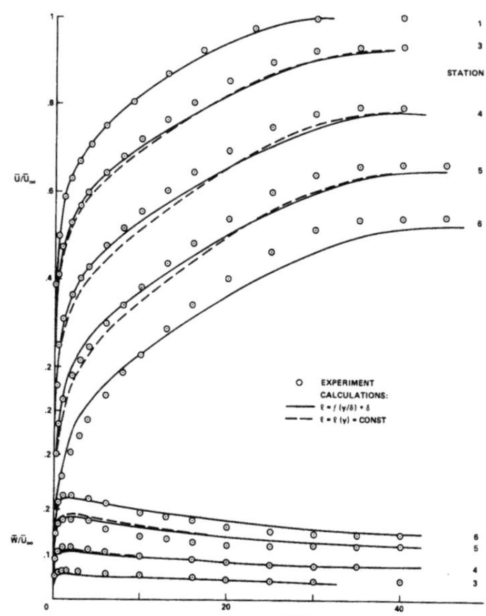

Fig.6. Comparison of measured and computed mean velocities (Dechow)

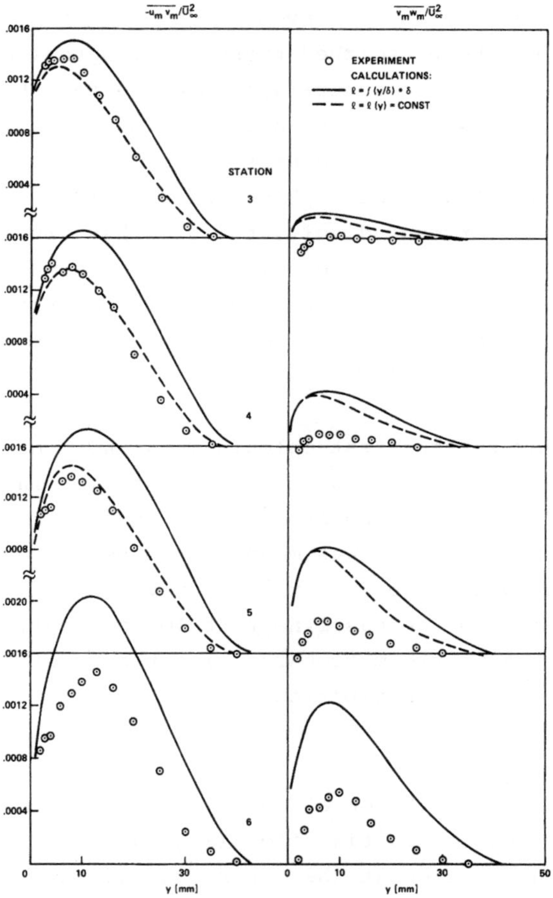

Fig.7. Comparison of measured and computed Reynolds shear stresses (Dechow)

are compared with the experimental data. Although at station 3 the peak value of the $\overline{u_m v_m}$ profile was only slightly higher than in the experiment, the maximum was 20% too high at station 5 and 25% too high at station 6. The measured Reynolds stresses $\overline{v_m w_m}$ were not well represented by the computations but were considerably overestimated. From the oil-flow pattern of Ref. 2 we extracted several spanwise distributions of the wall turning-angle. We found good overall agreement, even in the strong crossflow region, except very close to separation (x = 500 mm and z < 65 mm).

In additional computations, we investigated the sensitivity of the predictions to the prescribed turbulent length-scale distribution. Approximating the outer-layer measurements by a fixed length scale $\ell_e = 2.4$ mm we obtained the results shown by broken lines in Figs. 6 and 7. The profiles of the mean-velocity component $\bar{U}$ indicated considerably smaller velocities in the near-wall region. Correspondingly, the flow had already separated ahead of station 6. The magnitude of the Reynolds shear stress $\overline{u_m v_m}$ was considerably reduced, and the profiles were close to the experimental data. The crosswise shear stress only indicated a minor sensitivity.

As investigated by further test calculations (not shown), we found that the predicted near-wall flow quantities in the region close to separation significantly depended on the prescribed pressure distribution, which was measured to be larger at the wall than at the outer edge for $x \gtrsim 350$ mm. When using the latter data together with the baseline turbulence model, the flow sepatated farther downstream compared with the results obtained with the wall pressure distribution.

Experiment of Müller [3]

In this experiment, a 3DTBL on a flat plate was generated by laterally deflecting the initial flow (Fig. 8). The static pressure distribution measured at the outer edge of the boundary layer is shown in Fig. 9. We obtained a good representation of the pressure field, except for $z > 500$ mm, by approximating the measurements by polynomial functions of fourth order. At five z-coordinates the distribution of $c_p(x)$ was evaluated and then, at arbitrarily prescribed x-coordinates, the crosswise distributions $c_p(z)$ were computed. To save computer time and because we did not encounter problems in calculating two-dimensional flows or those with moderate crossflow, we started the numerical simulations at $x = 140$ mm instead of at $x = 0$. At each crosswise grid point the smoothed velocity profile of station B5 was normalized by the magnitude of the measured outer-edge flow and was assumed to be collateral to the outer-edge velocity vector. With this starting procedure we obtained velocity profiles at $x = 200$ mm, which differed less than 3% from the measurements

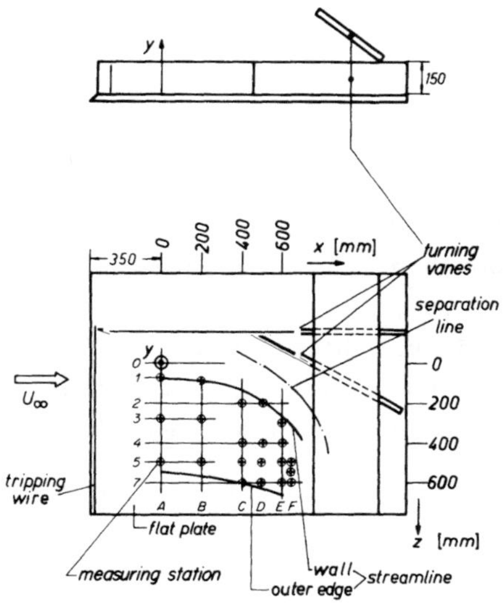

Fig.8. Experimental test setup (Müller)

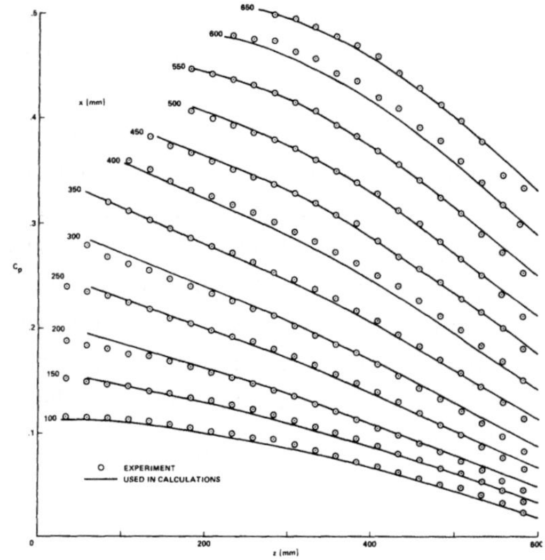

Fig.9. Measured and smoothed pressure distribution (Müller)

(see, e.g., profile B5 in Fig. 10). The lateral boundaries of the computational domain were defined by z_{max} = 600 mm and by

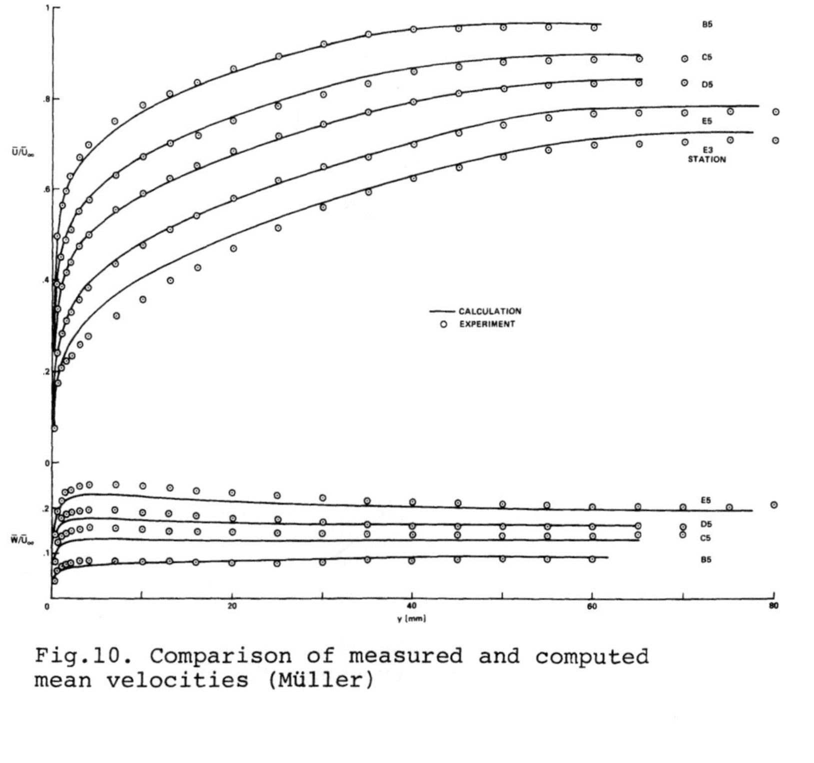

Fig.10. Comparison of measured and computed mean velocities (Müller)

a wall streamline starting at $x = 140$ mm and $z = 84$ mm, while x_{max} was prescribed with 650 mm.

In Figs. 10 to 13 the measured downstream development of the flow is compared with the computed one obtained with the baseline turbulence model. For all stations with $z = 500$ mm (row 5) the profiles of the mean-velocity component $\bar{U}$ were in close agreement with the measurements (Fig. 10), and the peaks in the profiles of the $\bar{W}$ component were underpredicted by about 10%. Close to the separation line, at stations D2 and E3, the agreement was noticeably worse. The computed and measured Reynolds

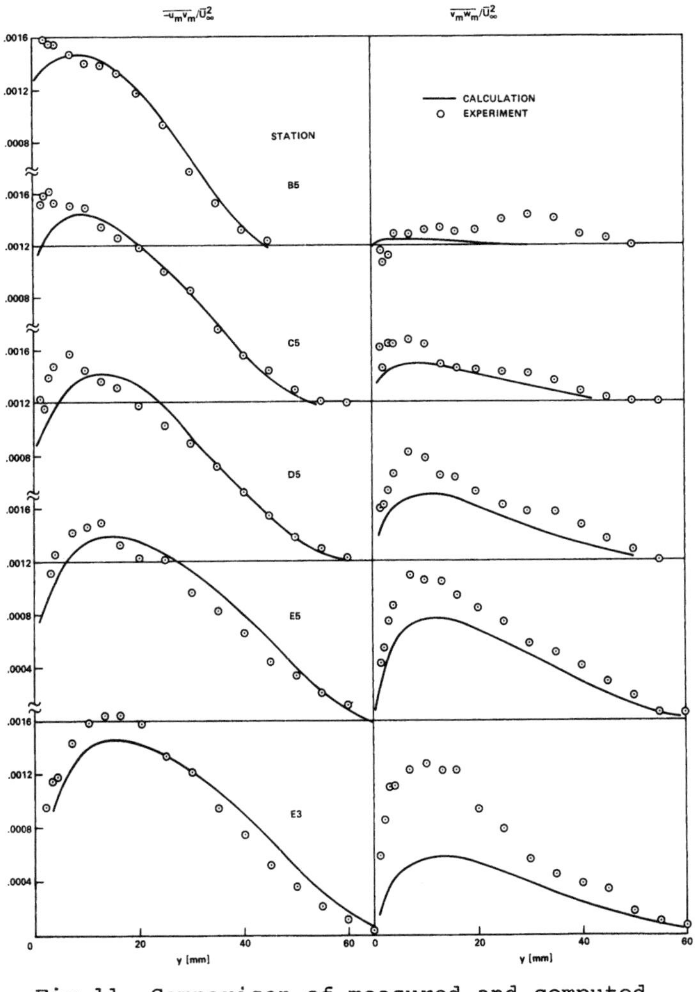

Fig.11. Comparison of measured and computed Reynolds shear stresses (Müller)

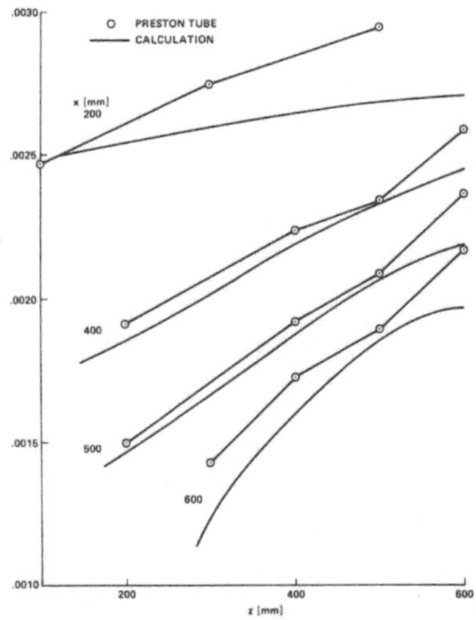

Fig.12. Comparison of measured and computed skin-friction coefficients (Müller)

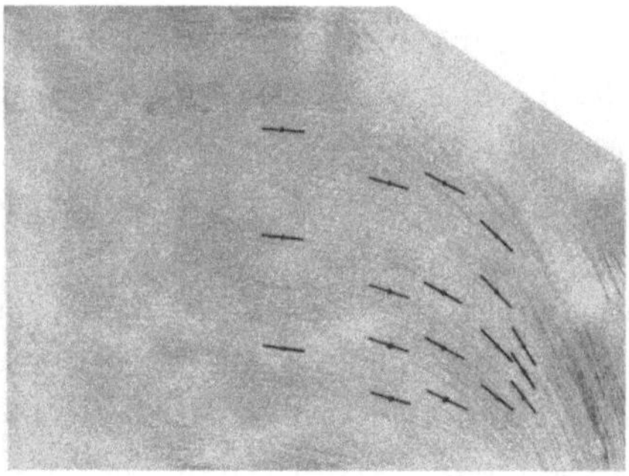

Fig.13. Comparison of oil-flow pattern with wall-streamline directions computed at measuring stations (Müller)

stresses $\overline{u_m v_m}$ and $\overline{v_m w_m}$ are compared in Fig. 11. Generally, good overall agreement was obtained for the streamwise component. The deviations for $y \lesssim 5$ mm at the upstream stations were probably caused by experimental inaccuracies. The profiles of the measured shear stress $\overline{v_m w_m}$ developed peaks at $y \approx 7$ mm which were not obtained from the computations and were underestimated by as much as 30% at station E5 and by 50% at station E3. In the outer layer, the comparison yielded satisfactory results.

The computed skin-friction coefficients based on the total wall shear stresses are displayed in Fig. 12. For $x > 200$ mm the agreement was good except for $z > 500$ mm, where the outer-edge pressure distribution had been prescribed too high. At $x =$ 600 mm the differences between measurements and computations increased with decreasing distance from separation and were as high as 15% at z_{min}. In Fig. 13, the directions of the resultant wall shear stresses, computed at the measuring stations, are compared with the wall flow visualization. Excluding the results at stations close to separation and those for $z > 500$ mm, the agreement was good, particularly since the crossflow angles β_w were as high as 40°.

Concluding Remarks

Numerical simulations of the 3DTBL experiments of van den Berg and Elsenaar [1], Dechow [2], and Müller [3] were carried out to test the applicability of an algebraic mixing-length turbulence model in flows with strong crossflow. Generally, for the regions with moderate skewing ($\beta_W < 15°$) the flow-field calculations agreed well with the experimental data. For the highly three-dimensional parts of the flows, however, we found different degrees of agreement for each case.

The computed downstream development of the flow considered by van den Berg and Elsenaar failed to simulate the measurements. Separation was obtained only if the pressure distribution prescribed was larger than the one measured at the wall and if simultaneously the magnitude of the eddy viscosities was reduced compared with the baseline model; this was also true in the computation of Elsenaar et al. [13]. Under these conditions the results proved to be extremely sensitive to either modification; for example, changing the pressure distribution by a few percent yielded large changes in the computed flow field. We feel, that the problems encountered were mainly a result of the breakdown of the boundary-layer concept in the downstream part of the flow. Though in the experiment the diffuser cross section approximately varied linearly throughout the flow field, the longitudinal pressure gradients rapidly decreased beyond station 5. This upstream pressure effect was caused by separation and by the flow blockage at the downstream end of the diffuser. Additionally, pressure variations normal to the wall probably influenced the development of this flow. For resolving these elliptic flow-field effects, predicting the location of separation and ultimately testing turbulence models, a Navier-Stokes prediction method would be necessary.

Our numerical simulation of Dechow's experiment yielded reasonable overall agreement with the measured mean-flow profiles and wall quantities even in the strong crossflow region with turning angles of about 30° between wall- and outer-edge flow direction. However, the flow field close to separation was found to be sensitive to the pressure distribution prescribed, as well as to the turbulence model used. In the three-dimensional flow, the measurement indicated normal pressure gradients, which impaired the comparison with solutions of the boundary-layer equations. Additionally, in the downstream part of the flow similarity laws, as developed for two-dimensional mean flow and turbulence fields, were no longer valid. Under these conditions we could not extract detailed information about turbulence modeling.

In Müller's experiment, which was performed in a wind tunnel with an open test section of 1 m ϕ, most measuring stations were located at sufficiently large distances away from separation in order to ensure that the normal pressure gradients were negligible. The results of the numerical simulation for this flow indicated a minor dependency on the accuracy of the prescribed pressure distribution and enabled us to study the applicability of the turbulence model used. Except for the region closest to separation, the flow development, computed with a baseline turbulence model, compared favorably with the measurements, even at the measuring stations farthest downstream with large skewing ($\beta_W > 40°$). In contrast to the other calculations, the results were found to be insensitive to the outer-layer mixing length and suggested the applicability of an algebraic eddy-viscosity model, as developed for two-dimensional boundary layers. In order to fully understand the results obtained, further investigations with more sophisticated turbulence models, whicn take into account the experimentally observed anisotropy of the turbulence field or the normal Reynolds stress diffusion, will be necessary.

The present study again demonstrates the lack of well-documented experiments that are suitable to guide both the development of computational methods for 3DTBLs and closure assumptions. Because pressure forces play a dominant role in flows like those investigated here, future experiments should accurately map out the pressure distribution and also ensure that the boundary layer concept is valid. On the basis of such experiments we expect progress in turbulence modeling for 3DTBLs as well as in the development of prediction methods for engineering-type applications.

References

1. van den Berg, B.; Elsenaar, A.: Measurements in a three-dimensional incompressible turbulent boundary layer in an adverse pressure gradient under infinite swept wing conditions. NLR TR 72092 U, 1972.

2. Dechow, R.: Mittlere Geschwindigkeit und Reynoldsscher Spannungstensor in der dreidimensionalen Wandgrenzschicht vor einem stehenden Zylinder. In: Strömungsmechanik und Strömungsmaschinen, Mitteilungen des Institutes für Strömungslehre und Strömungsmaschinen, Heft 21 (1977), Univ. Karlsruhe.

3. Müller, U. R.: Measurement of the Reynolds stresses and the mean-flow field in a three-dimensional pressure-driven boundary layer. To be published in J. Fluid Mech. (1982).

4. East, L.E.: Computation for three-dimensional turbulent boundary layers. FFA TN AE-1211, 1975.

5. Marvin, J.G.: Turbulence modeling for computational aerodynamics. AIAA Paper 82-0164, 1982.

6. Michel, R.; Quémard, C.; Durant, R.: Application d'un schemá de longeur de mélange á l'étude des couches limites turbulentes d'équilibre. ONERA TN 154, 1969.

7. Cebeci, T.: Calculation of three-dimensional boundary layers II. Three-dimensional flows in Cartesian coordinates. AIAA J. 13 (1975) 1056.

8. Rotta, J.C.: A family of turbulence models for three-dimensional thin shear layers. In: 1st symposium on turbulent shear flows, Pennsylvania State U., 1977.

9. Krause, E.; Hirschel, E.H.; Bothmann, Th.: Differenzenformeln zur Berechnung dreidimensionaler Grenzschichten. DLR FB 69-66, 1969.

10. Elsenaar, A.; Boelsma, S.H.: Measurements of the Reynolds stress tensor in a three-dimensional turbulent boundary layer under infinite swept wing conditions. NLR TR 74095 U, 1974.

11. Krause, E.: Strive for accuracy-improvement of predictions. Comp. and Fluids 8 (1980), 31.

12. Müller, U.R.: On the accuracy of turbulence measurements with inclined hot wires. To be published in J. Fluid Mech. (1982).

13. Elsenaar, A.; van den Berg, B.; Lindhout, J.P.F.: Three-dimensional separation of an incompressible turbulent boundary layer on an infinite swept wing. AGARD-CP-168, 1975.

A Three-Dimensional Law-of-the-Wall Including Skewness and Roughness Effects

I.L. RYHMING and T.K. FANNELOP

Fluid Mech.Lab. and Div.of Aero-Gasdynamics
ETH
Lausanne, Switzerland

NTH
Trondheim, Norway

Summary

A new mixing length formulation including velocity profile skewness in 3D boundary layer flows is proposed leading to an anisotropic behaviour of the eddy viscosities in the streamwise and cross-flow directions. The consequence of this new concept has been evaluated by calculating the boundary layer flow over an infinite swept wing by a finite difference method and comparing with available experimental data. Furthermore a new 3D law-of-the-wall has been derived including skewness as well as wall roughness effects. The ability of the new model to reproduce the skewing angle close to the wall has also been evaluated in this context. The new formulation is seen to give improved representation of some features of 3D boundary layer flows in comparison with the conventional isotropic eddy viscosity formulation.

1. Introduction

The law-of-the-wall in 2D turbulent boundary layers is well established from experimental results as well as on theoretical grounds. As turbulent motion is always threedimensional it has been suggested that this law for the mean flow can be extended with reasonable accuracy also to 3D flows.[1] Twodimensional boundary layer methods based on mixing length concepts are remarkably successful[2], and have therefore been adapted by many also for 3D problems. Hence Fannelop and Humphreys[3] replaced the 2D wall shear stress with the total shear stress which is employed both in the definition of the skinfriction velocity and in van Driests damping factor of the mixing length. The eddy viscosity was furthermore assumed proportional to the resultant rate of strain. This extention to 3D flow is the simplest possible which does not necessitate further physical postulates. The use of streamline coordinates is useful in studying this and an alternate extension to be considered in what follows. It was noted in Ref.3 that the proposed expression for the eddy viscosity takes into account the direction of the velocity vector in the boundary layer, but that its rate of change of direction, i.e. the skewing is not included. Hence, the proposed model may be inaccurate if the velocity vector is skewed to a large extent. van den Berg[4] has considered the details of the boundary layer flow over an infinite swept wing, and concludes that the skewing is important in this type of problem.

In several independent experimental investigations concerning 3D boundary layer flows, a reduced "effective eddy viscosity" has been measured in the cross-flow direction. In comparison with the streamwise value, the reduction is of the order of 30-40%, indicating an anisotropic behaviour of the eddy viscosity An attempt to establish a theoretical foundation for the anisotropic concept has been published recently by Rotta[5] based on considerations of pressure strain correlations. Schneider[6] has developed a model which in the end depends on a known empirical ratio T between the two eddy viscosity components, i.e. $0<T<1$. It is questionable, however, whether a fixed constant value of T suffices across the boundary layer from the wall to the edge and at all streamwise stations including the region immediately upstream of separation. In Ref.3 the NLR experimental case[7] was recalculated, whereby the cross-flow eddy viscosity was reduced by 40% according to the measurement of East[8], leading to an improved agreement with the experiments.

As separation is approached, the cross-flow becomes large and skewness effects increasingly important. In what follows we will formulate an eddy viscosity law which includes the effect of skewness. Based on this law we will extend the 2D law-of-the-wall to 3D and compare the result to van den Berg's extended law[4] which does not include skewness. We will furthermore investigate the effects of roughness on the 3D flow near the wall. The modifications of the 2D law due to roughness are well established, but the consequences for flow in 3D have not yet been explored.

2. Effects of skewness on eddy viscosity

The physical interpretation of Prandtl's 2D eddy viscosity relation is that fluid element (lumps), which move about as a result of the turbulent state, retain their momentum over a limited distance denoted the mixing length. In 3D flows, however, the dynamic effects of the rate of change of flow direction lead to an apparent anisotropy in the eddy viscosity law because the vertical momentum exchange which forms the core of Prandtl's mixing length assumption, is influenced not only by the rate of change of magnitude of the local velocity, but also by its rate of change of direction. The mathematical formulation is simplified in streamline coordinates.

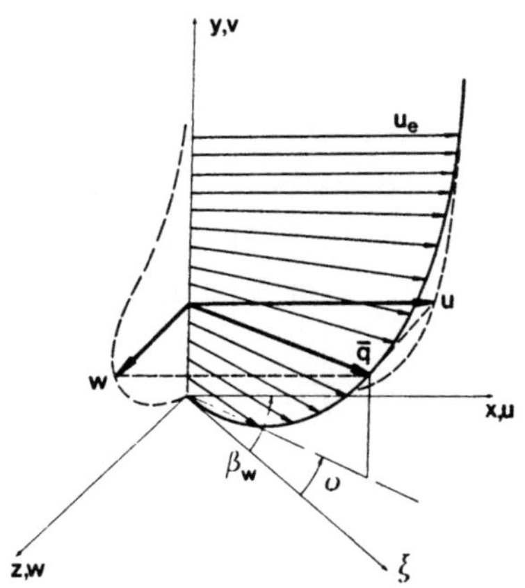

Fig.2.1. Coordinates and velocity components

Fig.2.1. indicates a skewed velocity profile with components u,w along the x- and z-axis. The x-axis is aligned with the projection on the surface of the external (inviscid) streamline, and the z-axis is normal to the x-direction. The coordinate y is normal to the wall. At a given distance y from the wall, the velocity vector $\bar{q}=\bar{q}(x,y,z)$. The rotation of this vector is most conveniently described with reference to the wall streamline, which forms the angle β_w with the x-axis. Hence,

$$\bar{q} = |\bar{q}| \cos(\beta_w - \varphi)\bar{e}_x + |\bar{q}| \sin(\beta_w - \varphi)\bar{e}_z \tag{2.1}$$

On noting that $\bar{q} = \bar{q}(x,y,z;\varphi)$ and that $\varphi = \varphi(x,y,z)$ the partial derivative of $\bar{q}$ with respect to y becomes

$$\frac{\partial \bar{q}}{\partial y} = \frac{\partial \bar{q}}{\partial y}\Big|_{\varphi} + \frac{\partial \bar{q}}{\partial \varphi}\frac{\partial \varphi}{\partial y} \tag{2.2}$$

On carrying out this operation on (2.1) and writing the result in terms of the components (u,w) we obtain

$$\frac{\partial u}{\partial y} = \frac{\partial u}{\partial y}\Big|_{\varphi} + w\frac{\partial \varphi}{\partial y}, \quad \frac{\partial w}{\partial y} = \frac{\partial w}{\partial y}\Big|_{\varphi} - u\frac{\partial \varphi}{\partial y} \tag{2.3a,b}$$

where the terms proportional to $\frac{\partial \varphi}{\partial y}$ express the rate of skewing. Prandtl's equations for the shear stress components are now generalized in such a way that their rate of skewing dependence is postulated as being directly proportional to the ratio of the skewed velocity vector rate of strain and the rate of strain of the corresponding non-rotated velocity vector. Consequently no rate of skewing effect arises in a 2D profile (w=0 and $\beta_w=\varphi=0$) or in a non-skewed 3D profile with $\varphi=0$ but $\beta_w \neq 0$. Hence we write

$$\tau_x = \varepsilon_x \frac{\partial u}{\partial y}, \quad \tau_z = \varepsilon_z \frac{\partial w}{\partial y} \tag{2.4a,b}$$

$$|\bar{\tau}| = (\tau_x^2 + \tau_z^2)^{\frac{1}{2}} = \rho \ell^2 \left[(\frac{\partial u}{\partial y})^2 + (\frac{\partial w}{\partial y})^2\right] \tag{2.5}$$

where ε_x and ε_z are taken as

$$\varepsilon_x = \rho \ell^2 \left[(\frac{\partial u}{\partial y})^2 + (\frac{\partial w}{\partial y})^2\right]\left[(C_x \frac{\partial u}{\partial y})^2 + (C_z \frac{\partial w}{\partial y})^2\right]^{-\frac{1}{2}} C_x \tag{2.6a}$$

$$\varepsilon_z = \rho \ell^2 \left[(\frac{\partial u}{\partial y})^2 + (\frac{\partial w}{\partial y})^2\right]\left[(C_x \frac{\partial u}{\partial y})^2 + (C_z \frac{\partial w}{\partial y})^2\right]^{-\frac{1}{2}} C_z \tag{2.6b}$$

and where the factors C_x and C_z are defined by

$$C_x = \left| \frac{\partial u}{\partial y} \left(\frac{\partial u}{\partial y}\bigg|_{\varphi} \right)^{-1} \right| = \left| \frac{\partial u}{\partial y} \left(\frac{\partial u}{\partial y} - w \frac{\partial \varphi}{\partial y} \right)^{-1} \right| \qquad (2.7a)$$

$$C_z = \left| \frac{\partial w}{\partial y} \left(\frac{\partial w}{\partial y}\bigg|_{\varphi} \right)^{-1} \right| = \left| \frac{\partial w}{\partial y} \left(\frac{\partial w}{\partial y} + u \frac{\partial \varphi}{\partial y} \right)^{-1} \right| \qquad (2.7b)$$

For the 2D or non-skewed profile, C_x and C_z are both equal to one, but for a skewed profile C_x and C_z become functions of position with numerical values different from one. Hence, an anisotropic eddy viscosity behaviour is obtained. From (2.6) and (2.7) it is seen that for the profile type shown in Fig.2.1, ε_z is decreased and ε_x is slightly increased in comparison with the isotropic values obtained for a non-skewed profile. In (2.6a,b) ℓ will be assumed equal to the distance from the wall; $\ell=\kappa y$ where κ denotes von Karman's constant. Equations (2.6a,b) represent the basis of our proposed anisotropic eddy viscosity formulation expressed in terms of the velocity components u and w and the skewing angle φ . The full expression for ε_x and ε_z in equation (2.6a,b) can be applied with only minor modifications in many existing finite difference methods based on eddy viscosity models for solving the 3D boundary layer equations. We will consider the results obtained from one such method, that of Fannelop-Humphreys[3)], in a later section. An extended law-of-the-wall, different from that of van den Berg[4)] can also be derived from (2.6a,b). We will consider this possibility in Section 4.

3. The 3D law-of-the-wall for smooth walls

In van den Berg's derivation[4)] of an extended law-of-the-wall, the 2D law (or "simple law" in van den Berg's terminology) is used to generate approximations to the 3D flow field in the inner layer via the equations of motion. Only those terms considered by van den Berg to be leading terms were retained, and the order of magnitude of the terms dropped was not given. A systematic expansion procedure shows that the terms not included are $0(\beta_w{}^2)$ or equivalently $0(w^2/u^2)$, i.e. the small cross-flow assumption is implicit in the analysis.

The basic assumption in van den Berg's as well as the present analysis is that the velocity near the wall can be expressed in terms of the components

$$u = u_{\tau_x} f(y^+) \quad , \quad w = u_{\tau_z} f(y^+) \qquad (3.1a,b)$$

where (u_{τ_x}, u_{τ_z}) are the components of the skinfriction velocity vector with magnitude $u_\tau = (\tau_w/\rho)^{\frac{1}{2}}$, $y^+ = yu_\tau/\nu$, and $f(y^+)$ is the simple law function. By following van den Berg the velocity components according to (3.1a,b) are substituted into the equations of motion for the 3D turbulent boundary layer to obtain improved values of the shear stress components τ_x and τ_z by integration. To simplify the result we will transform the variables to a wall-oriented coordinate system where the streamwise component ξ is aligned with the wall shear stress vector and ζ is normal to ξ. The transformation of the coordinates is thus specified as follows, see Fig.3.1.

$$\xi = x \cos \beta_w + z \sin \beta_w \ , \ \zeta = - x \sin \beta_w + z \cos \beta_w \qquad (3.2a,b)$$

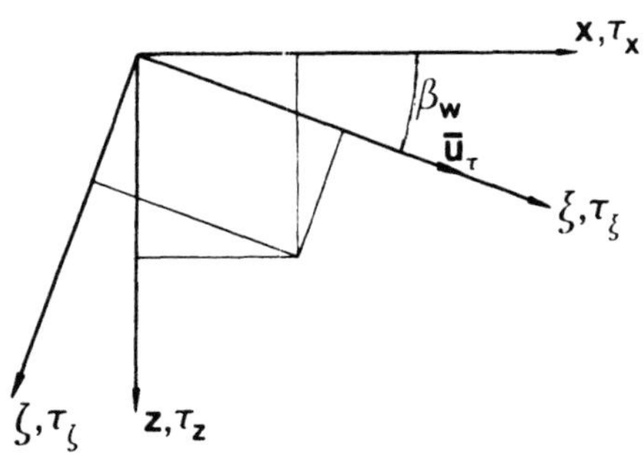

Fig.3.1. Wall coordinates

By restricting the attention to the y-axis only ($\xi=\zeta=0$) the transformation formulas for the derivatives become

$$\frac{\partial}{\partial x} = \cos\beta_w \frac{\partial}{\partial \xi} - \sin\beta_w \frac{\partial}{\partial \zeta} , \quad \frac{\partial}{\partial z} = \sin\beta_w \frac{\partial}{\partial \xi} + \cos\beta_w \frac{\partial}{\partial \zeta}$$

Furthermore

$$u_{\tau_x} = u_\tau \cos\beta_w , \quad u_{\tau_z} = u_\tau \sin\beta_w , \quad u_{\tau_\xi} = u_\tau , \quad u_{\tau_\zeta} = 0,$$

where u_τ and β_w are both functions of position.
The dependent variables are also transformed, e.g. τ_x and τ_z are expressed in terms of τ_ξ and τ_ζ (components of τ in the ξ and ζ directions) as follows:

$$\tau_x = \tau_\xi \cos\beta_w - \tau_\zeta \sin\beta_w, \quad \tau_z = \tau_\xi \sin\beta_w + \tau_\zeta \cos\beta_w \qquad (3.3a,b)$$

We obtain then the following relations:

$$\frac{\tau_\xi}{\tau_w} \cos\beta_w - \frac{\tau_\zeta}{\tau_w} \sin\beta_w = \cos\beta_w + \frac{\nu}{\rho u_\tau^3} \frac{\partial p}{\partial x} y^+ + \frac{\nu}{u_\tau^2} \cos\beta_w \frac{\partial u_\tau}{\partial \xi} I_2$$
$$- \frac{\nu}{u_\tau} \sin\beta_w \frac{\partial \beta_w}{\partial \xi} I_2 - \frac{\nu}{u_\tau} \cos\beta_w \frac{\partial \beta_w}{\partial \zeta} I_1 \qquad (3.4)$$

$$\frac{\tau_\xi}{\tau_w} \sin\beta_w + \frac{\tau_\zeta}{\tau_w} \cos\beta_w = \sin\beta_w + \frac{\nu}{\rho u_\tau^3} \frac{\partial p}{\partial z} y^+ + \frac{\nu}{u_\tau^2} \sin\beta_w \frac{\partial u_\tau}{\partial \xi} I_2$$
$$+ \frac{\nu}{u_\tau} \cos\beta_w \frac{\partial \beta_w}{\partial \xi} I_2 - \frac{\nu}{u_\tau} \sin\beta_w \frac{\partial \beta_w}{\partial \zeta} I_1 \qquad (3.5)$$

where

$$I_1 = f \int^{y^+} f\,dy' - \int^{y^+} f^2 dy', \quad I_2 = \int^{y^+} f^2 dy' \qquad (3.6a,b)$$

Here the terms proportional to $\frac{\partial \beta_w}{\partial \zeta}$ are neglected by van den Berg because I_1 is numerically very small, and the terms proportional to $\partial\beta_w / \partial\xi$ do not appear in his analysis in the same way as they do here. They give a contribution only to τ_ζ due to symmetry in the present analysis. The following parameters related to the outer flow are now introduced:

$$\alpha_x = \frac{\nu}{\rho u_\tau{}^3}\frac{\partial p}{\partial x} \quad , \quad \alpha_z = \frac{\nu}{\rho u_\tau{}^3}\frac{\partial p}{\partial z} \tag{3.7a,b}$$

The corresponding parameters in the (ξ,ζ)-system are

$$\alpha_\xi = \alpha_x \cos\beta_w + \alpha_z \sin\beta_w \quad , \quad \alpha_\zeta = -\alpha_x \sin\beta_w + \alpha_z \cos\beta_w \tag{3.8a,b}$$

Furthermore, the parameters β_ξ, γ_ξ and γ_ζ are defined as follows

$$\beta_\xi = \frac{\nu}{u_\tau{}^2}\frac{\partial u_\tau}{\partial \xi} \quad , \quad \gamma_\xi = \frac{\nu}{u_\tau}\frac{\partial \beta_w}{\partial \xi} \quad , \quad \gamma_\zeta = \frac{\nu}{u_\tau}\frac{\partial \beta_w}{\partial \zeta} \tag{3.9a,b,c}$$

On substituting these parameters into (3.4) and (3.5) and solving for τ_ξ, τ_ζ we obtain finally

$$\frac{\tau_\xi}{\tau_w} = 1 + \alpha_\xi y^+ + \beta_\xi I_2 - \gamma_\zeta I_1 \tag{3.10}$$

$$\frac{\tau_\zeta}{\tau_w} = \alpha_\zeta y^+ + \gamma_\xi I_2 \tag{3.11}$$

These are now the complete expressions with no deletions or simplifications.
The skewing angle φ is obtained from the velocity components, i.e.

$$\tan(\beta_w - \varphi) = \frac{w}{u} \quad \text{or} \quad \tan\varphi = -\frac{W}{U} \; , \; W < 0 \tag{3.12a,b}$$

where U and W are the velocity components in the (ξ,ζ)-system, see Fig.3.2.

By writing the conventional mixing length formulas for 3D-flows in wall coordinates we obtain ($C_x = C_z = 1$ in (2.6a,b))

$$\tau_\xi = \rho\kappa^2 y^2 \left[\left(\frac{\partial U}{\partial y}\right)^2 + \left(\frac{\partial W}{\partial y}\right)^2 \right]^{\frac{1}{2}} \frac{\partial U}{\partial y} \qquad (3.13)$$

$$\tau_\zeta = \rho\kappa^2 y^2 \left[\left(\frac{\partial U}{\partial y}\right)^2 + \left(\frac{\partial W}{\partial y}\right)^2 \right]^{\frac{1}{2}} \frac{\partial W}{\partial y} \qquad (3.14)$$

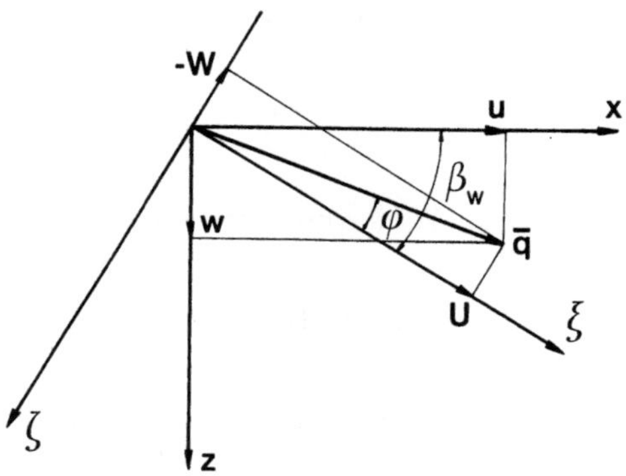

Fig.3.2. Velocity components and skewing angle referred to the wall coordinate system; $\bar{q}$ denotes the projection of the velocity vector on the surface.

By solving (3.13) and (3.14) in terms of $\frac{\partial U}{\partial y}$ and $\frac{\partial W}{\partial y}$ we get

$$\frac{\partial U}{\partial y} = \frac{\tau_\xi}{\kappa y \rho^{\frac{1}{2}}} (\tau_\xi{}^2+\tau_\zeta{}^2)^{-1/4} \cong \frac{\tau_\xi{}^{\frac{1}{2}}}{\kappa y \rho^{\frac{1}{2}}} \left[1 - \frac{1}{4}\left(\frac{\tau_\zeta}{\tau_\xi}\right)^2 + \ldots \right] \qquad (3.15)$$

$$\frac{\partial W}{\partial y} = \frac{\tau_\zeta}{\kappa y \rho^{\frac{1}{2}}} (\tau_\xi{}^2+\tau_\zeta{}^2)^{-1/4} \cong \frac{\tau_\zeta}{\kappa y (\rho\tau_\xi)^{\frac{1}{2}}} \left[1 - \frac{1}{4}\left(\frac{\tau_\zeta}{\tau_\xi}\right)^2 + \ldots \right] \qquad (3.16)$$

From these expressions the velocity components can be obtained either exactly by numerical integration or approximately by expanding first the right hand sides as indicated. By following the latter alternative we get on substituting for τ_ξ and τ_ζ from (3.10) and (3.11) and subsequent integration

$$\kappa U^{+} = \kappa \frac{U}{u_{\tau}} = \ln y^{+} + \tfrac{1}{2}\alpha_{\xi} y^{+} + \tfrac{1}{2}\beta_{\xi} \int^{y^{+}} I_{2}(y^{+'})^{-1} dy^{+'}$$

$$- \tfrac{1}{2}\gamma_{\zeta} \int^{y^{+}} I_{1}(y^{+'})^{-1} dy^{+'} + 0(\beta_{w}^{2}) \ln y^{+} + \text{Const}; \qquad (3.17)$$

$$\kappa W^{+} = \kappa \frac{W}{u_{\tau}} = \alpha_{\zeta} y^{+} + \gamma_{\xi} \int^{y^{+}} I_{2}(y^{+'})^{-1} dy^{+'} + 0(\beta_{w}^{2}) \ln y^{+}$$

$$+ \text{Const}; \qquad (3.18)$$

This is van den Berg's result[4)] when we interpret his parameters α_x, α_z etc as representing the gradients in the wall coordinate directions. The error estimate included here indicates a possible problem in cases involving large cross-flows. In the experiments used to verify the theoretical skewing angle, values as high as $\varphi \sim 12^{\circ}$ occur. In such cases it may be worthwhile to work with the exact relations given in (3.15) and (3.16) rather than the expansion indicated for small cross-flows, see section 7

4. The 3D law-of-the-wall including skewing effects

To derive an alternate law-of-the-wall which includes the effects of skewing, we have to consider the full stress components given by (2.4a,b). From (3.12a) we derive through differentiation

$$\frac{\partial \varphi}{\partial y} = \frac{1}{u^{2}+w^{2}} \left(-u \frac{\partial w}{\partial y} + w \frac{\partial u}{\partial y} \right) \qquad (4.1)$$

By substituting this expression in (2.3a,b) we get

$$\left.\frac{\partial u}{\partial y}\right|_{\varphi} = \frac{u^{2}}{u^{2}+w^{2}} \left(\frac{\partial u}{\partial y} + \frac{w}{u} \frac{\partial w}{\partial y} \right) , \quad \left.\frac{\partial w}{\partial y}\right|_{\varphi} = \frac{uw}{u^{2}+w^{2}} \left(\frac{\partial u}{\partial y} + \frac{w}{u} \frac{\partial w}{\partial y} \right) \qquad (4.2)$$

By using these expressions in the definitions of C_x and C_z and by solving the resulting equations (2.4a,b) in terms of $\partial u/\partial y$ and $\partial w/\partial y$ we obtain

$$\frac{\partial w}{\partial y} = \left[\frac{w|\tau_z|}{u\,\tau_x}\right]^{\frac{1}{2}} \frac{\partial u}{\partial y} \operatorname{sign} \tau_z \tag{4.3}$$

$$\frac{\partial u}{\partial y} = \frac{\tau_x^{\frac{1}{2}}}{\rho^{\frac{1}{2}}\kappa y} \left[1 + \left(\frac{\tau_z}{\tau_x}\right)^2\right]^{1/4} \left[1 + \frac{w|\tau_z|}{u\,\tau_x}\right]^{-\frac{1}{2}} \tag{4.4}$$

where τ_x and τ_z are given by (3.3a,b) in which τ_ξ and τ_ζ in turn are given by (3.10) and (3.11), respectively. Note that (4.3) and (4.4) implies a correct behaviour in the proximity of the wall, since as $y \to 0$ we have

$$\lim_{y\to 0} \frac{w}{u} = \tan \beta_w = \frac{\tau_z}{\tau_x} = \frac{\partial w}{\partial y} \Big/ \frac{\partial u}{\partial y}$$

Furthermore, (4.3) and (4.4) show an explicit dependence on the skewing angle φ exhibited by the factor w/u in the right hand sides. From (4.3) and (4.4) the velocity components are obtained by numerical integration. Unfortunately, first order results cannot be found in terms of known functions. A comparison between the new and the previous results are given in section 7.

5. The 3D law-of-the-wall including roughness effects

Experimental information on threedimensional turbulent boundary flows with roughness is nearly nonexistant and moreover, it is difficult to interpret the data[9]. The law-of-the-wall for rough walls can be written

$$\frac{u}{u_\tau} = f(y^+) - g(k^+) \tag{5.1}$$

The velocity shift $\Delta u^+ = g(k^+)$ will for y^+ sufficiently large be proportional to $\ln k^+$ where k^+ is related to the relevant roughness height k, i.e. $k^+ = ku_\tau/\nu$. For the sake of generality, we will assume that k varies in both the x and z directions. In order to deduce the threedimensional law which is the natural consequence of (5.1), we express our velocity components as follows

$$u = u_{\tau_x}(f - g), \quad w = u_{\tau_z}(f - g) \qquad (5.2a,b)$$

In evaluating the inertia terms in the boundary layer equations we will assume that g is small in comparison with f so that terms $0(g^2)$ can be neglected. In order to simplify the result we will again transform all terms to a coordinate system aligned with the wall streamline. Furthermore the symbols $\Delta\tau_\xi$ and $\Delta\tau_\zeta$ are used to denote the additional terms due to roughness effects so that for the rough wall, the total shear stress components are given by

$$(\tau_\xi)_r = \tau_\xi + \Delta\tau_\xi \; , \quad (\tau_\zeta)_r = \tau_\zeta + \Delta\tau_\zeta \qquad (5.3a,b)$$

After evaluation of the additional inertia terms and transformation to the (ξ,ζ)-system, we obtain

$$\frac{\Delta\tau_\xi}{\tau_w} = - \beta_\xi(2g+k^+ \frac{dg}{dk^+})(I_3+I_4)-\gamma_\zeta gI_4-\lambda_\xi k^+ \frac{dg}{dk^+} (I_3+I_4) \quad (5.4)$$

$$\frac{\Delta\tau_\zeta}{\tau_w} = - \gamma_\xi 2g\, I_3 \qquad (5.5)$$

where the new parameter λ_ξ is defined by

$$\lambda_\xi = \frac{\nu}{u_\tau} \frac{1}{k^+} \left.\frac{\partial k^+}{\partial \xi}\right|_{u_\tau} \qquad (5.6)$$

and where the integrals denoted by I_3 and I_4 are defined as follows

$$I_3 = \int^{y^+} f dy' = \frac{1}{\kappa} y^+(1+\ln y^+); \; I_4 = \int^{y^+} y \frac{df}{dy'} dy' = \kappa^{-1} y^+$$

(5.7a,b)

Here we have used the same approximation as van den Berg[4]. The integrals have been evaluated from the nominal y=0 which for a rough wall is not strictly correct. But the error involved will for terms of order g be negligible, or of order $O(g^2)$. For the leading terms it is essential in the integration of U and W, to define a lower limit which will reproduce the given velocity shift g in equations (5.2a,b).

It appears from equations (5.4) and (5.5) that additional stresses will be present only for nonzero gradients in pressure or roughness. This is to be expected as without such gradients the flow would remain twodimensional. Of perhaps greater interest is the presence of terms growing linearly (or as $y^+ \ln y^+$) with distance from the wall. This is distinctly different from the accepted twodimensional hypothesis, that the variation of the velocity profiles for rough wall and smooth wall boundary layers are the same at sufficiently large distances from the wall. We will use the symbols ΔU^+_r and ΔW^+_r to denote the additional terms due to roughness effects, in the 3D law-of-the-wall. These terms should be added to those derived previously for smooth walls, i.e. to equations (3.17) and (3.18)

$$\kappa \Delta U^+_r = -\kappa\, g \;-\; \frac{1}{\kappa}\left[\tfrac{1}{2}\beta_\xi\left(2g + k^+\frac{dg}{dk^+}\right)y^+(1+\ln y^+) + \tfrac{1}{2}\gamma_\zeta g y^+ + \tfrac{1}{2}\lambda_\xi k^+ \frac{dg}{dk^+}\, y^+(1+\ln y^+)\right] \tag{5.8}$$

$$\kappa \Delta W^+_r = -\frac{2}{\kappa}\, g\gamma_\xi y^+ \ln y^+ \tag{5.9}$$

6. The combined effects of skewness and roughness

The effects due to the interaction of roughness and skewness are not obvious. Roughness as such would tend to inhibit cross-flow due to an increase of transverse wall friction. But roughness also decreases the streamwise velocity component near the wall which in turn makes the transverse pressure gradient more effective in producing (increased) cross-flow. In order to find which one of these two opposing effects would be more important, it is necessary to carry out an analysis.

The additional shear stress terms have already been evaluated, i.e in (5.4) and (5.5), and these terms should be added to τ_ξ and τ_ζ given by (3.10) and (3.11), respectively, and the resulting expression substituted into (3.3a,b) for τ_x and τ_z. By using the so complemented expressions for τ_x and τ_z in (4.3) and (4.4), the first order combined effects of skewness and roughness can be obtained. To establish the magnitude of the cross-flow in terms of the skewing angle φ, it is necessary to consider a physically realistic case. This will be done in the following section.

7. Results and discussion

To test the new eddy viscosity model developed above, a series of calculations have been performed some of which will be reported here. First the boundary layer flow over an infinite swept wing corresponding to the NLR testcase[7)] has been calculated by the finite difference method described in Ref.3 where the anisotropic eddy viscosities were used. Among the very many variables describing this flow the most critical and difficult to predict is the evolution of β_w - the wall flow angle on the wing surface. Fig.7.1 shows β_w according to measurements and calculations using the new and the conventional (isotropic) eddy viscosity model. A substantial improvement is obtained with the new model.

Next, Fig.7.2 shows the result of the new 3D law-of-the-wall in comparison with the van den Berg extended law. Station 5 of the NLR test case is used and the parameter values of α_ζ and γ_ξ are those given in Ref.4. The parameter value of γ_ξ was evaluated in Ref.4 directly from experimental data points. Since differentiation of measured distributions is often inaccurate, a new set of parameter values at station 5 has been calculated using the finite difference code. As a result a clear improvement is obtained, see Fig.7.3.

The anisotropic eddy viscosity model is seen to produce a slightly increased W^+ in the entire range of y^+ values considered. This is to be expected from the discussion in section 2. Furthermore, the stress direction differs from the strain direction by 5.5^o at the maximum of W^+. This appears to be the right magnitude in comparison with known measurements in 3D turbulent boundary layer flows. Finally,to illustrate the effect of roughness on the law-of-the-wall, a standard sandgrain roughness of g=2 has been applied to the conditions specified for the NLR test case giving the result shown in Fig.7.4. As can be expected the roughness decreases the cross-flow, expressed here by an increase in the skewing angle φ. However, when skewing is considered the skewing angle is reduced to an intermediate value between the rough and the smooth wall result.

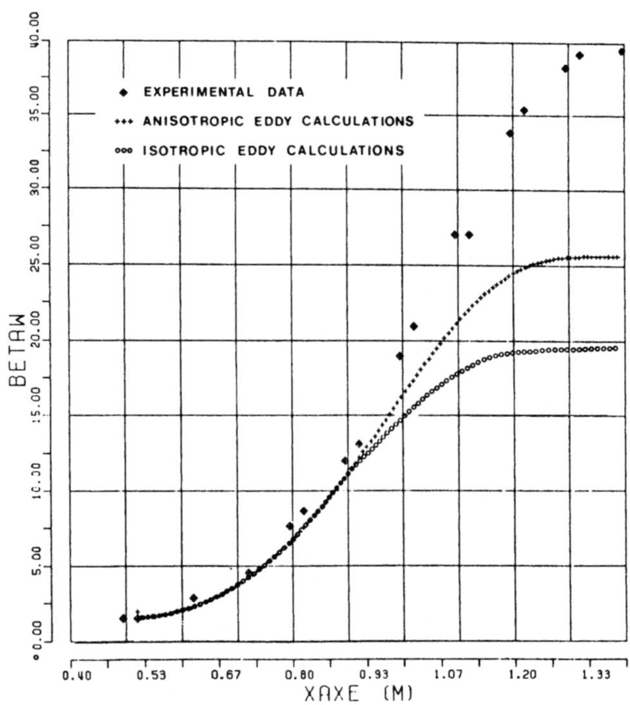

Fig.7.1. The wall angle evolution in the NLR test case according experiment and calculations using isotropic and anisotropic eddy viscosity.

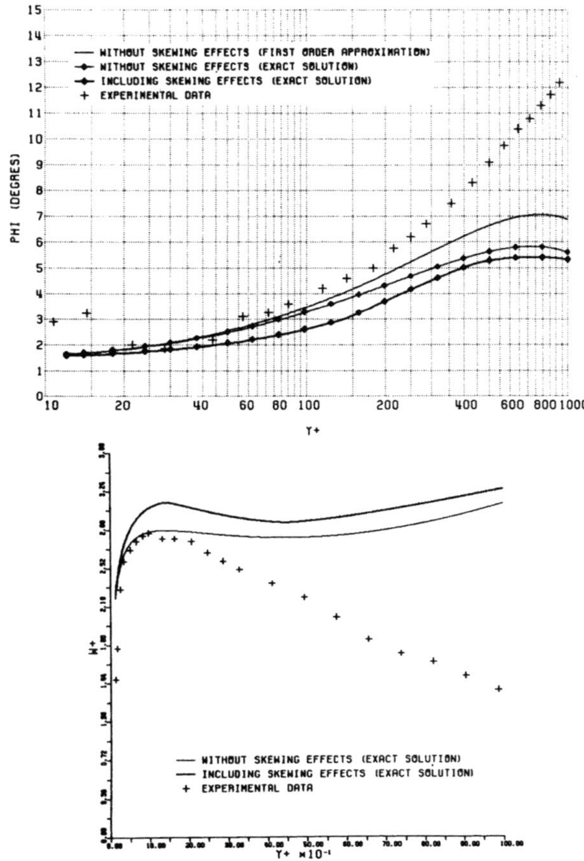

Fig.7.2. Skewing angle φ , and cross-flow velocity W^+ as functions of y^+ in station 5, NLR test case; $\alpha = -5.04 \times 10^{-3}$, $\gamma = 13.1 \times 10^{-6}$

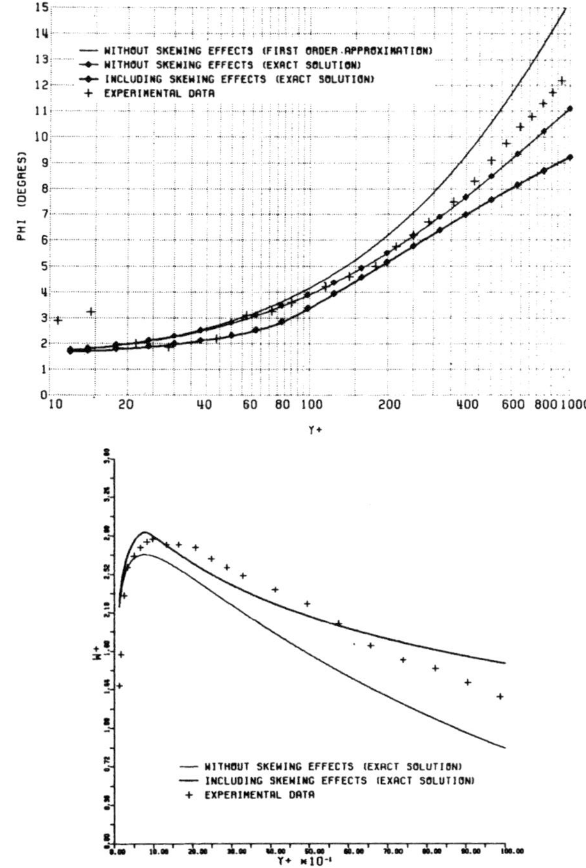

Fig.7.3. Skewing angle φ ,and cross-flow velocity W^+ as functions of y^+ at station 5, NLR test case; $\alpha = -5.38 \times 10^{-3}$, $\gamma = 9.96 \times 10^{-6}$

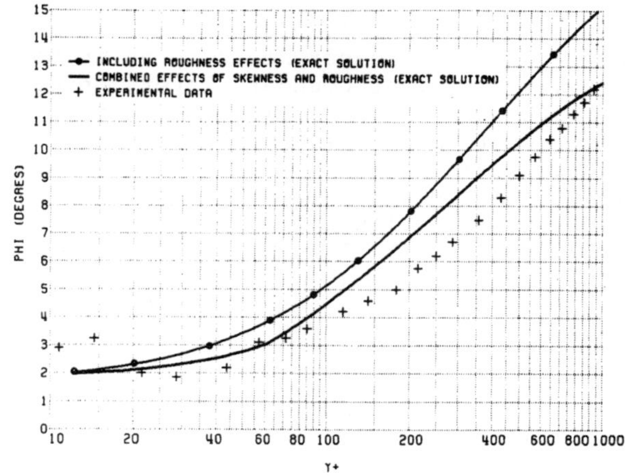

Fig.7.4. Effect of roughness on the NLR test case; sandgrain roughness $g = \Delta u^+ = 2$. The experimental points are those of the smooth wall at station 5. $\alpha_\zeta = -5.38 \times 10^{-3}$, $\gamma_\xi = 9.96 \times 10^{-6}$; "Exact" refers to the method of integration, i.e. the Runge-Kutta method.

Acknowledgements

The authors are indebted to Dr.Alice Nakkasyan and Mr.Vincent Bichet of the Fluid Mech.Lab, ETH, Lausanne, for their aid in performing the finite difference and Runge-Kutta calculations reported in this paper.

References

1) Nash,J.F. and Patel,V.C.: "Three-Dimensional Turbulent Boundary Layers", SBC Technical Books, Atlanta (1972)
2) Smith, A.M.O. and Cebeci,T.: "Numerical solution of the turbulent-boundary-layer equations", MC Donnell Douglas Rept. No DAC 33735 (1967)
3) Fannelop, T.K. and Humphreys, D.A.: "The solution of the laminar and turbulent threedimensional boundary layer equations with a simple finite difference technique", The Aeronautical Research Institute of Sweden, Rept.126 (1974)
4) van den Berg, B.: "A three-dimensional law of the wall for turbulent shear flows", J.Fluid Mech, vol 70, part 1, pp 149-160, (1975)
5) Rotta, J.C.: "Eine theoretische Untersuchung über den Einfluss der Druckscherkorrelationen auf die Entwicklung dreidimensionaler turbulenter Grenzschichten", DFVLR-FB79-05 (1979)
6) Schneider, G.R.: "Die Berechnung der turbulenten Grenzschicht an schiebenden Flügel unundlicher Spannweite mittels eines

dreidimensionalen Mischungswegmodells", DLR-FB77-73 (1978)
7) van den Berg, B. et al: "Measurements in an incompressible three-dimensional turbulent boundary layer, under infinite swept-wing conditions, and comparison with theory", J.Fluid Mech., vol 70,part 1, pp 127-148 (1975)
8) East, L.F.: "Measurements of the turbulent boundary layer on a slender wing", RAE TR 23141 (1974)
9) Krogstad, P.A.: "Calculation of Boundary Layers on Rough Surfaces", SINTEF Rept STF 15 A 81014, NTH, Trondheim.(1981).

Numerical Study of Corner Flows

D. ARNAL and J. COUSTEIX

ONERA/CERT
2 avenue Edouard Belin
31055 TOULOUSE CEDEX, FRANCE

SUMMARY

This paper is devoted to a numerical study of flows which develop along the intersection line of two planes. Attention is focused on three different geometries : symmetrical corner, straight duct and wing-body junction. Since they are characterized by the presence of longitudinal vortices, such flows constitute a good tool for testing turbulence models. It will be shown that neither an algebraic model nor an eddy viscosity model is able to predict satisfactorily all considered cases.

1. INTRODUCTION

Flows along corners formed by two intersecting walls exist in a great number of engineering applications, such as wing-body junctions or non-circular diffusers. In all of these situations, a transverse circulatory motion (secondary flow) develops and greatly alters the main flow. In fact, there are two mechanisms for the generation of secondary flows : one is due to Reynolds stress gradients, the other is induced by lateral skewing of a shear layer. In this paper, the first mechanism will be studied in the cases of a symmetrical corner flow and of a straight duct flow of square cross-section ; the decay of secondary flows generated by the second mechanism will be analyzed in the case of a simplified wing-body junction. For each case, various turbulence schemes have been used and will be compared.

2. GEOMETRY AND EQUATIONS

The considered geometry is given in Fig. 1. The corner is formed by two planes intersecting at right angles ; the x-axis is parallel to the external velocity U_e. The leading edges of the planes are not necessarily located at the same streamwise position, so that the thicknesses of the two merging boundary-layers can be different.

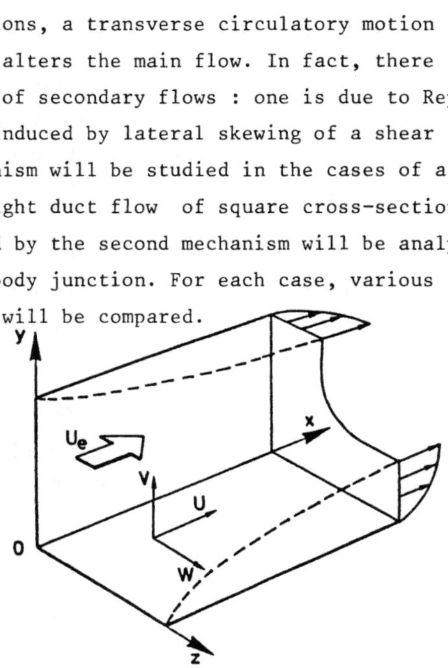

Fig. 1-Geometry and notations

This work was sponsored by the Société Technique des programmes Aéronautiques under GRANT STPA n° 81 95 008.

The equations to be solved are :

$$\frac{DU}{Dt}=-\frac{1}{\rho}\frac{d\bar{P}}{dx}+\frac{\partial}{\partial y}\left(\nu\frac{\partial U}{\partial y}-\overline{u'v'}\right)+\frac{\partial}{\partial z}\left(\nu\frac{\partial U}{\partial z}-\overline{u'w'}\right) \quad (1)$$

$$\frac{D\omega}{Dt}=\omega\frac{\partial U}{\partial x}+\omega_y\frac{\partial U}{\partial y}+\omega_z\frac{\partial U}{\partial z}+\nu\left(\frac{\partial^2\omega}{\partial y^2}+\frac{\partial^2\omega}{\partial z^2}\right)+\frac{\partial^2}{\partial y\partial z}\left(\overline{v'^2}-\overline{w'^2}\right)+\left(\frac{\partial^2}{\partial z^2}-\frac{\partial^2}{\partial y^2}\right)\overline{v'w'} \quad (2)$$

$$\frac{\partial^2 W}{\partial y^2}+\frac{\partial^2 W}{\partial z^2}=\frac{\partial\omega}{\partial y}-\frac{\partial^2 U}{\partial x\partial z} \quad (3)$$

$$\frac{\partial^2 V}{\partial y^2}+\frac{\partial^2 V}{\partial z^2}=-\frac{\partial\omega}{\partial z}-\frac{\partial^2 U}{\partial x\partial y} \quad (4)$$

with $\omega=\frac{\partial W}{\partial y}-\frac{\partial V}{\partial z} \qquad \omega_y=\frac{\partial U}{\partial z}-\frac{\partial W}{\partial x} \qquad \omega_z=\frac{\partial V}{\partial x}-\frac{\partial U}{\partial y}$ (5)

These equations are the streamwise momentum equation, the equation for the streamwise vorticity component and two POISSON equations for the secondary velocities V and W. A set of assumptions /1/ leads to a parabolized system in the x-direction. To solve the elliptic equations (3) and (4), a double sweep procedure is used.

In the ω-equation, the first term on the right-hand side describes the "vortex stretching" ; the two following terms describe the production of "PRANDTL's first kind secondary flows" (skew-induced secondary flows). The last two terms are responsible for maintaining "PRANDTL's second kind secondary flows" (stress-induced secondary flows).

3. CLOSURE RELATIONSHIPS

3.1. Eddy viscosity and algebraic stress models

Two turbulence models have been considered.

The first model (eddy viscosity model) constitutes an extension of BOUSSINESQ's assumption : $\overline{u_i'u_j'}=\frac{2}{3}\delta_{ij}k-\nu_t\left(\frac{\partial U_i}{\partial x_j}+\frac{\partial U_j}{\partial x_i}\right)$

δ_{ij} is KRONECKER's symbol, k the turbulent kinetic energy and ν_t the eddy viscosity. By neglecting the x-derivatives, the stresses appearing in the longitudinal momentum equation are given by :

$$-\overline{u'v'}=\nu_t\frac{\partial U}{\partial y} \qquad -\overline{u'w'}=\nu_t\frac{\partial U}{\partial z} \quad (6)$$

and the vorticity production terms are expressed as :

$$\overline{v'^2}-\overline{w'^2}=-2\,\nu_t\left(\frac{\partial V}{\partial y}-\frac{\partial W}{\partial z}\right)$$

$$\overline{v'w'}=-\nu_t\left(\frac{\partial V}{\partial z}+\frac{\partial W}{\partial y}\right)$$

In the second model, the Reynolds shear stresses are given by algebraic relations deduced from simplified forms of the transport equations for $\overline{u_i'u_j'}$ (algebraic model) /2/. $\overline{u'v'}$ and $\overline{u'w'}$ have the same expression as previously, with $\nu_t=C_\mu\frac{k^2}{\varepsilon}$ (7) and $C_\mu=\frac{1}{C_1}\left(\frac{4}{15}+\frac{-6C_2^2-12C_2+2}{33\,C_1}\right)$, but the expressions of the other stresses are very different :

$$\overline{v'^2} - \overline{w'^2} = \frac{4}{C_1}\frac{(4C_2-1)}{11}\frac{k}{\varepsilon}\left(\overline{u'v'}\frac{\partial U}{\partial y} - \overline{u'w'}\frac{\partial U}{\partial z}\right)$$

$$\overline{v'w'} = \frac{2}{C_1}\frac{(4C_2-1)}{11}\frac{k}{\varepsilon}\left(\overline{u'v'}\frac{\partial U}{\partial z} + \overline{u'w'}\frac{\partial U}{\partial y}\right) \qquad (8)$$

ε represents the isotropic dissipation rate.

This model contains two constants, C_1 and C_2. With $C_2 = 0.25$, no second kind secondary flow can be generated. Calculations show /1/ that, with $C_2 = 0.20$, vortices appear near the conrner, but the circulatory motion is in the opposite sense to that observed in experiments. Better results are obtained with $C_2 = 0.30$; this value has been used for all computations. The recommended value for C_1 lies between 1 and 2 /2/. For the symmetrical corner flow (§4) and for the wing-body junction flow (§6), C_1 is set equal to 1.5, which gives $C_\mu = 0.149$. In fact, this value of C_μ leads to underestimated levels for k, at least in the straight duct flow. To obtain a more accurate description of the turbulent kinetic energy profiles, it is necessary to take $C_\mu \simeq 0.08$.

3.2. Calculation of k, ε and ν_t

Two different methods have been checked :

a) Mixing length - k is deduced from (7) and ε is assumed to be equal to the production rate :

$$k = \sqrt{\frac{\nu_t \varepsilon}{C_\mu}} \quad \text{and} \quad \varepsilon = -\overline{u'v'}\frac{\partial U}{\partial y} - \overline{u'w'}\frac{\partial U}{\partial z} \qquad (9)$$

ν_t is expressed by a classic mixing length relationship :

$$\nu_t = F^2 l^2 \left[\left(\frac{\partial U}{\partial y}\right)^2 + \left(\frac{\partial U}{\partial z}\right)^2\right]^{1/2}$$

where l is the mixing length :

$$\frac{l}{\delta} = \mathcal{L} \tanh\left[\frac{0.41}{\mathcal{L}}\frac{d}{\delta}\right]$$

$\mathcal{L}$ represents the asymptotic mixing length level when d becomes large. In the absence of other specifications, $\mathcal{L}$ will be set equal to 0.085.

The thickness δ has the following expression :

$$\delta = (y\delta_y + z\delta_z)/(y+z)$$

where δ_y (or δ_z) represents the boundary-layer thickness for y = a (or z = b); a and b are defined below (§3.3.). For symmetrical cases, $\delta_y = \delta_z = \delta_\infty$ and $\delta = \delta_\infty$ (see Figure 2).

The length d is calculated from BULEEV's formulation :

$$d = 2yz/((y^2+z^2)^{1/2} + y + z)$$

b) Transport equations for k and ε - Another solution for calculating $\nu_t \sim \frac{k^2}{\varepsilon}$ is to solve the differential equations for k and ε (HANJALIC and LAUNDER /3/) :

$$\frac{Dk}{Dt} = -\overline{u'v'}\frac{\partial U}{\partial y} - \overline{u'w'}\frac{\partial U}{\partial z} - \varepsilon + \frac{\partial}{\partial y}\left(\frac{\nu_t}{\sigma_k}\frac{\partial k}{\partial y}\right) + \frac{\partial}{\partial z}\left(\frac{\nu_t}{\sigma_k}\frac{\partial k}{\partial z}\right) \qquad (10)$$

$$\frac{D\varepsilon}{Dt} = -C_{\varepsilon 1}\frac{\varepsilon}{k}\left(\overline{u'v'}\frac{\partial U}{\partial y} + \overline{u'w'}\frac{\partial U}{\partial z}\right) - C_{\varepsilon 2}\frac{\varepsilon^2}{k} + \frac{\partial}{\partial y}\left(\frac{\nu_t}{\sigma_\varepsilon}\frac{\partial \varepsilon}{\partial y}\right) + \frac{\partial}{\partial z}\left(\frac{\nu_t}{\sigma_\varepsilon}\frac{\partial \varepsilon}{\partial z}\right) \qquad (11)$$

The values given to $C_{\varepsilon 2}$, σ_k and σ_ε are respectively 2, 1 and 1.3. When the ε-equation is applied to the logarithmic region of a two-dimensional boundary-layer, the following relation arises :

$$C_{\varepsilon 1} = C_{\varepsilon 2} - 0.41^2/(\sigma_\varepsilon \sqrt{C_\mu}) \qquad (12)$$

3.3. Near-wall regions treatment - Boundary conditions

The preceding equations are valid only in fully turbulent regions. When a wall is approached, the eddy viscosity ν_t is multiplied by F^2, where F represents a damping function deduced from VAN DRIEST's formulation :

$$F = 1 - \exp\left(\frac{-l}{10.7\,\mu}\sqrt{\tau\rho}\right) \quad , \quad \tau = (\mu + \rho\nu_t)\left[\left(\frac{\partial U}{\partial y}\right)^2 + \left(\frac{\partial U}{\partial z}\right)^2\right]$$

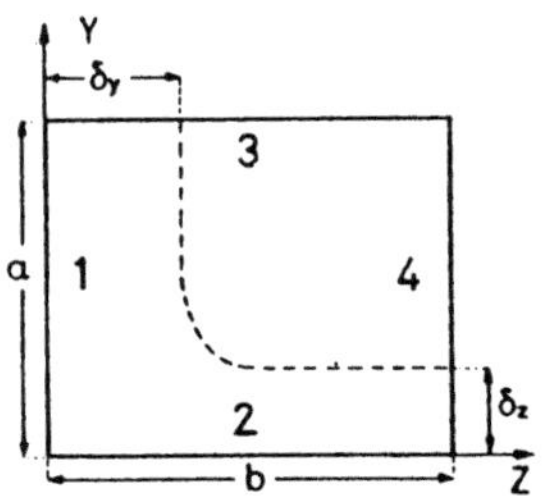

Fig. 2-Integration domain

In each (y, z) plane, the integration domain is rectangular (Figure 2). At solid boundaries 1 and 2, F = 0 and no-slip conditions are prescribed. The wall boundary conditions for ω are obtained by using the derivatives $\frac{\partial V}{\partial z}$ and $\frac{\partial W}{\partial y}$ calculated at y or z = 0 at the upstream station. When the turbulence Reynolds number becomes large, F tends towards unity ; the transport equations for k and ε are solved only for $F > 0.99$, with boundary conditions applied "near the wall" :

$$k = \left[\frac{\overline{u'v'}^2 + \overline{u'w'}^2}{C_\mu}\right]^{1/2}_{F=0.99}$$

$$\varepsilon = \left[-\overline{u'v'}\frac{\partial U}{\partial y} - \overline{u'w'}\frac{\partial U}{\partial z}\right]_{F=0.99} \qquad (13)$$

For the straight duct case, the integration domain remains unchanged when the streamwise distance increases, and a = b (duct half width). For the other two cases studied ("unbounded" corner flows), a = a(x) $\simeq$ 5 to 10 $\delta_z(x)$ and b = b(x) $\simeq$ 5 to 10 δ_y (x). The boundary conditions are :

$$\frac{\partial U}{\partial y} = \frac{\partial V}{\partial y} = \frac{\partial W}{\partial y} = \frac{\partial k}{\partial y} = \frac{\partial \varepsilon}{\partial y} = 0 \quad \text{and} \quad \omega = 0 \quad \text{along side 3}$$

$$\frac{\partial U}{\partial z} = \frac{\partial V}{\partial z} = \frac{\partial W}{\partial z} = \frac{\partial k}{\partial z} = \frac{\partial \varepsilon}{\partial z} = 0 \quad \text{and} \quad \omega = 0 \quad \text{along side 4}$$

These conditions are obvious for the square duct flow ; for the "unbounded" flow cases, the conditions $\frac{\partial V}{\partial y} = 0$ prescribed along boundary 3 can be replaced by V = 0 and the condition $\frac{\partial W}{\partial z} = 0$ imposed along boundary 4 can be replaced by W = 0, without important changes in the numerical results /1/.

4. SYMMETRICAL, UNBOUNDED STREAMWISE CORNER

The two intersecting surfaces are identical and semi-infinite. Experimental results /4/ /5/ indicate that the main feature of such a flow is the presence of stress-induced secondary flows. These secondary flows consist of two "vortices", the presence of which is shown by a distorsion of the isovels towards the corner (lines of constant streamwise velocity in a plane normal to the main flow).

In order to estimate the U, V and W fields in the initial (y, z) plane, CARRIER's formulation /6/, initially used for the laminar corner flow problem has been extended. This method is described in /1/. It can be pointed out that the two vortices responsible for the secondary flow are not included in the initial conditions.

ω is calculated from its definition. When the k-ε equations are solved, the initial values of these quantities are calculated by the relations (9), which are supposed to apply at each point where F $\geqslant$ 0.99.

Numerical results indicate that an accurate description of the anisotropy of turbulence is needed to reproduce the vortex-like pattern of the secondary flows. The eddy viscosity model gives rise to very small vorticity production terms and stress-induced secondary flows cannot develop /1/. For this reason, only numerical results obtained with the algebraic model are presented.

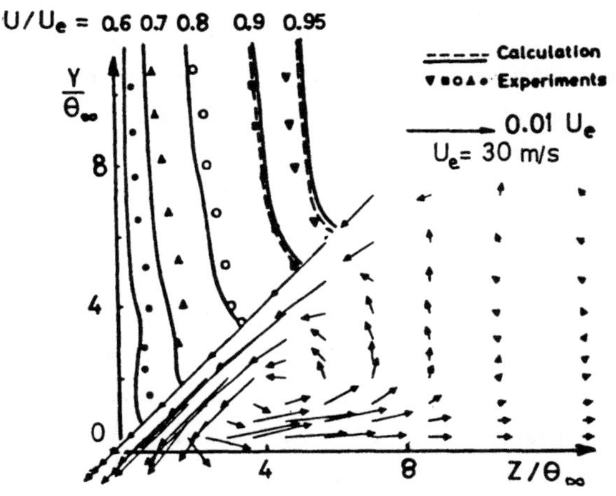

Fig. 3-Comparison with the experiments of MOJOLA and YOUNG —— *mixing length scheme* ---- *k-ε model*

Figure 3 allows a comparison between predicted and measured isovels (data of MOJOLA and YOUNG /4/, dP/dx = 0). Computations have been carried out by using the mixing length scheme and by solving the k-ε equations : both methods give similar results for the isovels, as well as for the secondary velocities. The overall distorsion of the experimental contours is well reproduced. Nevertheless, the calculation underestimates the extent of the distorted region. Figure 3 also shows one of the two vortices computed with the k-ε model.

5. STRAIGHT DUCT FLOW OF SQUARE CROSS SECTION

Due to symmetry considerations, calculations are performed in one quadrant only : $0 \leqslant y \leqslant a$, $0 \leqslant z \leqslant a$, where a represents the duct half width. As for the unbounded corner case, experimental results indicate the development of two vortices which are centered about the corner bisector in each quadrant.

The calculation starts at the duct inlet (x = 0), where the flow is supposed uniform : $U = U_b$ (bulk velocity). The secondary velocities, the vorticity, k and ε are set equal to zero. In the considered case (experiments of GESSNER et al. /7/), U_b = 15.2 m/s and a = 0.127 m.

Compared with the unbounded corner flow, the main numerical difference lies in the expression of $\frac{d\overline{P}}{dx}$, which is not a given function of the streamwise distance, but which must be calculated step by step so as to ensure the total mass conservation.

As for the unbounded corner flow, the eddy viscosity model is not adequate for calculating the development of secondary flows in a straight duct. Numerical results presented below are obtained with the algebraic model.

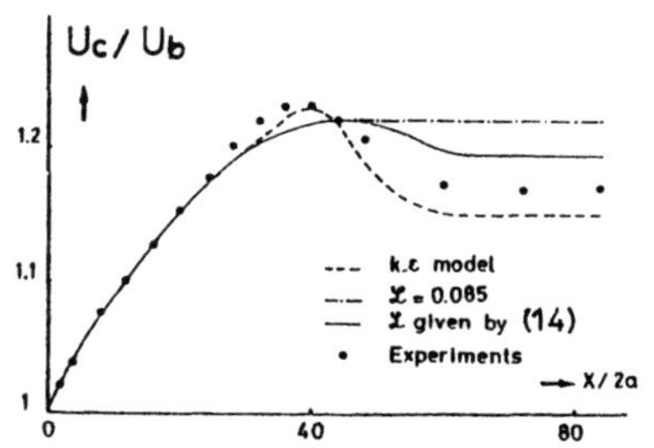

Fig. 4-Evolution of the axial velocity

Figure 4 shows the evolution of the axial velocity U_c made dimensionless with the bulk velocity. Experiments indicate a peak value in the region where the boundary layers start to merge ; at downstream stations, the flow is fully developed. When calculations are performed with a constant asymptotic mixing length level ($\mathcal{L}$ = 0.085), the "overshoot" does not appear. Better results are obtained by prescribing a streamwise evolution of $\mathcal{L}$:

$$\mathcal{L} = 0.1075 + 0.0143 \text{ Arctg}\left(\frac{x}{2a} - 50\right) \qquad (14)$$

With this expression, $\mathcal{L}$ exhibits a smooth evolution from 0.085 at the duct entrance to 0.13 for large values of x/2a. GESSNER and EMERY /8/ have developed calculations by using similar techniques. An important advantage of the k- ε model is to predict the peak value of the axial velocity without adjustment. Let us observe that the mixing length scheme is unable to predict the increase in the axial turbulence intensity when the flow develops.

The computed secondary velocities, indicated by arrows, and longitudinal iso-velocity lines are plotted in Figure 5 for two streamwise stations. Calculations are performed with the k- ε model.

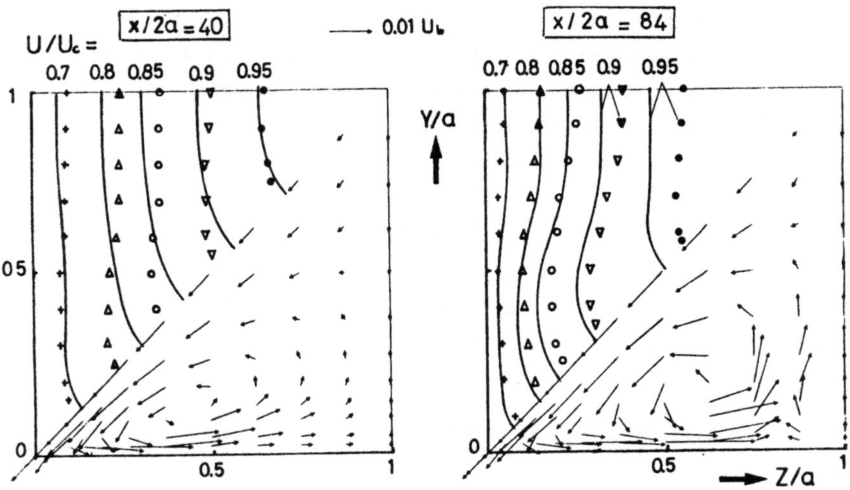

Fig. 5-Duct flow calculations : ✦ ▲ ○ ▼ ● *Experiments* —— *Calculations*

6. SIMPLIFIED WING-BODY JUNCTION

In corner flows or in duct flows without curvature, the terms describing the production of skew-induced secondary flows in the ω-equation are usually very small. On the other hand, in the case of the wing-body junction, first kind secondary flow will appear.

Such a configuration has been studied by SHABAKA /9/, with a near-zero pressure gradient. The wing is fixed to the wind tunnel floor and has a semi-elliptical leading edge followed by a section of constant thickness (Figure 6).

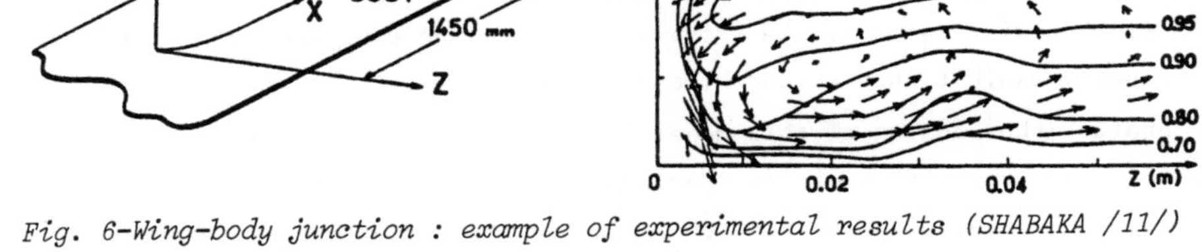

Fig. 6-Wing-body junction : example of experimental results (SHABAKA /11/)

Upstream from the leading edge, the body boundary-layer separation induces the formation of a strong horseshoe vortex ; as the flow accelerates, this vortex is stretched and forms two legs, each extending along one of the wing sides.

Detailed measurements are available only downstream from the end of the leading edge. The experimental results given in Figure 6 show that the first kind secondary flow looks like an anticlockwise rotary motion and is much larger than the stress-induced vortices measured in straight corners.

The slow decay of this skew-induced vortex has been computed by starting the calculation at the station corresponding to Figure 6. In the next figures, only numerical results obtained by the k- ε model are presented (the mixing length scheme gives nearly identical results). As was the rule for the STANFORD computations /10/, the initial conditions are the experimental fields of U, V, W (and then ω) and k ; the initial values of ε are obtained by writing $\varepsilon = C_\mu k^2/\nu_t$, ν_t being calculated from a mixing length scheme.

a) Algebraic model - Two values of C_2 have been tested : 0.25 and 0.30.

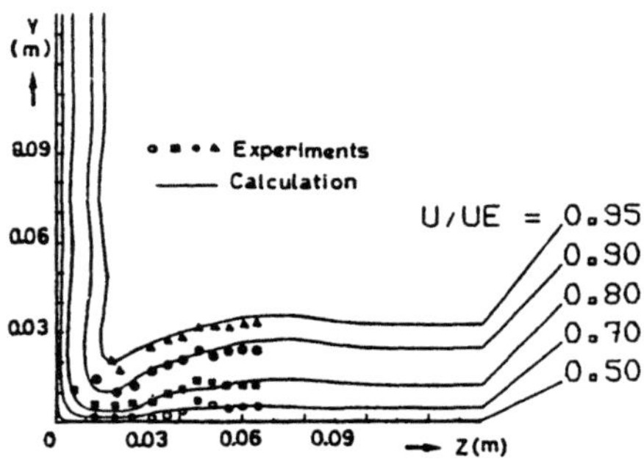

Fig. 7-Algebraic model : comparison with experiments

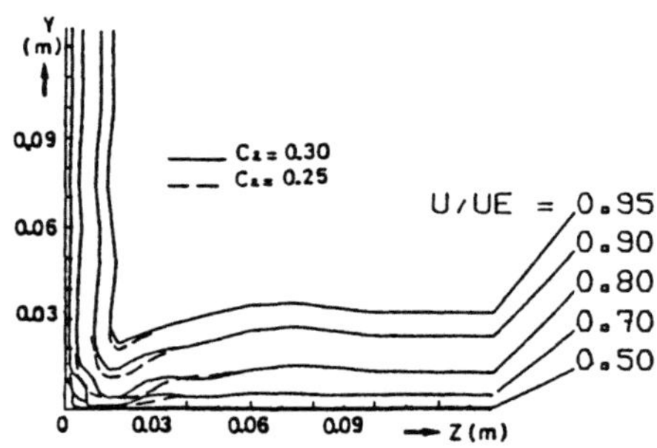

Fig. 8-Algebraic model : influence of C_2

Figure 7 shows a comparison between calculated and experimental isovels at x = 1.254 m, with C_2 = 0.25.

A comparison between calculations performed with the two values of C_2 is presented in Figure 8. The overall pattern of the isovels remains unchanged, but with C_2 = 0.30, two small vortices corresponding to stress-induced secondary flows are present near the corner ; their existence is indicated by a visible distorsion of the isovels. However, the experimental results do not clearly exhibit such second kind secondary flows.

Another shortcoming of this model is shown in Figure 9, which compares the secondary velocities fields. For z $>$ 4 cm, the calculated transverse velocity W has values which are too large and the vortex center is not correctly located.

b) Eddy viscosity model - Along the z-direction, the isovels are displaced towards the left, compared with the algebraic model (Figure 10). Impro-

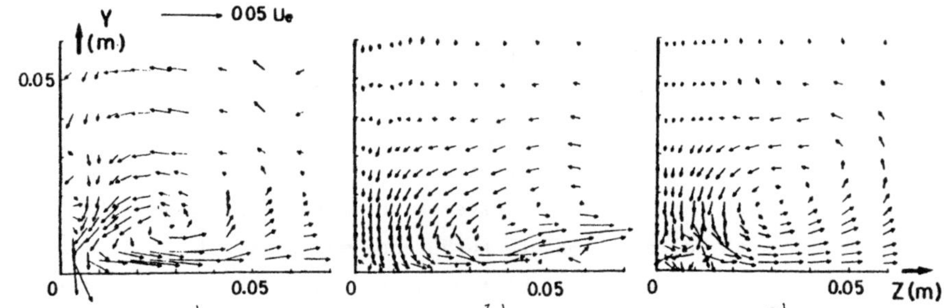

Fig. 9-Secondary velocities (x = 1.254 m) a) Experiments - b) Algebraic model (C_2 = 0.25) - c) Eddy viscosity model

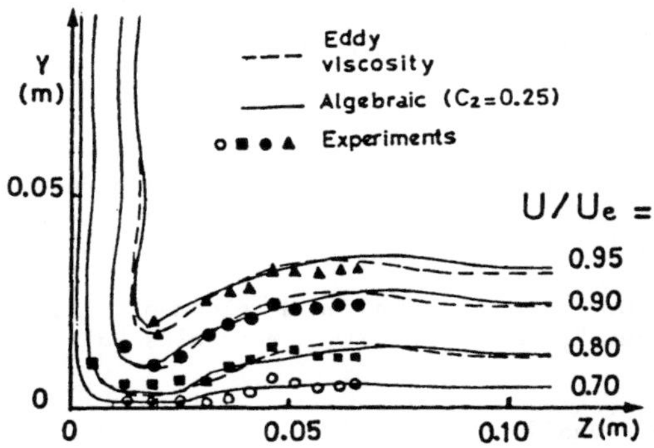

Fig. 10-Comparison between experimental and computed isovels

vements appear in the (V, W) field (Figure 9). The transverse velocity is in fairly good agreement with measurements, as well as the location of the vortex center. Comparisons with $\overline{u'v'}$-profiles confirm the superiority of the eddy viscosity model. These results are very close to those obtained by RODI et al. /10/.

These comparisons indicate that the skew-induced secondary flow has a strong effect on the vorticity turbulent production terms : the vortex induces Reynolds stresses which act in reducing the vortex intensity. This effect is not taken into account by the algebraic turbulence model, but is partially reproduced by the eddy viscosity model.

7. CONCLUSIONS

The most important conclusions can be summarized as follows :

In the case of external flows, such as unbounded corners or wing-body junction, the mixing length model and the k- ε model give similar results. The same remark is not valid for the internal flows ; for example, the flow development in a straight duct cannot be well reproduced by the mixing length model without empirical modifications.

When the vorticity production terms $\overline{v'^2} - \overline{w'^2}$ and $\overline{v'w'}$ are expressed by an algebraic model, stress-induced vortices appear in the calculations, and qualitatively good results are obtained. On the contrary, the eddy viscosity model is unable to predict second kind secondary flows.

In the case of the wing-body junction, Reynolds stresses set up by the skew-induced vortex reduce the vortex intensity. This slow attenuation is described by the terms $\frac{\partial W}{\partial y}$, $\frac{\partial W}{\partial z}$, $\frac{\partial V}{\partial y}$ and $\frac{\partial V}{\partial z}$ appearing in the eddy viscosity model. In the algebraic model, Reynolds stresses depend only upon U and the computed transverse velocity W remains too large.

The problem is to build a complete turbulence model including eddy viscosity and algebraic models for calculating a wide range of complex three-dimensional flows without adjustment from one case to another. A solution might be to solve Reynolds stress equations in which the terms $\frac{\partial U}{\partial y}$, $\frac{\partial U}{\partial z}$, $\frac{\partial V}{\partial y}$, $\frac{\partial V}{\partial z}$, $\frac{\partial W}{\partial y}$ and $\frac{\partial W}{\partial z}$ would be retained. Possibly, these equations might be simplified by neglecting the convection and diffusion terms.

REFERENCES

/1/ ARNAL D., COUSTEIX J. : "Turbulent flow in unbounded streamwise corners" Proceedings of the Third Symposium on Turbulent Shear Flows, University of CALIFORNIA, DAVIS, September 9-11, 1981

/2/ LAUNDER B.E., REECE G.J., RODI W. : "Progress in the development of a Reynolds stress turbulence closure", Journal of Fluid Mechanics Vol. 68 Part 3, 1975, pp. 537-566

/3/ HANJALIC K., LAUNDER B.E. : "A Reynolds stress model of turbulence and its application to thin shear flows", Journal of Fluid Mechanics, Vol.52 Part 4, 1972, pp. 609-638

/4/ MOJOLA O.O., YOUNG A.D. : "An experimental investigation of the turbulent boundary-layer along a streamwise corner", AGARD CP-93, Agard Symposium on Turbulent Shear Flows, 1972

/5/ BRAGG G.M. : "The turbulent boundary-layer in a corner", Journal of Fluid Mechanics, Vol. 36, Part 3, 1969, pp. 485-503

/6/ CARRIER G.F. : "The boundary-layer in a corner", Quart. Appl. Math. 4, 1947, pp. 367-370

/7/ GESSNER F.B., PO J.K., EMERY A.F. : "Measurements of developing turbulent flow in a square duct", Proceedings of the Symposium on turbulent shear flows, University Park, PENNSYLVANIA, April 18-20, 1977

/8/ GESSNER F.B., EMERY A.F. : "A length-scale model for developing turbulent flow in a rectangular duct", Journal of Fluids Engineering, June 1977

/9/ SHABAKA I.M.M.A. : "Turbulent flow in a simulated wing-body junction", Ph. D. Thesis, Aeronautics Dept., Imperial College London, ENGLAND, 1979

/10/ The 1980-81 AFOSR-HTTM STANFORD Conference on Complex Turbulent Flows-Comparison of computation and experiment, September 14-18, 1981 - STANFORD University (Preprints)

Three-Dimensional Boundary-Layer Calculations in Design Aerodynamics

E.H. Hirschel

Messerschmitt-Bölkow-Blohm GmbH
UFE122
Postfach 80 11 60
8000 München 80
Federal Republic of Germany

Summary

The results of a calculation of the three-dimensional, turbulent, compressible boundary layer on the fuselage of a transport airplane are discussed. The data were found by an integral method. The interpretation of the results at the base of the fuselage is made in detail. With the help of the $|\tau|$-minimum criterion a separation region is found.

Introduction

Methods for the computation of three-dimensional turbulent boundary layers are in such a state of development today that an application to aerodynamic design problems is possible. With test computations as for instance organized by the EUROVISC-Working Party on Three-Dimensional Shear Layers [1, 2] an acceptable level of confidence has been established. Of course the problems in design aerodynamics often are so complicated that only approximate answers can be found with boundary-layer computations. Severe problems may arise regarding the choice of initial data, the location of the line of laminar-turbulent transition, separation bubbles, shock-boundary-layer interactions and so on. The interpretation of the boundary-layer results for design purposes is a further problem, especially if separation is indicated. Three-dimensional boundary layers may separate in quite different ways, and, moreover, a breakdown of the integration process may not necessarily be due to actual separation [3].

Three-dimensional boundary-layer computations in design aerodynamics require besides integration methods for the boundary-layer equations methods for the definition of the surface boundary-layer coordinates and the accompanying metric tensor, as well as methods for the evaluation of the results of the boundary-layer computation (e.g. displacement thickness, skin-friction lines, etc.). Recently means have been developed to this end (for instance [3]) so that boundary layers on quite general body surfaces can be treated.

In the present paper the results of a calculation of the three-dimensional, turbulent, compressible boundary layer on the fuselage of a transport airplane are discussed. Special consideration is given to the flow at the rear end of the fuselage where separation due to its upsweep occurs. Unfortunately no experimental data besides some vague free-flight measurements of the boundary layer in that area are available. The overall results, however, appear acceptable, especially in view of results from other fuselage configurations where comparisons were possible. Of course, only a detailed comparison with experimental data could reveal the weak points of the solution.

The boundary-layer method used for the study is the integral method of COUSTEIX [4] in the form of COUSTEIX-AUPOIX. In this method a mixing length model is used together with similarity solutions to find families of both mainflow and crossflow profiles. The surface is treated as insulated. The continuity equation is applied as entrainment equation. The method is formulated for arbitrary boundary-layer coordinates without the assumption of small crossflow. The original method has been modified to take into account the non-alignment of the Reynolds stress and the velocity gradient by choosing a factor $0 \leq T \leq 1$.

The concepts used for the construction of the coordinates, the computation of the metric tensor, transformation etc. in the present study are those given in [3], which are also used to evaluate the results of the integral method [5, 6].

Geometry, Coordinates

The geometry of the fuselage is given in Fig. 1. All cross-sections except for those in the nose region are circular. The Cartesian reference coordinate system has its origin in the nose of the fuselage. The boundary-layer coordinates are surface oriented

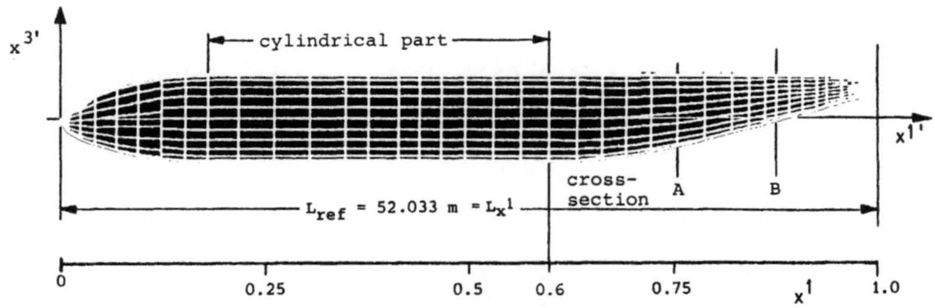

Fig. 1. Panel model of transport airplane fuselage

cross-section coordinates [3], which are orthogonal in the axisymmetric parts of the fuselage and non-orthogonal in the other parts, Fig. 2. Details and the metric tensor are given in [7].

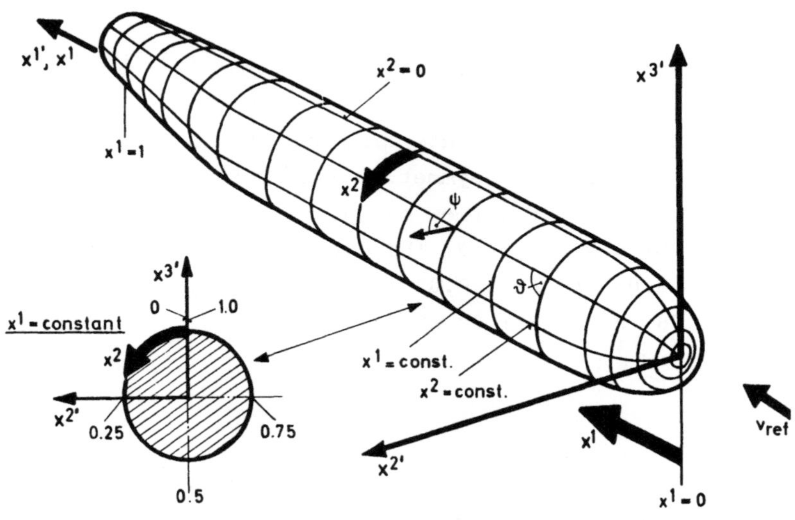

Fig. 2. Schematic of coordinates (Cartesian coordinates: $x^{i'}$ ($i' = 1, 2, 3$); boundary-layer coordinates: x^{α} ($\alpha = 1, 2$))

Flow Parameters, Initial Data, External Flow Field

The free-stream Mach number for the case presented is $M_\infty = 0.8$. The Reynolds number is $Re = 7.06 \cdot 10^8$ with the reference length $L_{ref} = 52.033$ m. Approximate initial data were prescribed at $x^1 = 0.08$ well behind the nose region, but in still accelerated inviscid flow. The inviscid outer flow field was computed with the MBB Panel Method [8] at zero angle of attack of the fuselage. Compressibility was accounted for by the Göthert rule.

In Fig. 3 the external inviscid velocity $v_e = |\underline{v}_e|$ is given as function of the surface parameters x^α. Fig. 4 shows the angle ψ_e

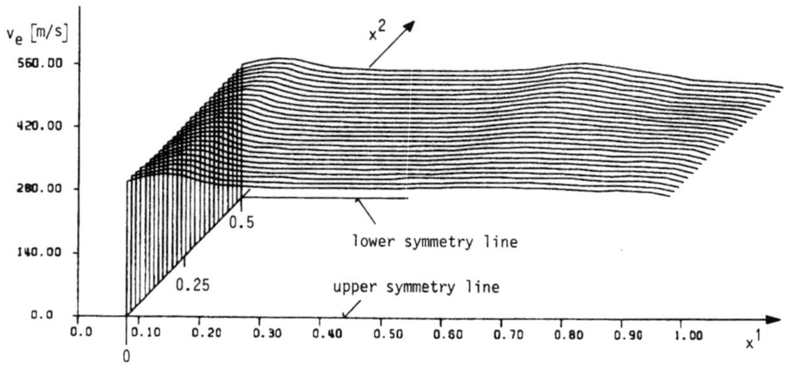

Fig. 3. External inviscid velocity $v_e(x^1, x^2)$ *)

between the x^2 = const.-coordinate line and the inviscid velocity vector. Because the flow is symmetric with respect to the $x^{1'}$-$x^{3'}$-plane (Fig. 2) only the flow on the right half of the fuselage ($0 \leq x^2 \leq 0.5$) was computed. The data were prescribed pointwise with $\Delta x^1 = 0.03$ and $\Delta x^2 = 0.02$ in the boundary-layer calculation domain $0.08 \leq x^1 \leq 0.98$; $0 \leq x^2 \leq 0.5$.

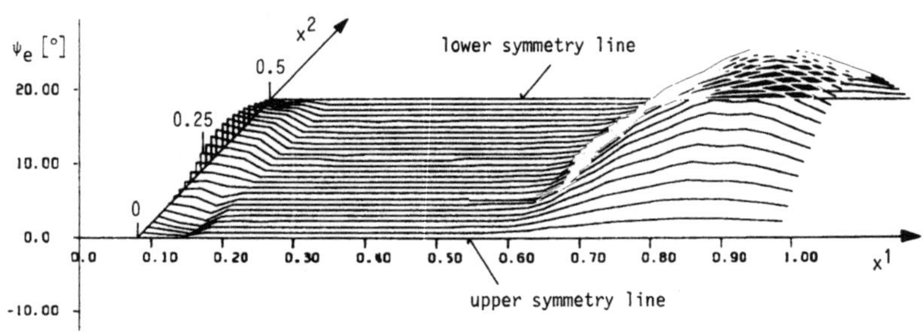

Fig. 4. Angle $\psi_e(x^1, x^2)$ between x^2 = const.-coordinate line and the vector $\underline{v}_e$ of the external inviscid flow

*) if not stated otherwise data are shown for right half of fuselage only

Results

The computed wall shear-stress coefficient $c_f = |\underline{c}_f|$ is given in Fig. 5. The value is nearly constant in the cylindrical part of the fuselage. Although the coefficient is referenced to the local

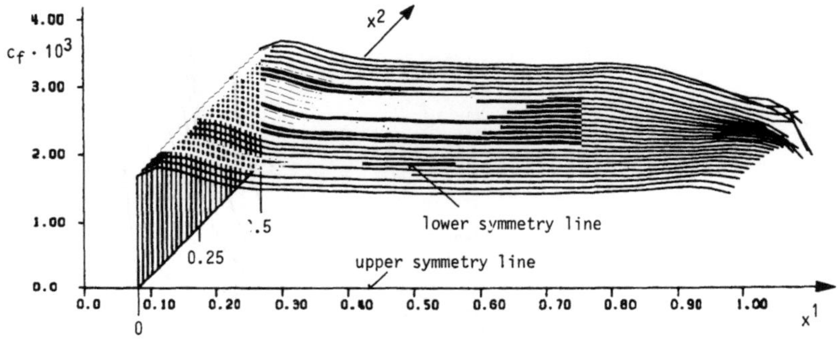

Fig. 5. Wall shear-stress coefficient $c_f(x^1, x^2)$

inviscid data $0.5\rho_e v_e^2$ which decrease slightly in that region the plateau cannot be explained. In the upper symmetry line for $x^1 >$ 0.6 the flow is divergent (Fig. 4) which leads to a rise of c_f and to a decrease of the boundary-layer thickness (see Figs. 7 and 8). In Fig. 6 the angle ψ_W between the x^2 = const.-coordinate line and the wall shear-stress vector behaves like the angle ψ_e in Fig. 4. However, at the rear end of the fuselage it is much more pronounced. Due to the upswept rear part of the fuselage the external flow is forced beneath the tail and the boundary layer follows on more curved streamlines.

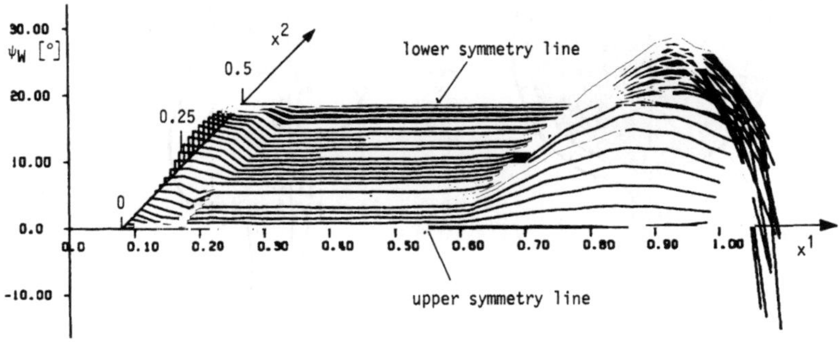

Fig. 6. Angle $\psi_W(x^1, x^2)$ between x^2 = const.-coordinate line and wall shear-stress vector $\underline{c}_f$

The boundary-layer thickness δ and the displacement thickness δ_1 Fig. 7, are other indicators of this flow behaviour. At the lower symmetry line both first rise slightly more than in the other regions. This is due to a slightly convergent flow already in the cylindrical part of the fuselage. In the rear part finally a very strong rise occurs because of the strongly convergent flow in

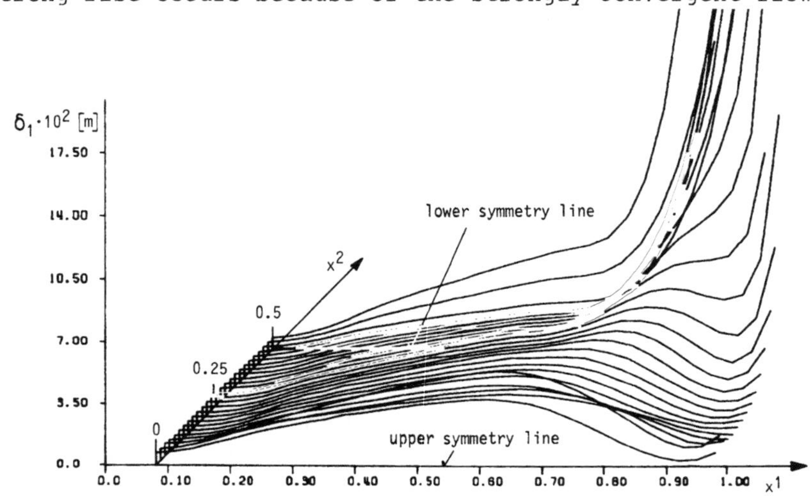

Fig. 7. Displacement thickness $\delta_1(x^1, x^2)$

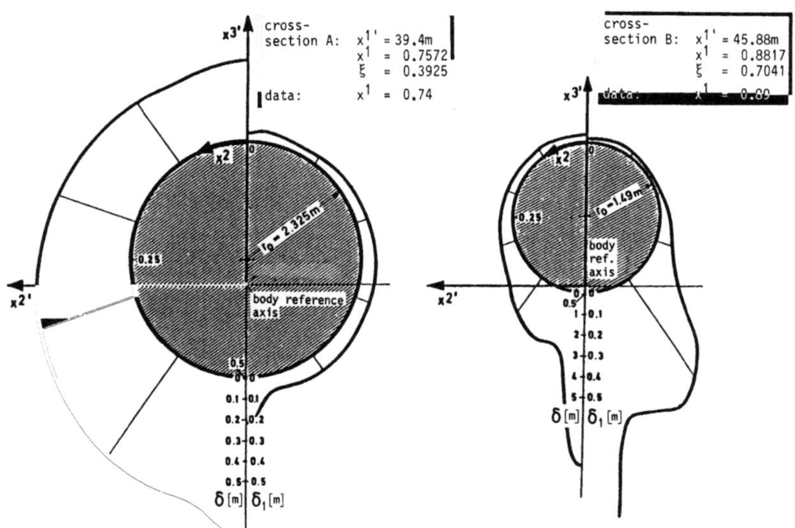

Fig. 8. Boundary layer thickness δ and displacement thickness δ_1 at cross-sections A and B

this region at the lower part of the fuselage. Fig. 8 shows δ and δ_1 in the two cross-sections A and B indicated in Fig. 1. In cross-section A the thickening in the lower symmetry line is well discernable, in section B the boundary layer obviously is near separation in this region (see next chapter).

The formal shape parameter $H_{tt} = \delta_{1t}/\delta_{2tt}$ in Fig. 9 is nearly constant all over the surface (t ≙ main-flow direction). At the lower symmetry line towards the end of the fuselage a strong rise occurs.

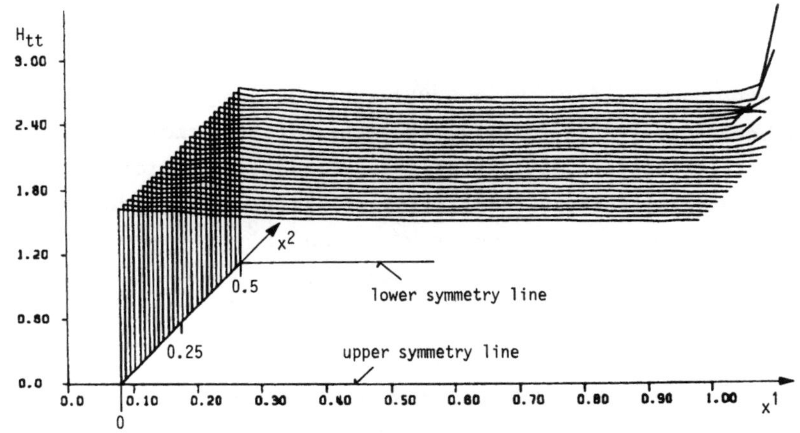

Fig. 9. Formal shape parameter $H_{tt}(x^1, x^2)$

Fig. 10 finally shows the cumulative, integrated friction forces on the surface. They are found from the computed wall shear-stress coefficients. The friction forces per surface element are

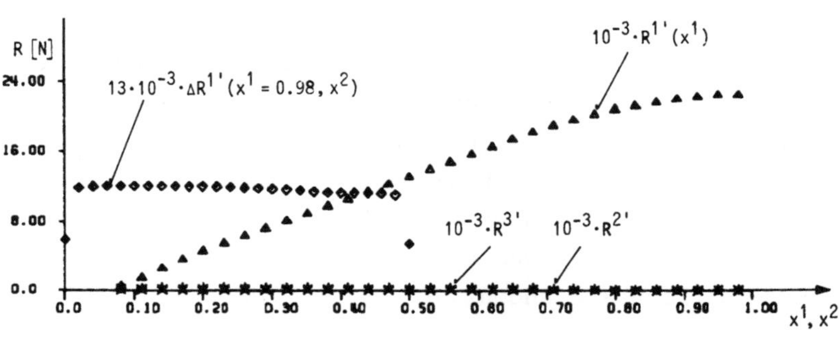

Fig. 10. Friction forces $R^{i'}(x^1)$ and $\Delta R^{1'}(x^1 = 0.98, x^2)$

$$\Delta R^{\alpha} = \tau_W^{\alpha 3}\, dF = \tau_W^{\alpha 3} \sqrt{a}\, dx^1 dx^2; \qquad (\alpha = 1, 2). \tag{1}$$

Here a is the determinant of the metric tensor of the surface coordinate system [3, 7]. The wall shear-stress $\tau^{\alpha 3}$ can be computed from c_f and ψ_W.

The components of the friction forces per surface element ΔR^{α} are then transformed into the Cartesian components $\Delta R^{i'}$ by

$$\Delta R^{i'} = \beta_{\alpha}^{i'}\, \Delta R^{\alpha}; \qquad (i' = 1, 2, 3), \tag{2}$$

where $\beta_{\alpha}^{i'}$ are the components of the base vectors of the surface coordinate system [3, 7].

The computed value of $R^{1'}(x^1 = 0.98) = 20450N$ compares well with the handbook value $R^{1'} = 20200N$. The cumulative friction forces in $x^{2'}$ and $x^{3'}$-direction are negligible. The distribution of $R^{1'}$ in circumferential direction $\Delta R'^{1}$ shows a decrease towards the lower symmetry line. The two points at $x^2 = 0$ and $x^2 = 0.5$ are referred to $\Delta x^2/2$ and therefore at a lower level.

Separation

A breakdown of the solution process occured in the region $0.92 \lesseqgtr x^1 \leq 0.98$; $0.3 \lesseqgtr x^2 \leq 0.5$. The questions are whether the boundary layer in reality is separating in this region and what kind of separation occurs. In the following the results presented in the preceding chapter are interpreted with regard to these questions.

First the results in this respect of a computation of the boundary layer on a 1:6 ellipsoid which was investigated experimentally are discussed. Fig. 11 shows the inviscid streamlines and the skin-friction lines on the ellipsoid at an angle of attack $\alpha = 5°$. The inviscid flow field was calculated for this case with the method [8] and the three-dimensional laminar boundary layer with the method [9]. The computed skin-friction lines in Fig. 13 show a strong convergence pattern at $x^1 > 0.5$; $x^2 \approx 0.18$ (the solution proceeded in the area only where the inviscid streamlines and the skin-friction pattern are shown).

At $x^1 \approx 0.25$ on the upper symmetry line lies the point where the

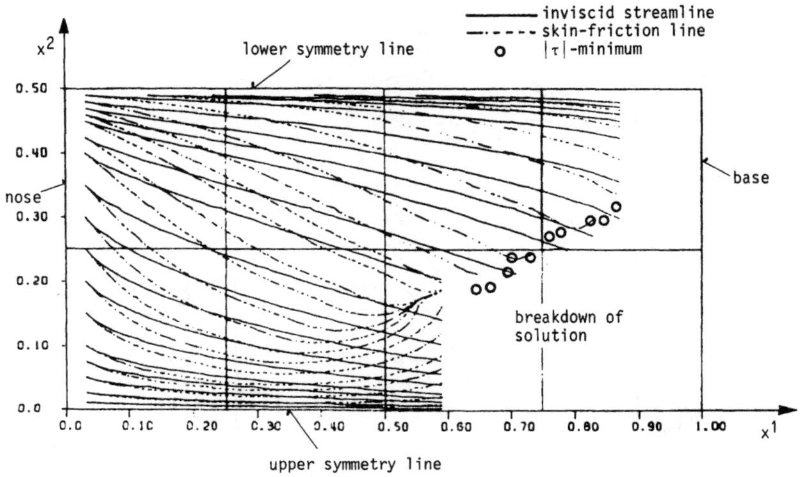

Fig. 11. Inviscid streamlines and skin-friction lines in the x^α-plane of a 1:6 ellipsoid at a.o.a. $\alpha = 5°$ (right half of ellipsoid)

general convergence starts (see also [10]). In Fig. 13 included are $|\tau|$-minimum-locations, which lie in the immediate vicinity of the separation line [11]. These points were found by applying the theory in [11] to the present results (in general with the theory [11] only separation lines of the type "local separation" [12] can be treated). The result, although with some scatter, agrees reasonably well with experimental data for the same case [13], which are not shown here.

In Fig.12a the computed energy-loss thickness δ_3 and the displacement thickness δ_1 are given. The curves show the typical bulging which appears in regions of skin-friction line convergence, and especially behind the convergence point where, at least for laminar flow, corners in the δ_1-contour are observed (see also [14]). In turbulent flow the bulging of δ_1 appears to be more pronounced, although no corners appear, and δ_1 becomes faster negative, Fig. 12b.

Thus the convergence of skin-friction lines, the appearance of a $|\tau|$-minimum line, the bulging of boundary-layer thickness contours appear to be an indicator of, at least, lines of "local separation". Singular points in the wall-shear stress vector field of course require completely vanishing wall-shear stress, which

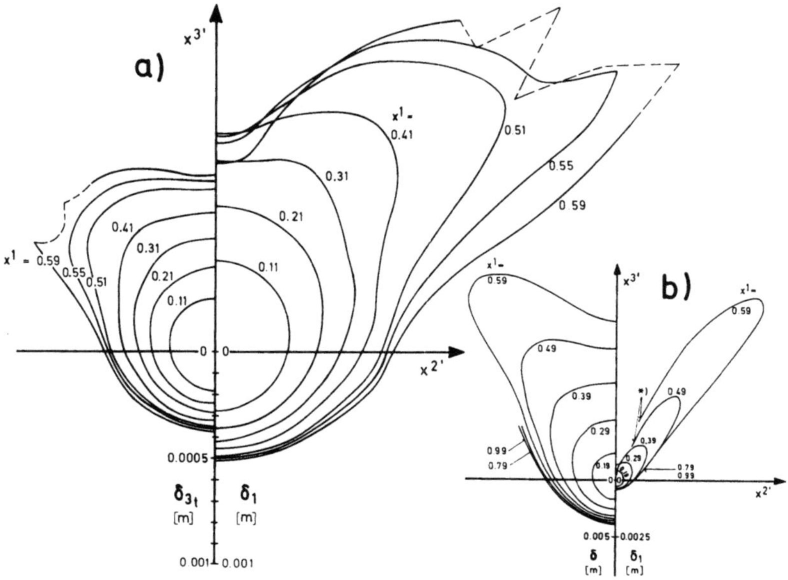

Fig. 12. a) Energy-loss thickness δ_3 and displacement thickness δ_1 at stations x^1 = const. of 1:6 ellipsoid at a.o.a. $\alpha = 5°$, laminar flow; b) boundary-layer thickness δ and displacement thickness δ_1 on cylindrical body at a.o.a. $\alpha = 2.5°$, turbulent flow [15]

here for the lower symmetry line appears to be approached at $x^1 \approx 0.98$, Fig. 5, accompanied by a steep rise of H_{tt}, Fig. 11. Of course, because neither local nor global interaction of the separating boundary layer with the inviscid outer flow is taken into

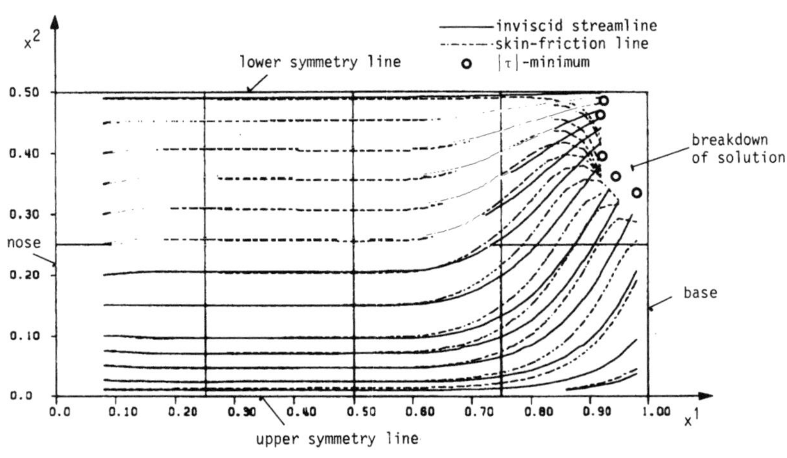

Fig. 13. Inviscid streamlines and skin-friction lines in the x^α-plane

account, locations of separation regions found in this way from boundary-layer computations are only approximate ones.

The breakdown of the solution for the transport-airplane fuselage appears in view of the above results to indicate separation. In Fig. 13 the skin-friction lines converge, an accompanying bulging of δ_1 is seen in Fig. 8. A $|\tau|$-minimum line resulted from application of the theory [11]. Figs. 8 and 13 seem to indicate a bubble-type separation straddling the lower symmetry line together with a "local separation" leading to a pair of lee-side-type vortices.

In Fig. 14 finally the inviscid streamlines and the skin-friction lines are given on the real configuration, with enlargements of the base region in Fig. 15.

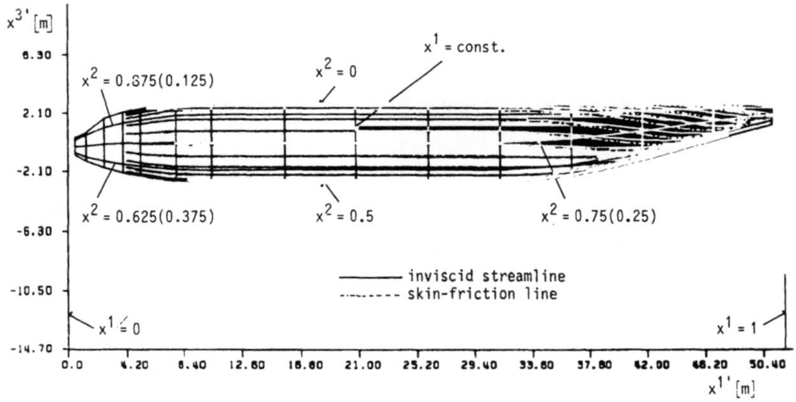

Fig. 14. Inviscid streamlines and skin-friction lines on the real configuration (side view)

Conclusions

The compressible turbulent boundary layer past a transport-airplane fuselage was computed. The results show the development of the solution up to the immediate vicinity of the base. Considerations of details of the solution together with an application of a separation criterion allows to identify the approximate location of a separation region. The discussion shows what results can be found from three-dimensional boundary-layer calculations

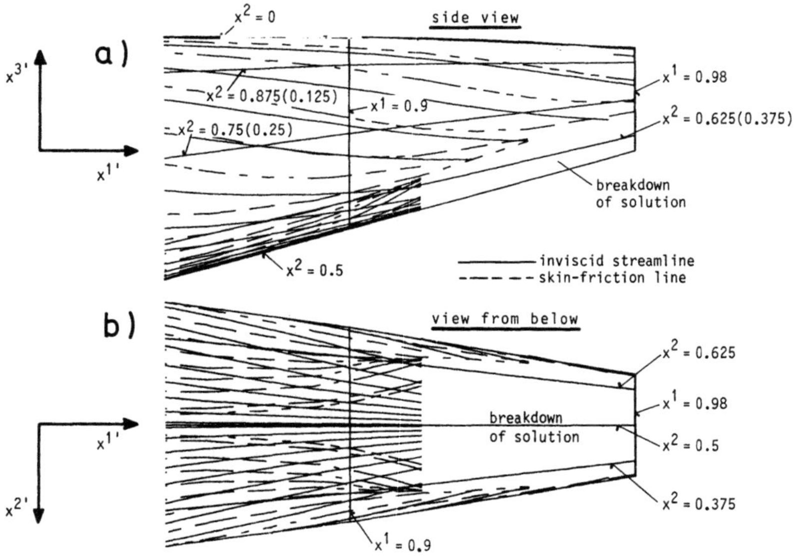

Fig. 15. Inviscid streamlines and skin-friction lines on the real configuration in the base region; a) side view, b) view from below

although many problems are not yet solved. From the point of view of the design aerodynamicist the, at least approximate, description of the local and global interaction in the separation region in order to arrive at force data for the complete configuration is the next necessary step.

References

1. Humphreys D.A.: Comparison of Boundary-Layer Calculations for a Wing: The May 1978 Stockholm Workshop Test Case. EFA TN AE-1522, Jan.1979, also AIAA J.19, No. 2, pp. 232-234 (1981).

2. Lindhout J.P.F.; van den Berg B.; Elsenaar A.: Comparison of Boundary-Layer Calculations for the Root Section of a Wing: The September 1979 Amsterdam Workshop Test Case. NLR MP 80028 U, 1981.

3. Hirschel E.H.; Kordulla W.: Shear Flow in Surface-Oriented Coordinates. Vieweg Verlag, Braunschweig 1981, Vol. 4 in Notes on Numerical Fluid Mechanics.

4. Cousteix J.: Theoretical Analysis and Prediction Methods for a Three-Dimensional Turbulent Boundary Layer. ESA TT-238 (1976).

5. Hirschel E.H.: Ein Verfahren zur Aufbereitung der Anfangs-, Rand- und metrischen Daten für die Berechnung dreidimensionaler Grenzschichten mit den COUSTEIX-Verfahren. MBB Bericht Nr. FE122/S/R/1530 (1981).

6. Hirschel E.H.: Ein Verfahren zur Berechnung von Wandschubspannungskomponenten, dreidimensionaler Verdrängungsdicke, äquivalenter reibungsfreier Quellenverteilung und Wandstromlinien aus den Ergebnissen der COUSTEIX-Verfahren. MBB-UFE-AERO-MT-471 (1981).

7. Hirschel E.H.: Boundary-Layer Coordinates on General Wings and Fuselages. To appear in ZFW 6 (1982), No. 3.

8. Kraus W.; Sacher P.: Das Panelverfahren zur Berechnung der Druckverteilung von Flugkörpern im Unterschallbereich. ZFW 21 (1973), pp. 301 - 311.

9. Cousteix J.: Solution de similitude et methode integrale de calcul des couches limites laminaires tridimensionelles sur paroi adiabatique. ONERA-DERAT, Document N° 1/5004 DN (1973), Note Technique.

10. Wang K.C.: Separation Patterns of Boundary Layer over an Inclined Body of Revolution. AIAA-Paper No. 71-130 (1971).

11. Hirschel E.H.; Kordulla W.: Local Properties of Separation Lines. ZFW 4 (1980), pp. 295 - 307.

12. Tobak M; Peake D.J.: Topological Structures of Three-Dimensional Separated Flows. AIAA-Paper No. 81-1260 (1981).

13. Meier H.U.: Personal communication of results to be published.

14. Stock H.W.: Laminar Boundary Layers on Inclined Ellipsoids of Revolution. ZFW 4 (1980), pp. 217 - 224.

15. Hirschel E.H.: Grenzschichtberechnungen an einem rotationssymmetrischen Rumpf mit dem COUSTEIX-Verfahren. MBB-UFE122-AERO-MT-472 (1981).

Turbulent Boundary Layers on Three-Dimensional Configurations

H.W. STOCK
W. SEIBERT

DORNIER GMBH
Postfach 1420
7990 Friedrichshafen

Abstract:

The Dornier integral method for calculating three-dimensional, turbulent, compressible boundary layers is described briefly. Computational results for flows along inclined ellipsoid of rotation are given and compared to available skin friction measurements. The boundary layer development on a ship hull is presented. Experimental results and computations of finite-difference solutions are shown for comparison. Finally the boundary layer development on road vehicle is discussed.

1. INTRODUCTION

For the computation of the development of three-dimensional compressible turbulent boundary layers an integral method is employed. The method uses a general surface oriented, curvilinear, non-orthogonal coordinates system. The boundary layer method requires as input the velocity outside the boundary layer in magnitude and direction and the configuration definition in a cartesian coordinate system.

2. SHORT DESCRIPTION OF THE METHOD

A detailed description of physical and numerical assumptions is given in Ref. [1].

2.1 Geometry treatment

The metric coefficients, which appear in the boundary layer equations written in a surface oriented, curvilinear non-orthogonal coordinate system (x, y) are computed in the following way.

After the choice of a suitable surface oriented coordinate system for a given configuration, grid points are generated given by $\sum_{i=1}^{n} x_i$ and $\sum_{j=1}^{m} y_j$. The identical surface points given in the x, y system are subsequently described in a reference cartesian coordinate system (X, Y, Z) via $\sum_{i=1}^{n} \sum_{j=1}^{m} X_{i,j}$, $\sum_{i=1}^{n} \sum_{j=1}^{m} Y_{i,j}$ and $\sum_{i=1}^{n} \sum_{j=1}^{m} Z_{i,j}$.

It is now possible to establish via a numerical procedure (a bicubic spline procedure was applied) the functional dependence between the coordinate systems such that one obtains for any point i,j in the computational domain

$$X_{i,j} = f_1 (x_i, y_j)$$
$$Y_{i,j} = f_2 (x_i, y_j)$$
$$Z_{i,j} = f_3 (x_i, y_j)$$

The metric coefficients appearing in the boundary layer equation are functions of f_1, f_2 and f_3.

2.2 Boundary layer treatment

Fig. 1 shows a schematic sketch of a three-dimensional velocity profile. That velocity profile, skewed in space, is fully described by the velocity profile u in the (x,z) plane and the velocity profile v in the (y,z) plane. The integral values in the boundary layer equations are based on the u and v profiles. It is almost impossible to describe u and v profiles by suitable profile families. Therefore the three-dimensional velocity distribution will be described by the profile U in the (t,z) plane (mainstream velocity profile) and by V in the (n,z)

plane (cross-flow velocity profile).
The t-direction is the direction of the resultant inviscid outer flow projected onto the surface and the n-direction is perpendicular to the t-direction. The angles appearing in fig. 1 are λ, α and β. λ is the angle between the positive x- and y-direction, α is the angle between the x- and the main stream direction and β (limiting streamline angle) the angle between the main stream and the limiting streamline direction or the resultant skin friction direction; β thus represents a measure for the three-dimensionality of the flow. Via the angles λ and α it is an easy matter of geometry to compute from integral quantities based an U, V profiles those in the x, y direction (u, v profiles).

Compressibility effects are taken into account by the Stewartson-Illingworth transformation. The turbulence model used, is based on the entrainment method of Head [2], which was developed for two-dimensional, equilibrium boundary layers. Based on the physics of the entrainment idea, it seems reasonably suited for the application in three-dimensional flows. The entrainment is done by the large eddies in the outer part of the boundary layer, which entrain inviscid fluid into the viscous layer. In that outer part the crossflow velocities are small, such that entrainment may be formulated based on variations in the mainstream profile.

An extension to Head's original method is used to account for non-equilitrium effects namely Horton's [3] lag-entrainment method.

2.3 Numerical method

Myring [4] has shown that the three-dimensional boundary layer integral equations are hyperbolic. A feature of hyperbolic systems of equations is that they have characteristic directions, where the two outermost characteristics define the zones of influence and dependence. The characteristics of the boundary layer equations approximately lie in any point of the x, y surface coordinate system within an angle bounded by the mainstream and the limiting streamline direction [4].

The x-direction is the marching direction for the integration, using are explicit scheme. To determine an approximate stability criterion for that scheme, the geometrical interpretation of the C.F.L.-rule is used, according to which the numerical domain of dependence must enclose the domain of dependence of the two outermost characteristic directions of the equations. To increase the acurracy of the method the Δx steps are decreased in regions where the gradients of the variables increase.

For the computation of the y-derivatives of the variables note is taken of the characteristic directions by using central, backward or foreward differences.

Initial conditions are supplied depending on the body configuration. For bodies like a fuselage, where the flow stagnates in one point, initial conditions are evaluated using the exact stagnation point solution. For a wing type body, where the flow attaches along a line, initial conditions are computed, assuming locally an equivalent flow along an infinite swept cylinder.

The problem of the boundary conditions is also configuration dependent. For bodies which have symmetry planes no boundary conditions are necessary. The variables are either symmetric (zero y-derivatives) or antimetric (y-derivatives are evaluated by inward differencing). For other configurations boundary conditions must be supplied along the side boundaries, where - speaking in physical terms - the flow is entering from outside the computational domain, on the other hand, where the flow is leaving the domain no conditions are needed.

3. RESULTS

3.1 Turbulent bounday layers on inclined ellipsoids

For the inviscid outer flow the potential flow pressure distribution was used. For not too large angles of attack, the potential flow pressure distribution and the experimentally observed one [5] is almost identical in the foreward part of the body ($x/a \leq 0.5$), Fig. 2. Under these circumstances calculated boundary layer quantities may be compared to experimental ones. For the geometry analytical expressions are employed to compute the metric coefficients.

Fig. 3 and 4 show calculated results for fully turbulent flow for the ellipsoid of thickness ratio F = 6 at $\alpha = 10^{\circ}$ for a Reynolds number of $Re_L = 7.2 \times 10^6$. The momentum thickness Θ_{11}, the displacement thickness δ_1^*, the skin friction c_f of the streamwise velocity profiles and the limiting streamline angle β are shown for 6 different stations ψ, which are indicated in the figure. (Θ is the circumferential angle, Θ=O windward, $\Theta=180^{\circ}$ leeward symmetry plane).

Station 6 is the last station where boundary layer data can be generated for the whole circumference. Downstream of station 6 a free vortex layer separation situation is predicted. Maximum values of the boundary layer thicknesses show up on the side of the ellipsoid, the maximum is moving towards the windward symmetry plane when going downstream. At the same time the skin friction c_f is going to vanish and the flow close to the surface changes rapidly its direction as seen by the drastic change in β in only a small range of Θ. Fig. 5 shows the development of the 3-D displacement thickness δ^* for an angle of attack of $\alpha = 5^{\circ}$. Downstream of station 8 a free vortex layer separation is predicted, all the vorticity contained in the two narrow sheets of curve 8 is shed into the free stream in form of a pair of counterrotating vortices.

The free vortex layer separation line is computed by integrating the flow only on the windward side (from the windward symmetry plane up to the free vortex layer separation line). Θ stations beyond the separation line are omitted. Fig. 6 shows for F = 6 and $\alpha = 5^{\circ}$ and 10° that line for fully laminar and fully turbulent flow.

Finally it is tried to compute some of the measured quantities [5]. Fig. 7 gives a sketch of the experimentally observed flow pattern, the shaded region indicates transitional flow. The position, where in the computation the laminar procedure [1] is switched to the turbulent one (no transitional boundary layer calculation), is shown, too. On the leeward side of the body the experimentally observed transition is close to that position, whereas on the windward side transition occurs far downstream of that position.

In comparing the calculated results with the measurements, reasonable agreement should be expected on the leeward side of the ellipsoid. Indeed, as shown in Fig. 8 for station $x/_a = 0.304$ the agreement on the leeward side for the total skin friction coefficient c_{f_t} with measurements is fairly good. Contrary on the windward side, the experimental values of c_{f_t} are by far larger than the calculated ones, a feature which should be expected if transition is fixed in the calculation far upstream of its experimentally observed location. The angle γ $(\alpha+\beta)$ in Fig. 8 gives the direction of the total skin friction, here again the agreement with measurements is fairly good.

3.2 Turbulent boundary layer on a ship hull

The turbulent boundary layer development is calculated on the ship hull of the "Larson test case" (Stockholm 1980). The configuration and the computational grid is shown in Fig. 9. The geometry program was used for the computation of the metric coefficients. Fig. 10 give the results for the momentum thickness Θ_{11}, the shape parameter H and the limiting streamline angle β plotted vs. $\frac{2G}{L}$ at different stations $\frac{2X}{L}$. ($\frac{2G}{L}$ is a circumferential coordinate, i.e. the local arc length from the keel line, G = 0). The computational results of Cebeci [6] and McLean [6] (finite difference methods) and the experiments [6] are shown, too. The calculations show reasonable agreement for the stations $\frac{2X}{L}$ = -0.744, 0.291 and 0.502. The maxima of Θ_{11} occur at about the same circumferential station $\frac{2G}{L}$. The maximum values of Θ_{11}, predicted by the different methods, are fairly in disagreement, which may be mainly due to the fact that each method uses a different number of circumferential stations. At the station $\frac{2X}{L}$ = 0.884 the integral method calculates a boundary layer flow, which is almost separated. Two maxima in H show up with H $\geq$ 2.0, which indicate flow on the verge of separation. The finite difference methods calculate flow separation for $\frac{2G}{L} \geq 0.12$, demonstrating clearly a drastic thickening of the boundary layer at $\frac{2G}{L} \sim 0.075$.

It seems difficult to compare conclusively, at least at the rear end of the ship, the different computational methods based on those results shown in Fig. 10 due to the following problems: 1. Any method uses a different computational grid. 2. There is no common treatment of the surface geometry and 3. no common treatment of the inviscid outer flow.Furthermore, it is well known, that for flows close to separation only minor differences in the metrics or in the outer flow treatment produce quite different results for the boundary layer development.

A comparison between the experimental [6] and theoretical results, at least at the rear end of the ship, is not possible: 1. As input to the computations the potential flow pressure distribution was given, not taking into account the strong viscous, inviscid interaction at the rear end of the ship. 2. The boundary layer in that location is extremely thick. Fig. 11 gives iso-lines of the velocity ratio U/U_∞ in the stations $\frac{2X}{L}$ = 0.947 and 0.981. Taking U/U_∞ = 0.9 as being somewhat representative for the boundary layer thickness, it is clearly seen that the thickness is of the same order of magnitude as a representative body diameter. This situation cannot be handled by either of the methods.

3.3 Turbulent boundary layer on a road vehicle

The inviscid irrotational flow around a road vehicle is calculated by a conventional panel method. The ground was simulated by mirroring the singularities. The configuration for the panel method input is shown in Fig. 14. Four stations X/L are indicated for which results will be shown. The Reynolds number for the boundary layer calculation is based on L = 4 m vehicle length and a velocity of 100 km/h.

Fig. 15 shows the inviscid pressure distribution for the four stations on the circumference. In the back part large gradients in circumferential direction of the pressure are produced at the junction of the roof and the side of the vehicle. Fig. 16 shows the results for the three-dimensional displacement thickness δ^*. At station X/L = 0.9326 δ^* shows a marked maximum at the junction (roof, side wall). Downstream of that station a free vortex layer separation is indicated. Fig. 17 gives the limiting streamline angle, which varies drastically over only a small distance in circumferential direction. The boundary layer development close to separation is similar to that on inclined ellipsoids.

Fig. 18 gives the pressure distribution in the symmetry plane on top of the vehicle. The measurements agree quite well with the inviscid pressure distribution except at the rear end of the vehicle. Two supplementary panel method computations were done, based on two different concepts to include viscous effects. One posibility to account for viscous effects is the displacement thickness concept, which brings up the necessity of repanelling the body surface. The second possibility is the equivalent source concept, whereby the viscous effects are simulated by sources on the original body surface. The advantage of that method is that the panelling of a configuration is done once for all in the inviscid-viscous iteration loops.

Fig. 18 shows clearly that the two concepts produce different pressure distributions as seen also in fig. 19 where C_p is plotted in the circumference at station X/L = 0.9326. The deviations in the computed pressure distribution show up in regions where the boundary layer is thick and rapidly growing. The idea came up that the two concepts will produce the same pressure distribution, if for the equivalent source concept the pressure is not evaluated on the original surface, but on the displaced surface. Results of that iteration method are not included, they are in progress.

REFERENCES

[1] STOCK, H.W.
Integral Method for the Calculation of Three-Dimensional Laminar and Turbulent Boundary Layers
NASA-TM-75320, 1978

[2] HEAD, M.R.
Entrainment in the Turbulent Boundary Layer
A.R.C. R & M 3152, 1958

[3] HORTON, H.P.
The Calculation of Compressible Turbulent Boundary Layers by the Entrainment History Method
Hawker Siddelay Rep. Researh/1048/H.P.H.

[4] MYRING, D.F.
In Integral Prediction Method for Three-Dimensional Turbulent Boundary Layer in Incompressible Flow R.A.E. Rep. 40147, 1970

[5] H.P. KREPLIN, H.V. MEIER
Experimental Determination of Wall Sheer Stress Vectors on an Inclined Prolate Spheroid
AFFOL-TR-80-3088
Proceeding of DEA-Meeting on Viscous and Interacting Flow Field Effects 1980

[6] L. LARSON
Stockholm Test Case on Turbulent Boundary Layers on a Ship Hull, Stockholm 1980

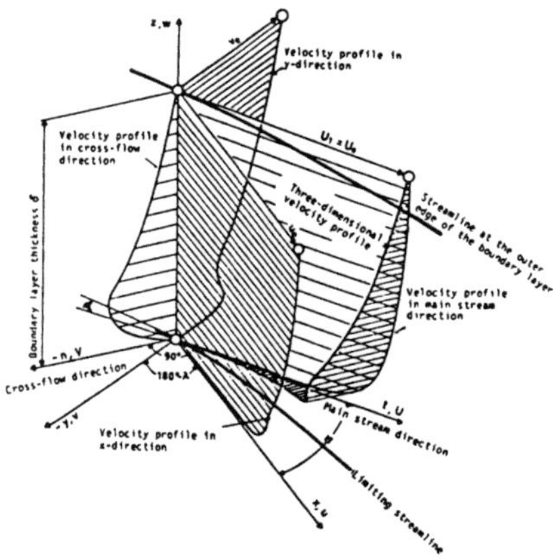

Fig. 1: Schematic sketch of velocity profiles in a three-dimensional boundary layer

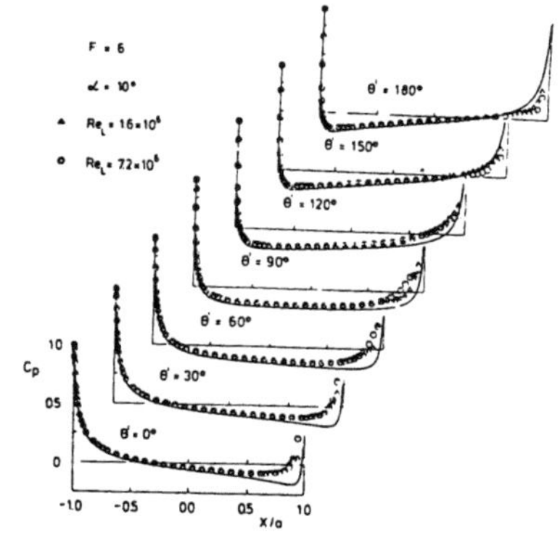

Fig. 2: Measured and calculated (potential flow) pressure distribution on an inclined ellipsoid

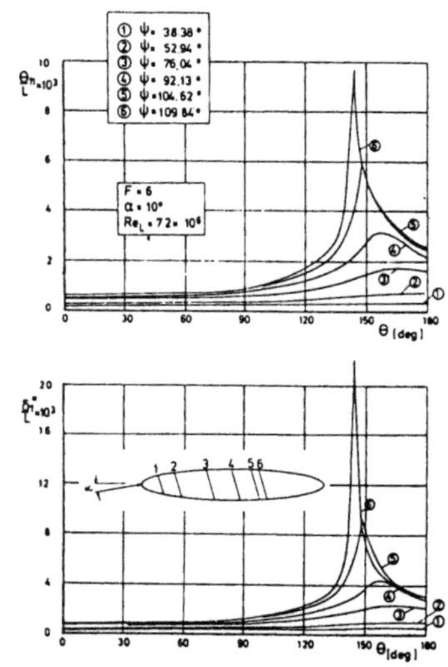

Fig. 3: Computed momentum and displacement thickness of the mainstream velocity profile on an inclined ellipsoid

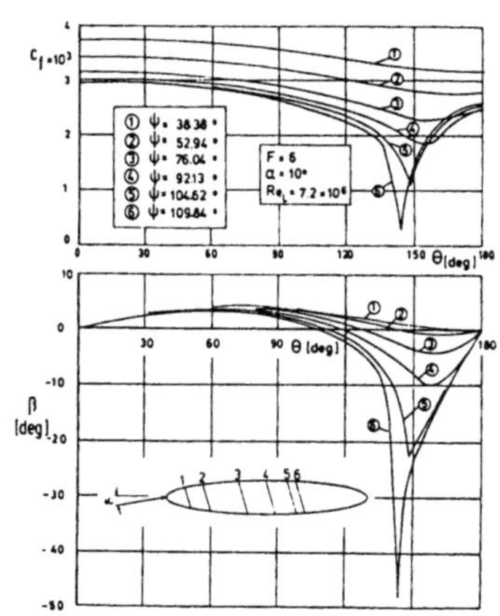

Fig. 4: Computed skin friction component in mainstream direction and limiting streamline angle

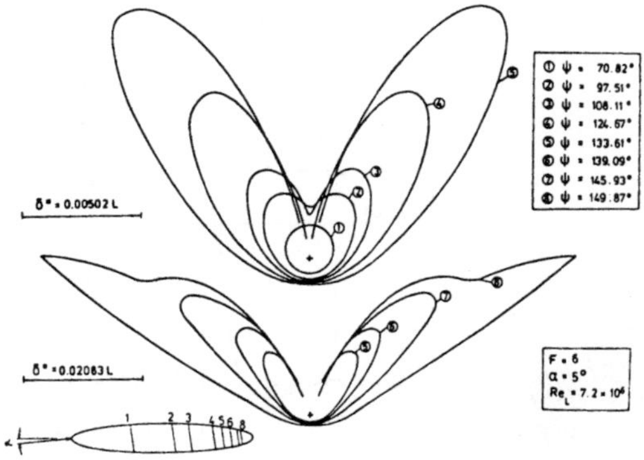

Fig. 5: Computed displacement thickness of the three-dimensional boundary layer

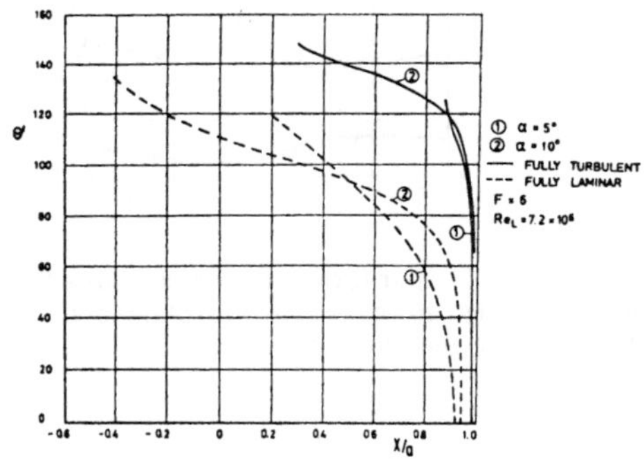

Fig. 6: Free vortex layer separation lines for fully laminar and fully turbulent flows on an inclined ellipsoid

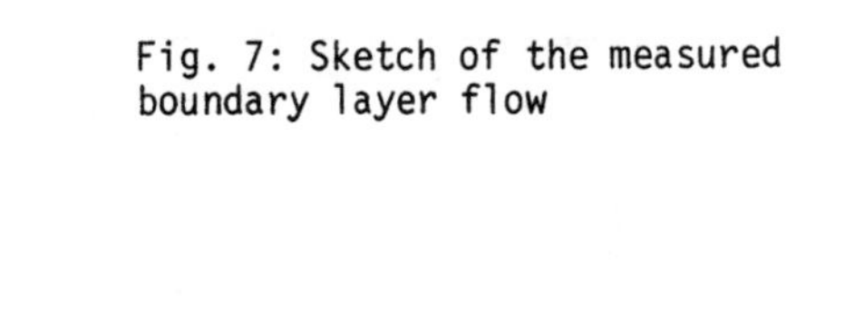

Fig. 7: Sketch of the measured boundary layer flow

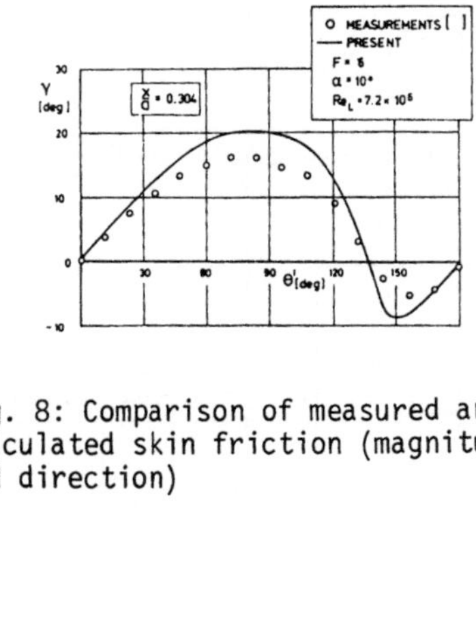

Fig. 8: Comparison of measured and calculated skin friction (magnitude and direction)

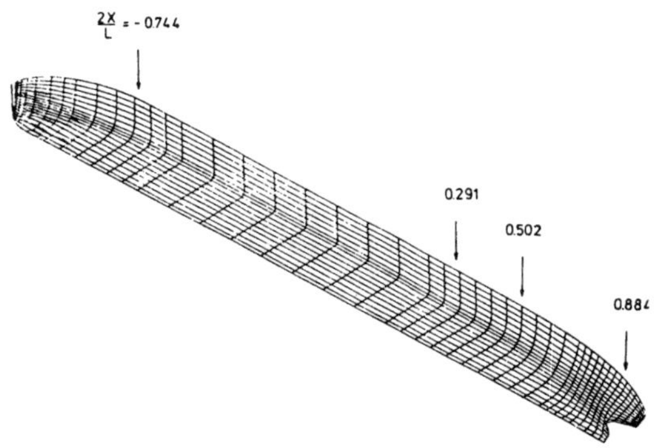

Fig. 9: Panel grid for the ship hull

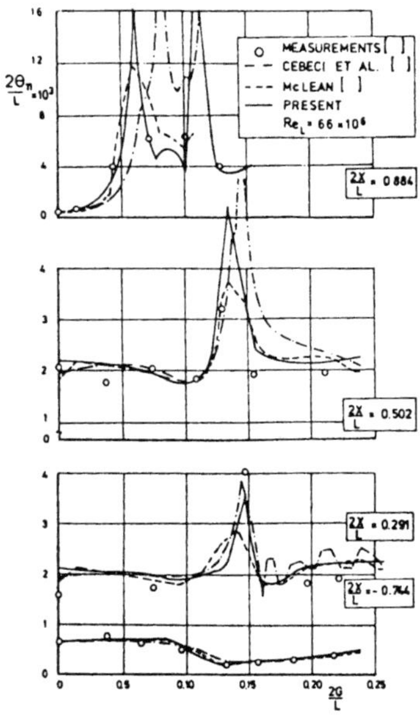

Fig. 10a: Comparison of the measured and calculated momentum thickness of the mainstream velocity profile

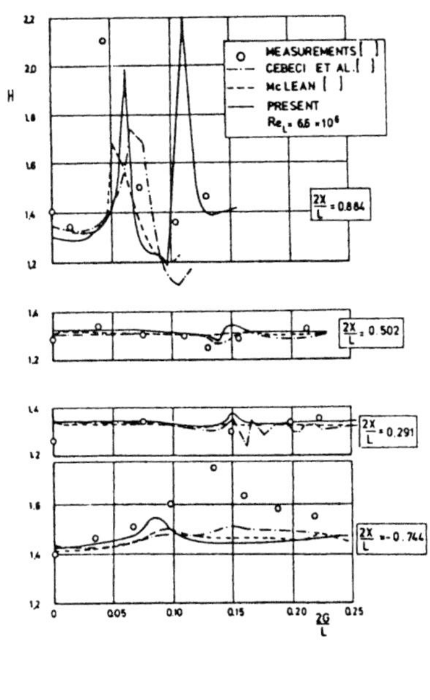

Fig. 10b: Continued, shape parameter of the mainstream velocity profile

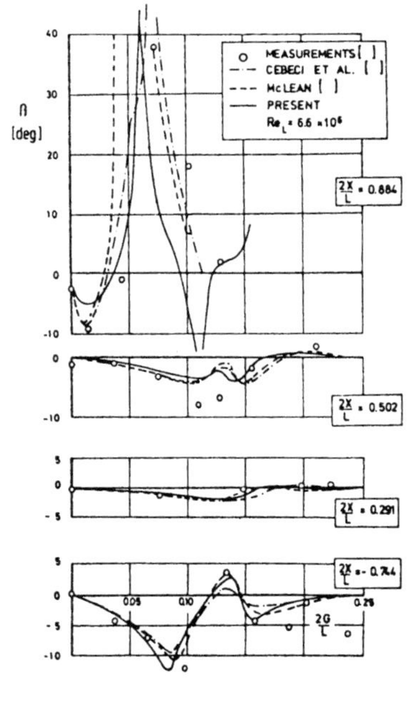

Fig. 10c: Continued, limiting stream line angle

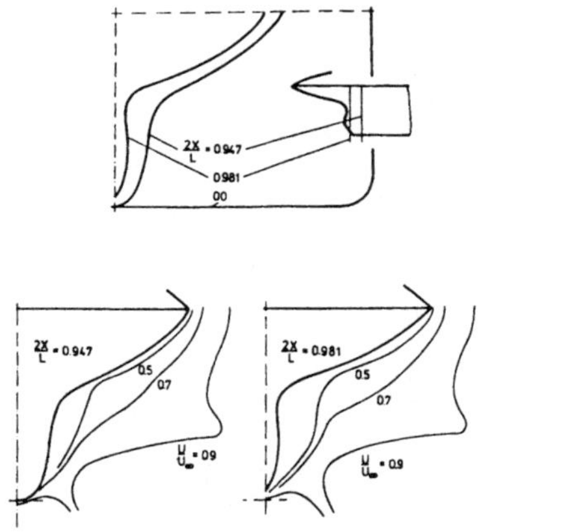

Fig. 11: Measured iso-velocity lines on the ship hull

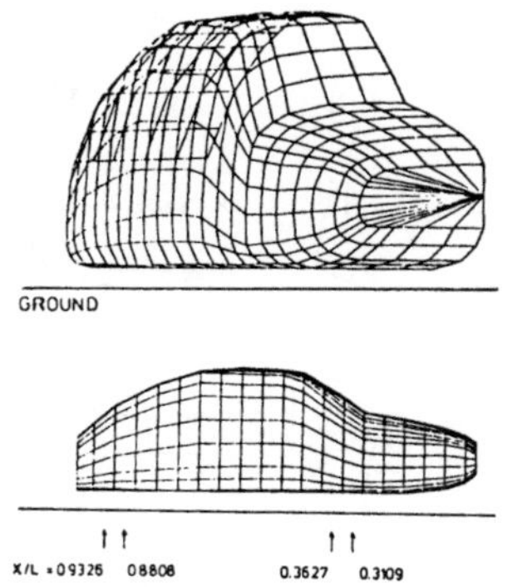

Fig. 12: Panel grid for the road vehicle

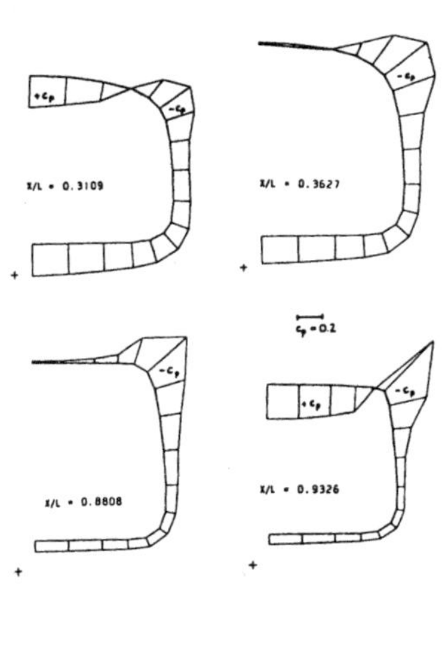

Fig. 13a: Pressure distribution at different x-stations on the road vehicle

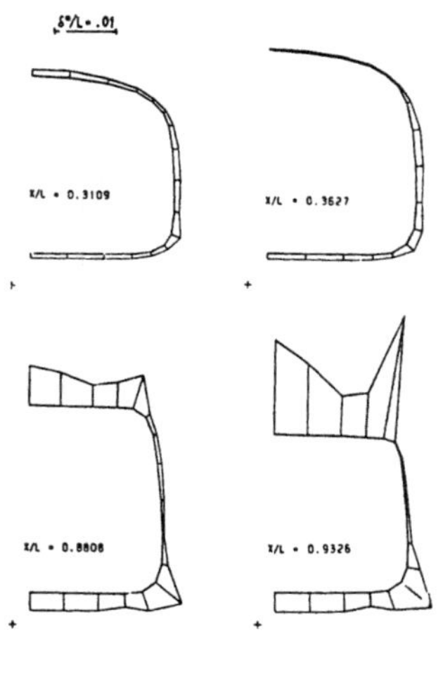

Fig. 13b: Continued, displacement thickness of the three-dimensional boundary layer

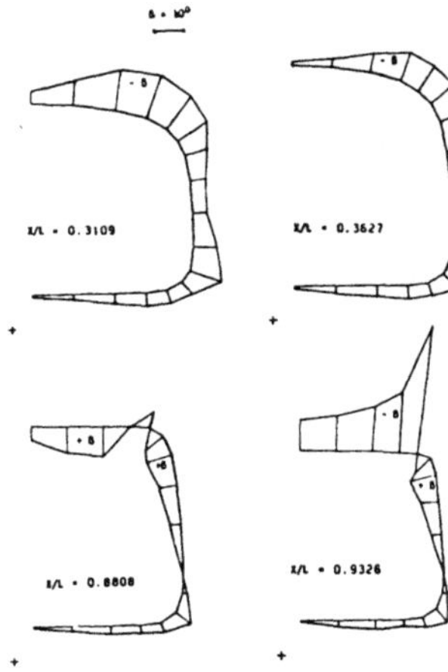

Fig. 13c: Continued, limiting stream line angle

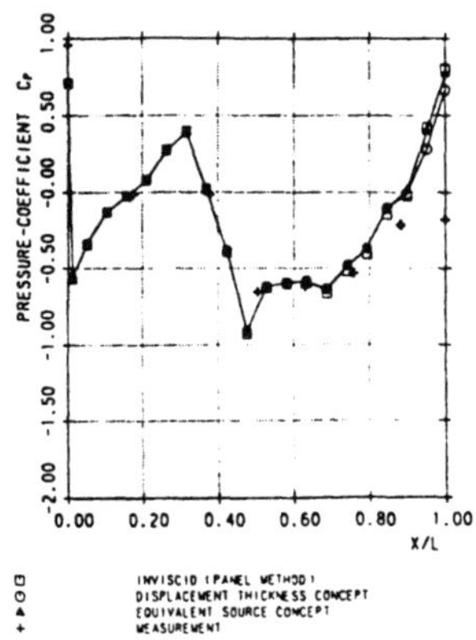

Fig. 14: Comparison of the measured and computed pressure distribution in the upper surface symmetry plane

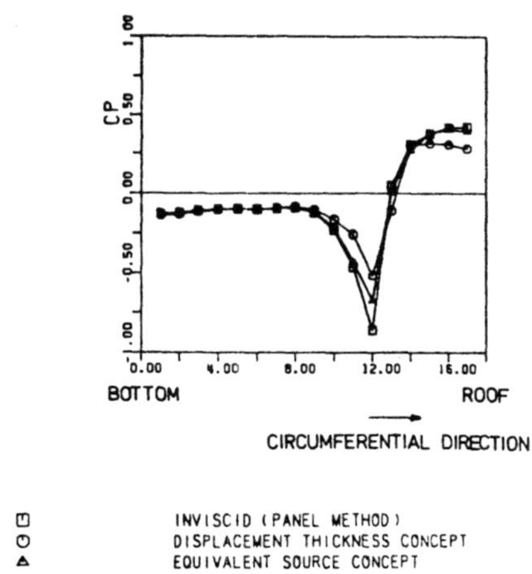

Fig. 15: Calculated pressure distribution at the station x/L = 0.9326

An Evaluation of the Final Discussion of the IUTAM Symposium on Three-Dimensional Boundary Layers

H.H. FERNHOLZ

Hermann-Föttinger-Institut für Thermo- und Fluiddynamik
Technische Universität Berlin
Straße des 17. Juni 135, D-1000 Berlin 12

Summary

The final discussion of the symposium was chaired by Krause (RWTH Aachen) and dealt with four main topics:

(1) Requirements for future experiments both in incompressible and in compressible flows,
(2) Flow separation in three-dimensional boundary layers,
(3) Measuring techniques and the evaluation of data,
(4) Turbulence modelling (disaster area) and numerical modelling.

Requirements for future experiments

Young (Queen Mary College London) stated three objectives for measurements in general: (a) to provide some basic ideas from which the physics of a flow can be understood, (b) to provide computors with information that can be built into models for computing purposes and to check assumptions that go into turbulence models. and (c) to provide very good data against which computors can check their calculation methods.
The extra dimension brought in by 3-D flows has presented major problems and the physics have not yet been properly understood. This means that we want to put all our efforts on a better understanding of the physics and on the assumptions for the turbulence models. Comparisons of the different calculation procedures should not yet be a primary objective.

Van den Berg (NLR Amsterdam) confirmed Young's views. In order to get more physical insight it was less important to measure the same quantities which we have measured so far in a still greater number of flow configurations but to try to get more information about those terms in the transport equations for the Reynolds stresses which were needed for closure.
Fernholz (TU Berlin) wanted to have those quantities (possibly

triple correlations as suggested by Gibson, Imperial College London) specified more distinctly which had first priority, apart from the full Reynolds stress tensor, and reminded the audience of an earlier suggestion by Bogdonoff (Princeton University) that the same experiment should be performed in two different wind tunnels by two experienced teams. He offered to serve as a clearing house both for suggestions for the type of the experiment and for the quantities and measuring techniques. As a preliminary proposal it was mentioned that the 3-D boundary layers should contain large crossflow (say $\beta_W > 25°$) and that initial and boundary conditions must be given in great detail. It would be preferable to have one experiment where the turbulence level close to separation was below 20 % as in the NLR experiment and one with a high turbulence level close to separation as in the Berlin 3-D boundary layer.
Bogdonoff (Princeton University) was overwhelmed with the ability of computors to calculate what he considered to be very complex shapes but stated his conviction that complex shapes will never give a picture of the real physics. So experiments should be designed for specific physics. As far as Reynolds stress is concerned it was not at all clear to him that this is the most important quantity which one can measure. If the flow is attached it appears as though it does not make any difference what turbulence model is used. Most of the calculations seem to do reasonably well. What may be more important than the turbulence model is a sensitivity study. For this the parameter is changed by some significant amount and one sees whether it makes any difference or not.
In the design of an experiment one is caught between a very important set of applications on one side and an experiment which is fundamentally important on the other, but we are trying to make too big a step between the simple experiment which gives an insight into the physics and the computation of an important practical problem, such as an aircraft (wing-body junction) or a ship. He was furthermore overwhelmed by the confidence of people who tackle problems in which the physics and the parameters are not clear. Parameter variation is in his opinion the key part of an experiment. We have too many experiments in which

there is too little variation of parameters. Another possibility is perhaps not to make such good experiments but to try to change parameters, e.g. a geometrical change or a fluid mechanical change, such as Reynolds number or Mach number variations. If computors are given some range of parameters and find that they do not get results at both ends of the parameter range it is obvious that their method does not have anything fundamental.

Another problem, the importance of which may not yet have been recognized enough, is whether the flow is steady or not. This must be checked because it may be another parameter which one should measure, because all computations are practically steady mean flow computations. Nobody has a high frequency computation. Experimentalists should not try to provide complex data sets but simple data sets from which a certain amount of fundamental physics could be obtained. One example of a fundamental property in three-dimensional flows which is not well understood is spanwise spreading. In general, there is no such thing as a sound base in turbulence and if this is so we lack a sound base from which to start and can be led all over the place. Bogdonoff finally suggested a definition of problem areas into those flows where computations can be done, because we know what is going on, and into these which one cannot compute, because the physics are not understood well enough.

Joubert (University of Melbourne) proposed a check list for the experimenter and for those who evaluate data:

(1) Check velocity profiles, particularly the law of the wall and the velocities that go into it, i.e. in three-dimensional boundary layers the effects of secondary flow.

(2) Check whether the inner, so called collateral layer curves or whether it is straight. Are the shear stresses lined up with the mean shear strain rate and what is their relation with the local velocity?

(3) Examine the principal assumptions on which calculation methods are based. They are often obscure because computors do not specify them clearly enough (e.g. isotropy in the inner layer).

(4) Make sure that more than one method is used for measuring skin friction at the wall. Reynolds stress models do not

work well in predicting skin friction.
He finally suggested surface roughness as another major variable that should be investigated in its effect on, say pressure driven boundary layers.
Fannelop (NTH Trondheim) reported that the three-dimensional boundary layer on a rough flat plate with a semi-infinite body had been investigated in his group and that no significant differences between the 3-D boundary layer on a smooth surface and on a rough surface had been found. No increase in cross flow had occurred and the main effect had been the increase in boundary layer thickness.
Bogdonoff (Princeton University) pointed out that in many ways it was easier to design an experiment in compressible than in incompressible flow. The tests were cleaner because a whole set of parameters could be defined clearly in supersonic flow and effects were very often larger. Instrumentation may, however, sometimes be more difficult. Nevertheless he hoped that high-speed work could be helpful for low speed work where effects were smaller. In general he thought that it should be much easier to work on three-dimensional flow than on two-dimensional flow because the "foot prints" were larger, things were smeared out and the gradients were much smaller. In his experience it is the strong gradients which give us most trouble.

Flow separation in three-dimensional boundary layers

Young (Queen Mary College, London) took up an earlier remark by Cousteix who had stated that he ran into difficulties with his computation when he got to points in the flow where influences from downstream were present. In Young's opinion this was probably a key feature of what we call separation lines, that is there are points where there are two sheets of streamlines which come from different areas. They approach each other and this implies that they must then form another sheet which leaves the surface. Now, whether it leaves the surface and moves off to a great distance or whether it leaves the surface and remains close to the surface is something which depends on the circumstances of the setup, but which probably leads to some confusion because we are a little too accustomed to two dimensions. The layer is in such circumstances that there is a great blow up and what follows is a rather chaotic mess. When one comes to three dimensions however, the sheet leaves the surface, the direction being in a plane normal to the line. Separation lines can therefore be associated with vortices, either shallow or very large. If shallow, their effect on the pressure distribution will not be large. All one is seeing from what happens on the surface is streamlines which turn gently along the separation line and everything - from an experimental point of view - is relatively smooth. From a computational point of view, however, a singularity occurs unless the downstream influence can somehow be taken into account. So three-dimensional separation need not necessarily be something which is unpleasant and very dramatic, it can be reasonably well organized but in each case it has to be examined on its merits.

Joubert (University of Melbourne) emphasized that separation in a three-dimensional flow is quite a different phenomenon from separation in 2-D flow and unless the flow pattern on the body surface can be drawn one does not understand what one is doing. The laws of critical point theory about the number of nodes, saddles and vortex points on the body must be obeyed and without knowing these laws flow patterns on a body should never ever be drawn.

Hirschel (MBB München) drew attention to the extreme sensitivity of computations to the pressure gradient $\partial p/\partial y$ normal to the wall. Since normal pressure gradients are almost inevitable in regions close to separation the only rational approach to model the flow is on the basis of the Navier-Stokes equations.

Lindhout (NLR/Enschede) thought that Hirschel's suggestion was a good means to get rid of the singularity mentioned by Cousteix.

Krause (RWTH Aachen) finally pointed out that if the results of calculations in three-dimensional boundary layers were seen in a historical perspective the progress was not too bad.

Measuring techniques and the evaluation of data

Fernholz (TU Berlin) distinguished two main types of flow where measuring techniques were relatively safe, and where they still had problems. If an experiment is done in attached flow, if the turbulence level is low and if one stays away 3 mm or more from the wall, then our measuring techniques may be called adequate. Problem areas are still the near wall region, regions with reverse flow or flow zones where the turbulence level is high. In his opinion wall pressure measurements and mean velocity measurements (the latter performed by Pitot tubes or hot wires) in flows without flow reversal (not even instantaneous reversal), were reliable. There was no reliable technique, however, for measuring the static pressure normal to the wall which would be an important piece of information for many three-dimensional boundary layers. Since there is no general proof yet that some form of the law of the wall is valid in all types of three-dimensional boundary layers we must rely on direct wall shear stress measurements as they have been performed by Pierce (VPI) and on calibrations of indirect skin friction meters by means of a direct technique. Apart from some LDV data measured by Simpson et al. (see JFM 1982) in the outer region of a 2-D boundary layer, there exists only hot-wire data of fluctuating quantities close to separation and hardly any in the near wall region. Data checks by different methods would be very important, for example by means of the pulsed-wire technique. As far as the flow angle is concerned cobra probes can be used away from the wall but for flow angles close to the wall hot wires seem to be the only means.

Pierce (Virginia Polytechnic Institute) informed about problems with signals from carefully calibrated probes caused by velocity and shear gradients close to the wall. He warned furthermore that even tiny misalignment may lead to appreciable errors if probes - as is usually done - have been calibrated in a uniform flow.

Bogdonoff (Princeton University) did not see the problem of measuring flow angles close to the wall as being so important since one could always extrapolate over the last millimeter to the angle given by the visualization of the surface flow pattern where the angle could be measured very accurately. Surface flow patterns could be generated without any trouble if the technique were applied with skill. Among others it means finding a carrier (e.g. alcohol or hydrocarbons of different volatility) for the lamp black which evaporates in the time scale which is correct for the specific facility. The mixture must stay liquid until equilibrium of the flow is reached and should then evaporate in a short time. As it evaporates, the layer gets very thin and markings are extremely fine. By using a sheet of "Scotch tape" a permanent record of the surface flow pattern can easily be obtained. Bogdonoff warned, however, that one could not tell from the surface flow pattern whether the flow is steady or unsteady.

In order to find out whether the flow is separated or not the well known vapor screen method is used in a localized way. For this purpose a small amount of volatile liquid is fed into the boundary layer through static pressure orifices, one at a time. With the right amount of liquid and by use of a laser a sheet of vaporizing material can be generated and gives an impression of what is happening in that region. With a high speed camera and the vapor screen one can get some idea of the dynamics of the flow, for example whether it is steady or whether cells roll up in vortices.

Bogdonoff finally urged the more frequent use of optics in 3-D flows but in order to do this new windows had to be put into the walls of supersonic windtunnels at the right place. Flow field visualization is urgently needed because only by its use can one get the idea that the observed bow shock is not that nice clean line which is seen on the surface but behaves in an unsteady way.

Gessner (University of Washington) commented on the problem of attempting to measure flow direction or Reynolds stresses in a skewed three-dimensional boundary layer, in which one cannot align the probe with the mean streamline. Which kind of probe or which type of response model does one tend to use to extract information on the flow field? Unfortunately, of the various techniques that have been proposed only a few construct a set of response equations and then follow it with a validation study (e.g. Acrivlellis who has only developed response models) in order to test how well posed or how ill conditioned the coefficient matrix may be. Other techniques such as that by Bissonnette & Mellor have validated their rotatable single wire technique only under non skewed flow conditions, i.e. in fully developed pipe flow and have not actually simulated a skewed flow condition.

Turbulence modelling and numerical modelling

Bogdonoff (Princeton University) stated that turbulence modelling was in a terrible state. He based his opinion on results of turbulence measurements published by the Princeton group over the last few years. A recently started research programme, for example, seems to show that gradients caused by rather weak shocks in turbulent high Reynolds number boundary layers modify the turbulence level a great deal. There is no turbulence model that has been proposed which could deal with this situation. With a 1 % turbulence level in the free stream upstream of a weak shock wave one measures 2 % downstream of the shock, with 6 % and 15 to 17 % in the boundary respectively. Once a boundary layer separates and becomes a free shear layer, the turbulence level goes up very quickly. Free shear layers seem to be much more sensitive than boundary layers. Beginning with a boundary layer of about 6 % turbulence level at separation, 16 % was found in the free shear layer after a small region of separated flow and 40 to 50 % at reattachment. These latter numbers have little meaning due to linearization of the signals but the effect is very large. So when the results of a calculation do not come out in the way they are expected to, may be it is due to the numerics which then have to be straightened out, but also one should not get the idea that the simple turbulence models which are around really have any fundamental universal solid foundations, they can be greatly in error.
There was agreement among those who took part in the discussion that there was no universal model and that the constants which had to be chosen for individual types of flows, e.g. shear or pressure driven 3-D boundary layers, were probably not even constants but functions.
Bogdonoff was then asked by Gessner for a statement on the

perspective of turbulence measurements in supersonic flows. Though he stated at the beginning that he was the wrong person to answer that question Bogdonoff made some interesting remarks along the lines of the question. At first no high accuracy was required but the new tools were very critical. Laser Doppler anemometry has been around for a decade and has cost at least a billion dollars in the U.S. alone. The amount of data is, however, very small and cost per data point is astronomical. Yet this was supposed to be the instrument that could do everything. Although results are now coming out, it is unlikely that they will answer our questions in the lower part of the boundary layer which we are especially interested in, nor in high speed flow because particles will not follow the large gradients nor in separated flows because the particles never get where one wants them to get.

Fortunately, however, new techniques are coming up, e.g. a resonant doppler system with some possibility of doing major flow visualization and with the potential to do turbulence work by using atoms or molecules as the tracers.

Fernholz (TU Berlin) confirmed the very unsatisfactory state of affairs of turbulence measurements in compressible boundary layers on the basis of the recent survey by Fernholz & Finley (AGARDograph 263).

Krause (RWTH Aachen) closed the discussion with the statement that the next ten years will confront us with the fact that we will have to throw away our nicely developed algorithmus if we get to use the next generation of computers and parallel processors. M o r e theoretical background from the mathematical side is needed and simple approaches based on physical concepts must be used more.

MIX
Papier aus verantwortungsvollen Quellen
Paper from responsible sources
FSC® C105338

If you have any concerns about our products,
you can contact us on
ProductSafety@springernature.com

In case Publisher is established outside the EU,
the EU authorized representative is:
Springer Nature Customer Service Center GmbH
Europaplatz 3, 69115 Heidelberg, Germany

Printed by Libri Plureos GmbH
in Hamburg, Germany